Time-Synchronized Control: Analysis and Design

Dongyu Li · Shuzhi Sam Ge · Tong Heng Lee

Time-Synchronized Control: Analysis and Design

Springer

Dongyu Li
School of Cyber Science and Technology
Beihang University
Beijing, China

Tong Heng Lee
Department of Electrical and Computer Engineering
National University of Singapore
Singapore, Singapore

Shuzhi Sam Ge
Department of Electrical and Computer Engineering
National University of Singapore
Singapore, Singapore

ISBN 978-981-16-3088-0   ISBN 978-981-16-3089-7 (eBook)
https://doi.org/10.1007/978-981-16-3089-7

This Springer imprint is published by the registered company Springer Nature Singapore Pte Ltd.
The registered company address is: 152 Beach Road, #21-01/04 Gateway East, Singapore 189721, Singapore

To parents, Huiying and Youning for their companionship in the journey, and to mentors for their guidance without reservation

Dongyu Li

To parents for their love and contentment in life, to Jinlan for her care to the families, and to friends for their supports

Shuzhi Sam Ge

To my family, and to sincere friends

Tong Heng Lee

Preface

In the literature of control science and engineering, we are more concerned with convergence of the states of any dynamic multi-input-multi-output systems, then pay attention to academically challenging and practically relevant finite-time convergence where finite-time control drives the states to converge within a certain time moment, and finally solve the important and critical issue in solving how each state element converges as demanded in many accurate, precise, and delicate operations in many applications including robotic autonomy, manufacturing operations, and exploration in oceans and space.

This monograph develops and introduces various fundamental and basics of time-synchronized control and its stability because of the demand for precise and dedicated operations in practice. Interconnections with finite-time control and sliding mode control are explored. The work forges ahead beyond the well-established notions and outcomes of standard "Finite-Time Stability", and investigates and develops results and outcomes for a newer (and in various important instances, more impactful) notion of "Time-Synchronized Stability". This work is oriented towards researchers, engineers, and students who dedicate themselves for the best control system design in practice yet with sound theoretical support by seamlessly integrating or fusing academic exploration and engineering regulations.

Specifically and importantly, under time-synchronized control, all the state elements of a target system converge to the origin exactly *at the same time*, called *time-synchronized convergence*. The associated time-synchronized stability is built on top of the well-known standard notion of finite-time stability, but adds to it the further notion of time-synchronized convergence. In this monograph, we underscore and discover important outcomes and properties associated with finite-time stability, and then define a notion and concept of a control system—*ratio persistence*. All of these together contribute to time-synchronized stability. Ratio persistence is an essential property that keeps each pair of the closed-loop system state elements being constant. Note that even the classical proportional control (P control) can appropriately enable a control system to be ratio persistent. Thus, P control can be seen as a very special type of time-synchronized control, where its synchronized settling time is $t \to \infty$. Of course herein, our explorations and attained outcomes are much more impactful beyond this in consequence. The results and outcomes here pertain to realizing this ratio persistence, and time-synchronized control with finite-time stability as well.

Time-synchronized control can be viewed as offering coordination of different states of the system in time for different applications in the sense that, as time passes, the system state elements converge to the equilibrium while always preserving a constant ratio among them. These unique nature and property make time-synchronized control highly attractive in providing not only time-synchronized convergence, but also more nice properties for this interesting time-synchronized convergence property attained such as a smooth system output, a short state trajectory, reduced energy consumption, improved convergence accuracy, among others.

The main content of this book includes time-synchronized stability and its relevant Lyapunov theorems, time-synchronized control design for different types of systems with practical considerations, and explorations in time-synchronized consensus.

In Chap. 2, time-synchronized stability, fixed-time-synchronized stability, and predefined-time stability as well as their respective Lyapunov-like theorems are proposed. Fixed- and predefined-time-synchronized stability are regarded as special types of time-synchronized stability.

In Chap. 3, quasi-continuous and continuous time-synchronized sliding modes are introduced to facilitate the achievement of time-synchronized stability. Their convergence rate, singularity issue, types of continuity are properly discussed.

In Chaps. 4–6, time-synchronized controllers are proposed for affine systems, Euler-Lagrange systems, and general MIMO systems under matched and unmatched disturbances, where time-synchronized or fixed-time-synchronized convergence is achieved. An important matter of singularity avoidance is suitably addressed. Further, we introduce the concept of "the least upper bound of synchronized settling time" to accurately estimate the synchronized settling time. An extension to predefined-time-synchronized control is further addressed.

In Chap. 7, the time-synchronized consensus problem is defined and solved. It shows that all the state elements of the multi-agent system are capable of achieving consensus time-synchronously when the graph theory is properly invoked with the proposed time-synchronized consensus controller.

In Chaps. 8 and 9, (practical) time-synchronized control is proposed for disturbed spacecraft under multiple practical constrains with hardware-in-loop experiments.

In Chap. 10, pertinent conclusions are drawn with interesting but challenging possible future directions.

Budding and Genesis

The genesis and mulling of this monography were very much inspired and motivated by many excellent scientific works including *Linear Controller Design: Limits of Performance* by S. Boyd and C. Barratt, *Optimal Control* by F. L. Lewis, D. Vrabie and V. L. Syrmos, *Sliding mode control: theory and applications* by C. Edwards, S. Spurgeon, and many excellent works presented at TCCT Workshop on Cooperative Control and Multi-Agent Systems under the leadership of Jie Chen, Beijing. As we investigated different control performance specifications for multi-agent systems, we further investigated the description and usage of two different versions of the signum function—the usual standard version and a norm-normalized version, i.e. one defined by the signum element and the other by the standard unit vector as illustrated in *Adaptive Neural Network Control of Robotic Manipulators* by S. S. Ge, T. H. Lee and C. J. Harris. We explored that this norm-normalized version had that special property (essentially of ratio persistence) to assure, what we had termed at that initial stage, "Simultaneous Arrival To Origin" (SATO) convergence (despite rather simple underlying mathematical analysis, that is actually, time-synchronized convergence). Motivated by this interesting property, we determined to seek a revisit of existing outcomes in "Finite-Time Stability" and "Sliding Mode Control" to explore SATO convergence in depth. In 2018, we had managed to successfully develop some early promising results along this line, where time-synchronized controls were designed to achieve time-synchronized convergence for simple first-order systems. In 2019, we further developed this work and managed to synthesize methodologies improving the outcomes, where more types of time-synchronized stability and more practical control systems were considered. With the journal manuscripts submitted, review comments came back which additionally benefited our work. It is indeed pleasant on how all the pieces have now fallen nicely in place; here this book/monograph is available to all who had expressed their great interest, and also to many others who, we hope and trust, will appreciate the great potential of the concepts and results here. We welcome interested excellent individuals to further push the boundary well beyond.

Please enjoy reading!

Beijing, China — Dongyu Li
Singapore, Singapore — Shuzhi Sam Ge
Singapore, Singapore — Tong Heng Lee

Acknowledgments

We are here to extend our gratitude towards colleagues and collaborators. We would also like to express our sincere appreciation to numerous valuable suggestions, comments, and multi-faceted supports to our understanding and formulation of the current research works as the foundation for this monograph, in particular, from (in alphabetical order of last names)

Rodrigo Aldana of University of Zaragoza,
Linda Bushnell of University of Washington,
Hui Cao of Xi'an Jiaotong University,
Xiao Chen of Shanghai Institute of Satellite Engineering,
Kun Cao of Nanyang Technological University,
Huixu Dong of National University of Singapore,
Shuo Ding of National University of Singapore,
Leonid Fridman of National Autonomous University of Mexico,
Yanning Guo of Harbin Institute of Technology,
Yufeng Gao of Harbin Institute of Technology,
Qinglei Hu of Beihang University,
Wei He of University of Science and Technology Beijing,
Zhimin Hou of National University of Singapore,
Wanyue Jiang of Qingdao University,
Frank L. Lewis of University of Texas at Arlington,
Jianwei Liu of Beihang University,
Shuai Liu of Shandong University,
Chuanjiang Li of Harbin Institute of Technology,
Yueyong Lv of Harbin Institute of Technology,
Xiaoling Liang of National University of Singapore,
Xian Li of Qingdao University,
Xing Liu of Northwestern Polytechnical University,
Guangfu Ma of Harbin Institute of Technology,
Jun Ma of Harvard University,
Lorenzo Marconi of University of Bologna,
Yongduan Song of Chongqing University,
Xiaodong Shao of Beihang University,

Keng Peng Tee of Agency for Science, Technology and Research (SG),
Yan Wu of Agency for Science, Technology and Research (SG),
Lihua Xie of Nanyang Technological University,
Haoyong Yu of National University of Singapore,
Wei You of Shanghai Institute of Satellite Engineering,
Haoyang Yang of Beihang University,
Jianying Zheng of Beihang University,
Wei Zhang of Shanghai Institute of Satellite Engineering,
Yanzhong Zhao of Chinese Academy of Sciences, among others.

Special thanks towards William Achauer, Anil Chandy, Jasmine Dou, Bhagyalakkshme from Springer, for their consistent and reliable support and comments in the process of publishing the book, and towards IEEE for granting us the permission to revisit materials from our publications.

Beijing, China — Dongyu Li
Singapore, Singapore — Shuzhi Sam Ge
Singapore, Singapore — Tong Heng Lee
June 2021

Contents

Acronyms and Notation

Acronyms

ISS	Input to state stable/stability
MIMO	Multiple-input multiple-output
SISO	Single-input single-output

Notation

$\mathbb{R}$	Field of real numbers
Σ	Summation
max	Maximum
min	Minimum
$\forall$	For all
$\in$	Belongs to
$>$ ($<$)	Greater (less) than
$\geq$ ($\leq$)	Greater (less) than or equal to
$t \to T$	t approaches T
$t \uparrow T$	t approaches T from below (in an increasing manner)
$t \downarrow T$	t approaches T from above (in an decreasing manner)
$\otimes$	Kronecker product
w.r.t.	With respect to
s.t.	Such That or Subject to
i.e.	The Latin phrase id est, meaning "that is"
e.g.	The Latin phrase exempli gratia, meaning "for example"
$\mathbb{R}^n$	The n-dimensional Euclidean space
I_n	The n-dimensional identity matrix
1_n	The n-dimensional vector with all entries being 1
$\vert \bullet \vert$	The absolute of a scalar $\bullet$
$\Vert \bullet \Vert$	The norm of a vector $\bullet$

$\dot{f}$	The first derivative of f with respect to time
$\ddot{f}$	The second derivative of f with respect to time
$\mathbf{D}^+ f$	The right-hand upper Dini derivative of f with respect to time
$\mathrm{diag}\{x_1, \cdots, x_n\}$	A diagonal matrix with diagonal elements x_1 to x_n
$H > 0$	A positive definite matrix H
$H \geq 0$	A positive semi-definite matrix H
X^T	The transpose of matrix/vector X
$\lambda_{\max}(A)/\lambda_{\min}(A)$	The maximum/minimum eigenvalue of matrix A
$A^{\dagger}$	The Moore-Penrose pseudo-inverse of matrix A
$\mathrm{vec}(A)$	The vectorization form of matrix A

Chapter 1
Introduction

Abstract This chapter briefly recalls and re-collects the history of finite-time control, and the budding and genesis of key ideas and developments in the further notion of time-synchronized control. Time-synchronized control is, in fact, a unique type of finite-time control; which already meets all the requirements of finite-time stability, and additionally drives all the system state elements to the equilibrium at the same time. The merit of this noteworthy and so-termed time-synchronized control consists in not only the synchronized property on the convergence time, but also the abilities to smooth the system output, shorten the state travel length, and reduce the energy consumption. Finally in the chapter, the overall book organization is shown in a summary of the content of the remaining chapters.

1.1 Finite-Time Control

With the developments of powerful sensors, actuators, control processors, and computer hardware, sophisticated control algorithms outperform simple ones that have sufficed in the past [9]. Finite-time control, as an important control systems methodology, had emerged and grown in interest in modern control systems design to cater to the growing requirements in a wide range of progressive control tasks; to address the high-precision and fast-response performance. As the notion implies, finite-time stabilization offers a control system convergence in finite-time [21]; under which, more specifically, the situation is such that before a certain time instant, all the state elements are assured to reach the origin (or a bounded neighborhood of the origin). Its wide applications include satellite stabilization [14, 54, 61], manipulator tracking [18, 24, 58], multi-agent consensus [10, 29, 39, 52], state estimation [5, 16, 42, 44], control of network systems [8, 29, 34, 55, 56, 62], control under practical constraints [23, 25, 46, 47, 59], among others.

Dating back to the 1950s in the initial instances of its development, the early idea of finite-time control likely first appeared in [28, 30, 31]. In the 1960s, the concept of finite-time control is further polished in [13, 53]. Again in the 1960s, another instance of optimal control design presented a similar concept of finite-time control

D. Li et al., *Time-Synchronized Control: Analysis and Design*,
https://doi.org/10.1007/978-981-16-3089-7_1

$$v_1(\|x\|) \leq V(t,x) \leq v_2(\|x\|), \tag{2.134}$$

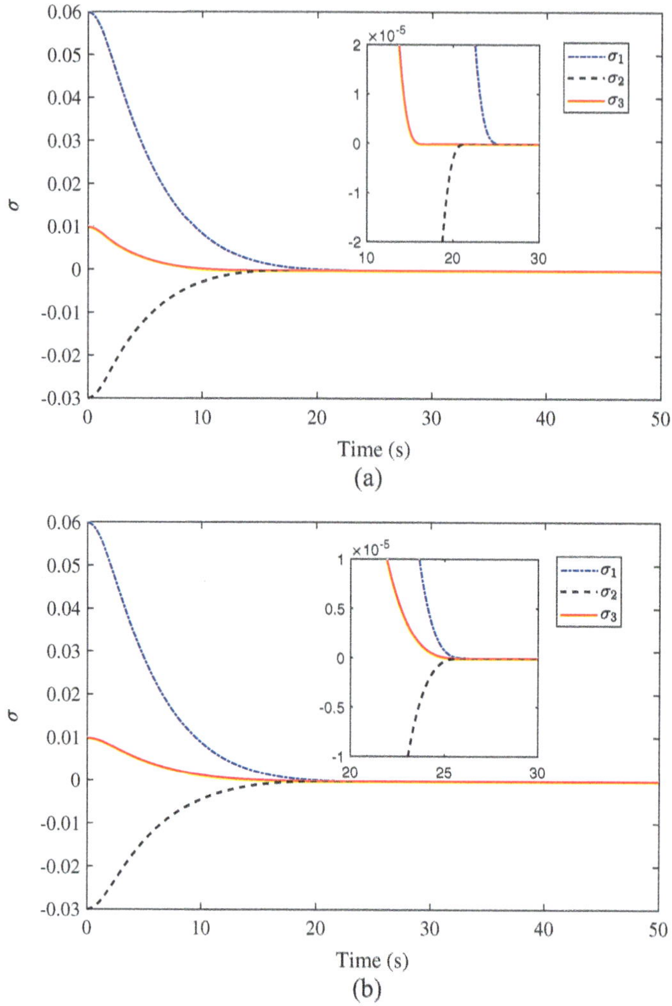

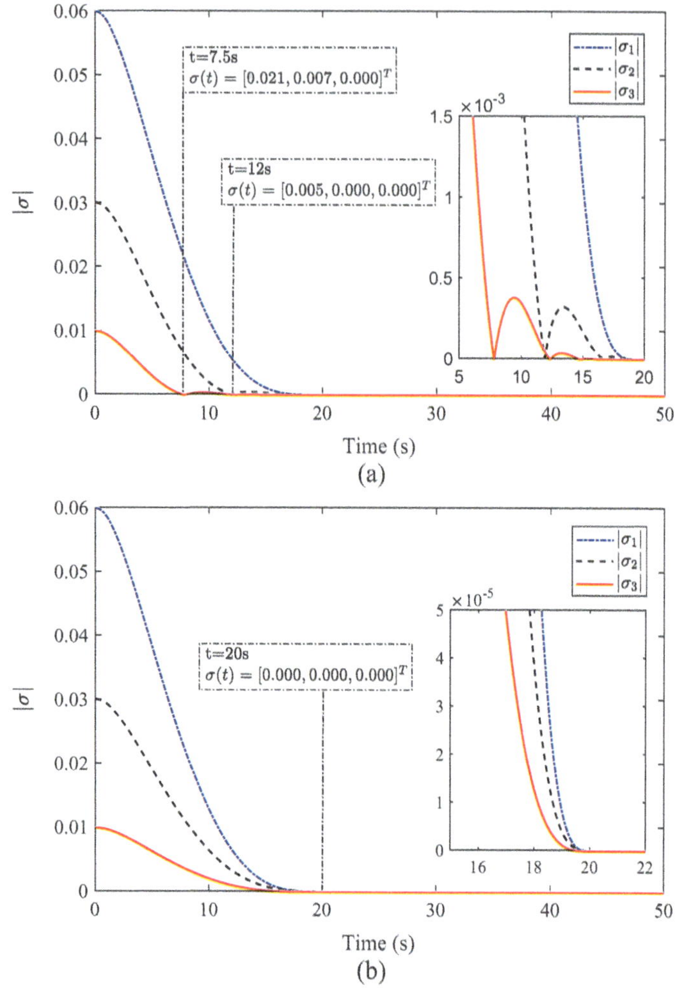

$$s=x_2+\alpha_1\text{sig}_{\text{n}}^{p_1}(x_1)+\beta_1\text{sig}_{\text{n}}^{g_1}(x_1), \tag{6.44}$$

$$u = -g_0^{-1}\left(\alpha_2 \mathrm{sig}_{\mathrm{n}}^{p_2}(s) + \beta_2 \mathrm{sig}_{\mathrm{n}}^{g_2}(s) + f_0 + J(x_1)\right), \tag{6.69}$$

$$\begin{aligned} J(x_1) = &\alpha_1 \rho_1 x_1 x_1^T \mathrm{sig}_{\mathrm{n}}^{p_1}(x_1) + \beta_1 \rho_1 x_1 x_1^T \mathrm{sig}_{\mathrm{n}}^{g_1}(x_1) \\ &+\alpha_1 \rho_2 \mathrm{sig}_{\mathrm{n}}^{p_1}(x_1) + \beta_1 \rho_2 \mathrm{sig}_{\mathrm{n}}^{g_1}(x_1). \end{aligned} \tag{6.70}$$

Substituting ρ_1 and ρ_2 (introduced in (6.54)–(6.55)) into (6.70) and noticing $x_1 x_1^T \mathrm{sig}_{\mathrm{n}}^{p_1}(x_1) = \mathrm{sig}_{\mathrm{n}}^{p_1+2}(x_1)$, we have

$$\begin{aligned} J(x_1) = &\alpha_1^2 p_1 \mathrm{sig}_{\mathrm{n}}^{2p_1-1}(x_1) + \beta_1^2 g_1 \mathrm{sig}_{\mathrm{n}}^{2g_1-1}(x_1) \\ &+\alpha_1 \beta_1 (p_1 + g_1) \mathrm{sig}_{\mathrm{n}}^{p_1+g_1-1}(x_1). \end{aligned} \tag{6.71}$$

Remark 6.6 For the switching law (6.51), the general sliding manifold $s_s = l_1 x_1 + l_2 \mathrm{sig}_{\mathrm{n}}^4(x_1)$ only takes place when $s^* \neq 0,\ \|x_1\| \le \varepsilon$. By selecting a sufficiently small ε, such that the trajectory of x_1 is governed by $s_s = \alpha_1 \mathrm{sig}_{\mathrm{n}}^{p_1}(x_1) + \beta_1 \mathrm{sig}_{\mathrm{n}}^{g_1}(x_1)$ rather than $s_s = l_1 x_1 + l_2 \mathrm{sig}_{\mathrm{n}}^4(x_1)$. Despite the fact that the general one does not arises when ε is small enough, the closed-loop convergence property in this case ($s^* \neq 0,\ \|x_1\| \le \varepsilon$) can still be explored. One may choose an input-to-state stable (ISS) Lyapunov function $V_{ISS} = x_1^T x_1 / 2$ associated with a Lyapunov gain

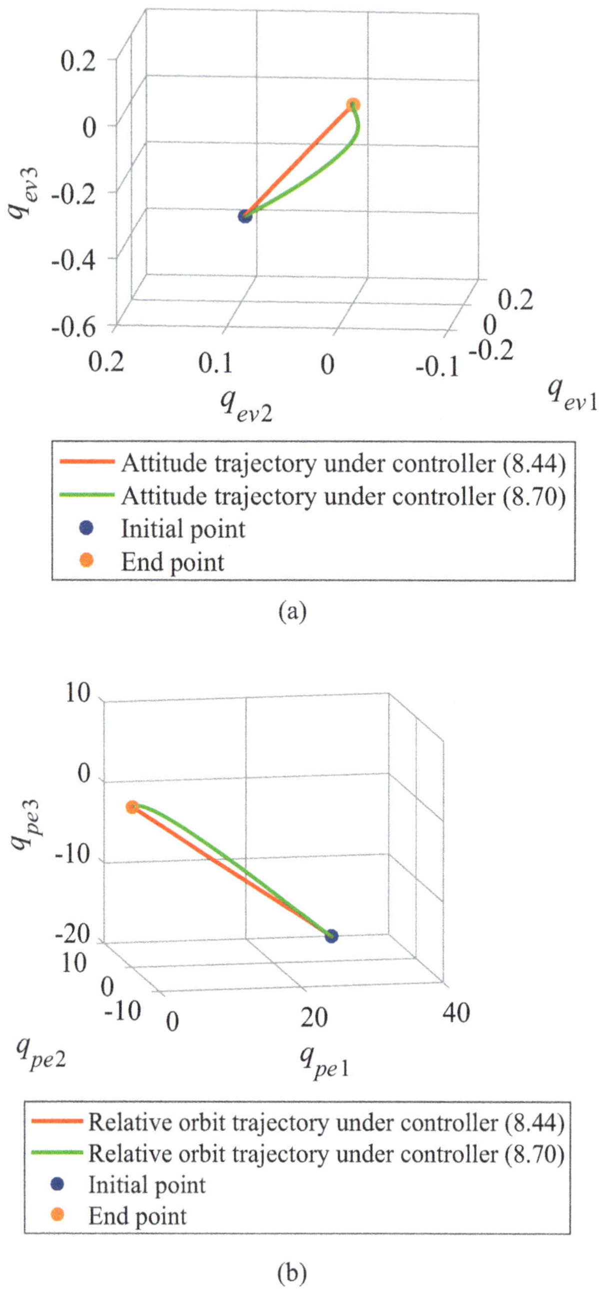

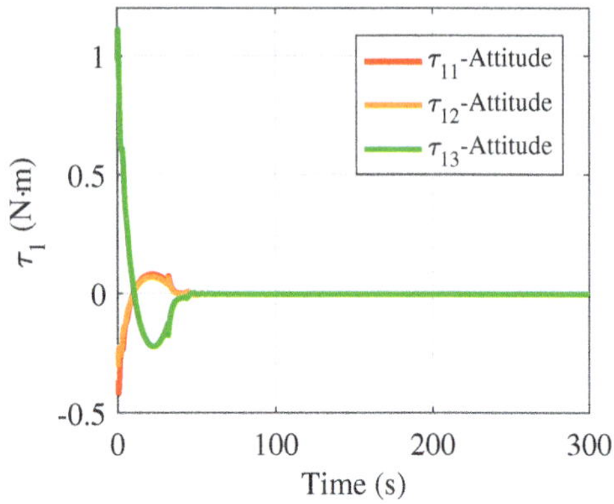

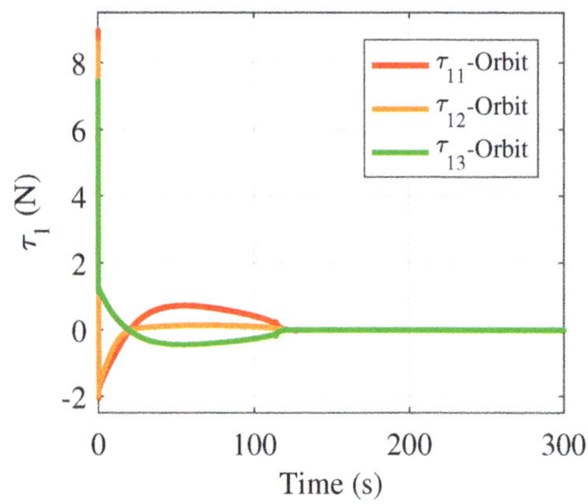

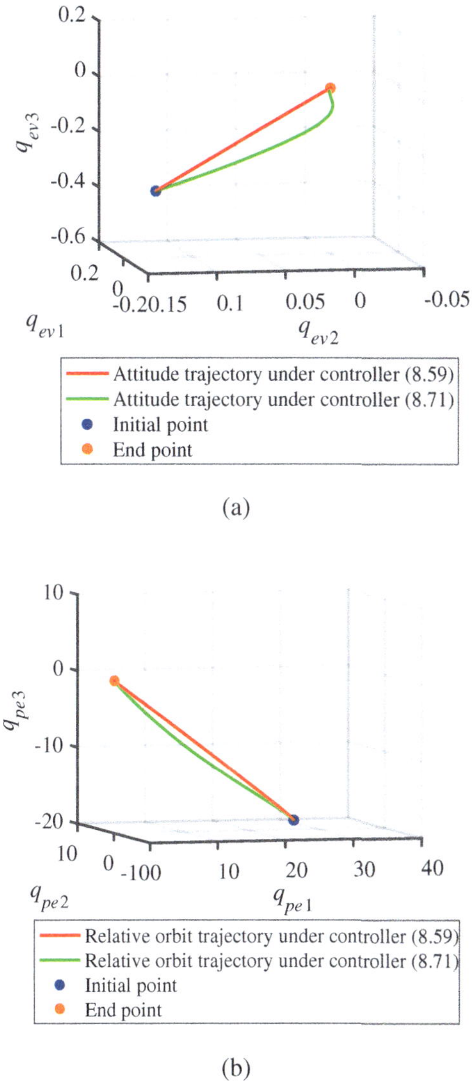

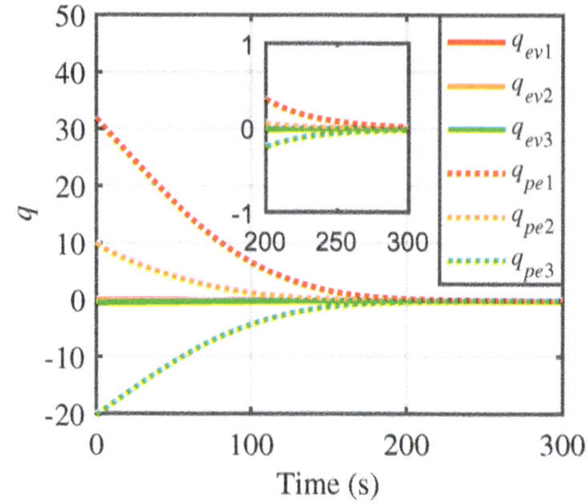

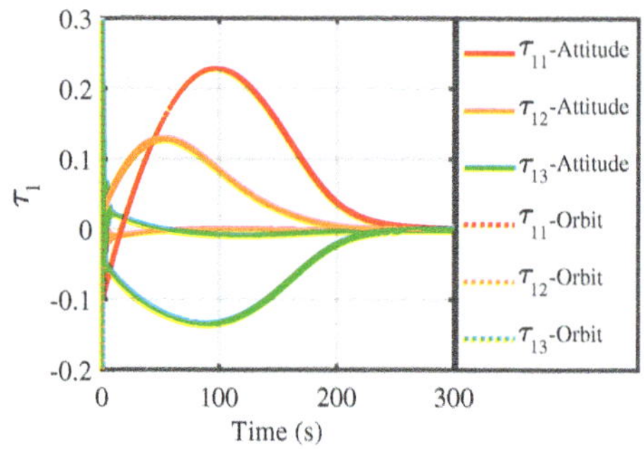

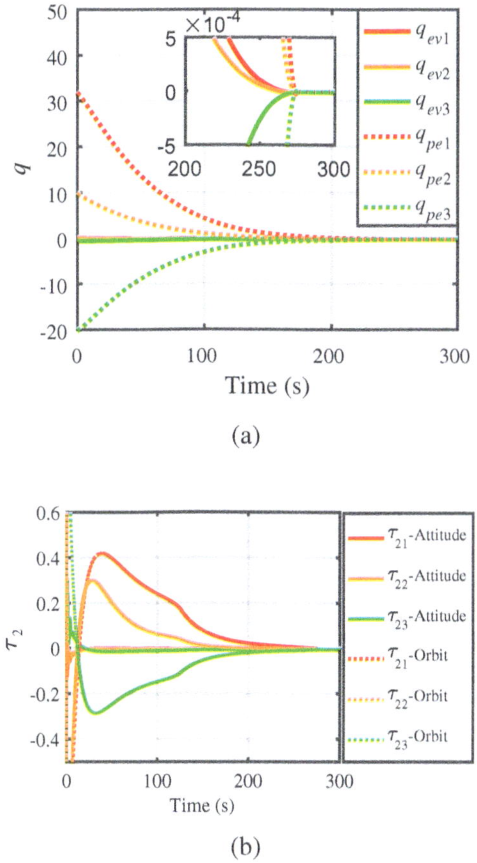

Considering $\|\varepsilon\| \le \varepsilon_{\max}$ and $\|\dot{\varepsilon}\| \le \varepsilon_{d\max}$, for $\forall t \ge T_z(z_0)$, we have

in [17], which was called the *Fuller effect* therein. We here briefly introduce the main essential elements in [17]. Consider thus a second-order system

$$\ddot{x} = u(x, \dot{x}), \tag{1.1}$$

where $x,\ u \in \mathbb{R}$ denote the system state and the control input under the constraints $|u| \leq 1$. Under the cost function $J = \int_0^\infty x^2 \mathrm{d}t$, the optimal scheme is

$$u = -\operatorname{sign}(s), \tag{1.2}$$

$$s = x + c\dot{x}|\dot{x}|, \tag{1.3}$$

where a positive constant c is chosen such that the system output crosses the surface $s = 0$ at a countably infinite number of points but located within a finite-time interval. From analysis, the time intervals will decrease as a geometric progression, contributing to a finite accumulation point called *Zeno behavior*: regarding a system jumps an infinite number of times in a finite amount of time. This further infers finite-time convergence since it has finite settling time (as also summarized in [50]).

To prove finite-time stability, one can find a Lyapunov function $V(x)$ satisfying

$$\dot{V}(x) + cV(x)^{\alpha} \leq 0, \tag{1.4}$$

where constants $c > 0$, $\alpha \in [0, 1)$. Then, we have $V(x) = 0$ for $\forall t \geq T(x_0)$, where x_0 is the initial system state, and the settling time is bounded by [6]

$$T\left(x_0\right) \leq \frac{V\left(x_0\right)^{1-\alpha}}{c(1-\alpha)}. \tag{1.5}$$

Another method to prove finite-time stability is given by first proving asymptotic stability and verifying homogeneity of the closed-loop system [7].

Finite-time control drives the system state to the origin within a bounded settling time subject to the initial state [21]. Fixed-time stability is then introduced, where the solution of a system is globally finite-time stable and the bounded settling time is regardless of any initial conditions [43]. To prove fixed-time stability, it requires a Lyapunov function $V(x)$ to satisfy a stronger condition,

$$\dot{V}(x) \leq -\left(\alpha V^{p}(x) + \beta V^{g}(x)\right)^{\chi}, \tag{1.6}$$

where positive constants α, β, p, g and χ satisfy $p\chi < 1$ and $g\chi > 1$. The corresponding settling time $T\left(x_0\right)$ is bounded as

$$T\left(x_0\right) \leq \frac{1}{\alpha^{\chi}(1-p\chi)} + \frac{1}{\beta^{\chi}(g\chi-1)}. \tag{1.7}$$

Different from finite-time control and other existing control techniques, the merit of fixed-time control lies in the capability of forcing the system state to the equilibrium in fixed time. This decisive characteristic allows a system engineer to manipulate the set-point value after a priori calculated settling time with disregard of initial states [2].

Recently, predefined-time control is proposed based on time-varying high-gain feedback, which likewise achieves global finite-time convergence with predefined settling time that explicitly selected as a control parameter [40, 60]. For example, given a first-order system $\dot{x} = u \in \mathbb{R}$, the following controller is capable of realizing predefined-time convergence,

$$u = \begin{cases} \frac{-k(e^x-1)}{e^x(t_f-t)}, & \text{if } 0 \le t < t_f, \\ 0, & \text{otherwise,} \end{cases} \tag{1.8}$$

where k is a positive control gain, and t_f is the predefined settling time. Basically, we have $u \to \infty$ as $t \to t_f$, which ensures convergence at t_f (the detailed proof is already provided in [40]).

1.2 Time-Synchronized Control

Before introducing time-synchronized control, we first briefly recall the history of finite-time control for the single-input single-output (SISO) system and the multi-input multi-output (MIMO) system.

For a SISO system, a simple relay controller is able to drive all the system state to the origin in finite time, i.e., a relative degree one sliding mode [57]. However, the extension of finite-time control from SISO systems to MIMO systems is non-trivial. In the MIMO case, instead of the selection and combination of multiple individual SISO sliding modes [51], elegant controllers are designed to drive all partial state elements to the equilibrium simultaneously by treating the state vector as a whole, the sliding mode control for multivariable uncertain systems [19, 20, 32], the unit vector control for multivariable systems [15] (Sect. 3.6, *The Unit Vector Approach*), the unit controller for MIMO affine systems in [49] (Sect. 3.5, *Unit Control*), the integral sliding mode in [22], the stabilization control of mechanical systems in [45], the quasi-continuous MIMO sliding mode in [33], to name a few.

In contrast with finite-time control, the design of fixed-time control and predefined-time control are more complicated due to the inherent complexity in their stability analysis. For the SISO system (or the MIMO system divided into multiple scalar SISO systems), fixed-time control and predefined-time control are explored and investigated [1, 27, 40, 43, 60]. Similar to the evolution of finite-time control, some recent approaches extend the fixed-time control to MIMO systems via super-twisting-like algorithms [3, 4, 41].

Actually, the performance of practical control systems not only stands on when the system state converges, but also depends largely on when and how each state element converges. Despite the superior convergence property of finite-time control (including fixed-time and predefined-time control), for some applications, standalone finite-time control is not sufficient, when a MIMO system is expected to reach all (or some of) the target values of the state elements *time-synchronously* (at the same time). It happens in practice when one requires a MIMO system (or a system with diversified/networked subsystems) to accomplish multiple *time-synchronized* actions. Considering a robotic hand, to stably grasp an object's surface/contour, its finger joints are to reach the target angles time-synchronously, otherwise the object may escape or slip during the manipulation. By treating a multi-agent system as a whole, its state elements are often required to reach desired locations time-synchronously in cooperative attack, transport, and formation missions [26]. Especially in the field of space offense and defense, it may be even catastrophic when some synchronous criteria are not satisfied during practical applications.

Motivated by the above, the time-synchronized control problem is defined. Naturally, time-synchronized control is usually designed for MIMO systems. It ensures that all the system state elements converge to the equilibrium at the same time, which has unique performance advantages including smoothing the system output trajectory, alleviating the chattering phenomenon, reducing the energy consumption and improving the convergence accuracy. In general, if we use a sliding mode to guarantee theoretically exact compensation of matched disturbances, the chattering phenomenon is impossible to be fully eliminated. However, this chattering could be alleviated when time-synchronized control is applied with the usage of proper sign functions. The concept of time-synchronized control was first put forward in public by the authors of this book together, in a plenary speech at the 2019 International Conference on Automatic Control (CACS 2019, Keelung, Taiwan), called "New Theoretical Developments in Fundamental Signum Functions Applicable in Control Systems", which reveals the time-synchronized control design of an extremely simple first-order system.

In [36], several effective controllers are proposed for different types of systems to achieve simultaneous-arrival-to-origin convergence introduced therein (which is, in fact, time-synchronized convergence). This approach revisits the sign functions well-applied in control design. It shows that when a norm-normalized sign function is properly invoked in the sliding-mode control, simultaneous-arrival-to-origin convergence can be expected. Next, to deeply articulate the time-synchronized convergence property and to formulate the stability associated, time-synchronized stability and its Lyapunov-like theorems are formally defined in [38]. Time-synchronized stability is built on top of finite-time stability. As introduced previously, finite-time stability requires all the system state elements to reach the equilibrium before a certain time instant ignoring how each state element converges. In contrast, time-synchronized stability requires all the system state elements (with non-zero initial values) to reach the equilibrium at exactly the same time. Similar to the various types of finite-time stability, special types of time-synchronized stability include: (1) fixed-synchronized stability where the upper bound of the synchronized settling time is invariant with any initial state; and (2) predefined-time-synchronized stability where the upper bound of the synchronized settling time is selected as a control parameter in advance. In summary, we have the following mapping:

$$\text{Finite}-\text{Time Stability} \Rightarrow \text{Time}-\text{Synchronized Stability}$$
$$\text{Fixed}-\text{Time Stability} \Rightarrow \text{Fixed}-\text{Fime}-\text{Synchronized Stability}$$
$$\text{Predefined}-\text{Time Stability} \Rightarrow \text{Predefined}-\text{Time}-\text{Synchronized Stability}$$
$$\text{Other Types of Finite}-\text{Time Stability} \Rightarrow \text{Other Types of Time}-\text{Synchronized Stability}$$

Time-synchronized and fixed-time-synchronized control schemes are also proposed in [38]. These control methods naturally reduce the energy consumption and provide the shortest output trajectory due to the inherent properties of ratio persistence. Ratio persistence is a critical component of the sufficient conditions of time-synchronized stability. In general, the solution of a system is ratio persistent if the ratio of the closed-loop state elements (non-zero) is time-invariant in forward time. Straightforwardly, since the ratio is constant, the ratio persistent trajectory is the shortest path between any initial state and the equilibrium, which is optimal in terms of its travel length. For example, if the system state is $\in \mathbb{R}^2$ or $\mathbb{R}^3$ Euclidean space, the output trajectory is a straight line. In addition, ratio persistence likewise reduces the energy consumption by aligning the direction of the control signal with the direction of the error vector.

Despite the establishment of time-synchronized stability in [38], the time-synchronized controllers therein are confined to the assumption that the system model is ideally accurate and suffers from no external disturbances. This inevitably degrades the time-synchronized performance in possible practical implementations. In [37], under external disturbances and model uncertainties that can be lumped into disturbances, robust time-synchronized controllers are proposed. Further, based on time-synchronized stability, for a time-synchronized stable system, its analytical solution is derived in [37], providing a quantitative way to explicitly preview and predesign the time-synchronized performance of the closed-loop system.

The discussion in [37, 38] points out that many known control schemes in the existing results (e.g. [15, 19, 22, 33, 45, 48, 49]) already meet or partially meet the Lyapunov conditions of time-synchronized stability. Although many of their numerical results may not directly reflect time-synchronized convergence, they have the potential to achieve such a convergence property after some modifications following the proposed theorems in [37, 38]. It evidently approves the feasibility and generality of the presented conditions of time-synchronized stability, as they widely exist in the state of the art literature although such convergence property and stability results are not studied there.

Despite the time-synchronized convergence attained in [37, 38], they study only one single MIMO system (under some practical considerations). In real-world applications, multi-agent missions depend greatly on how and when the networked system states converge, e.g., for cooperative missile attacks [26], synthetic-aperture radar satellite lineup [11], and spacecraft formation-flying (PROBA-3, EXO-S, SWIFT, etc. [12]), the networked systems are usually required to reach a target or to achieve a spacial configuration synchronously. An extension of time-synchronized control from one single system to the multi-agent system is addressed in [35]. The concept of time-synchronized consensus is formally defined and formulated, where all the elements of all the agent states achieve consensus at exactly the same time. It can be extended to time-synchronized formation control problems by forming a consensus-based spatial configuration synchronously.

1.3 Why Time-Synchronized Control

In this section, we use an extremely simple but representative time-synchronized controller to explain the rationale rather than dig deeply into the mathematical analysis.

For a vector $x = [x_1,\ x_2, \ldots,\ x_n]^T \in \mathbb{R}^n$, we define the classical sign function $\mathrm{sign_c}$ and the norm-normalized sign function $\mathrm{sign_n}$

$$\mathrm{sign_c}\,(x) \triangleq \left[\mathrm{sgn}\,(x_1),\ \mathrm{sgn}\,(x_2), \ldots,\ \mathrm{sgn}\,(x_n)\right]^T, \tag{1.9}$$

$$\mathrm{sign_n}\,(x) \triangleq \begin{cases} \frac{x}{\|x\|}, & x \neq 0, \\ 0, & x = 0, \end{cases} \tag{1.10}$$

where for $i = 1, 2, \ldots, n$,

$$\mathrm{sgn}\,(x_i) \triangleq \begin{cases} +1, & x_i > 0, \\ 0, & x_i = 0, \\ -1, & x_i < 0, \end{cases} \tag{1.11}$$

For a single-integrator system $\dot{x} = u$ in $\mathbb{R}^n$, where $x = [x_1,\ x_2, \ldots,\ x_n]^T \in \mathbb{R}^n$, we propose a time-synchronized controller $u_n = -\,\mathrm{sign_n}(x)$. The analytical solution of the closed-loop system is

$$x(t) = (\|x_0\| - t)\,\frac{x_0}{\|x_0\|},\quad t \leq \|x_0\|, \tag{1.12}$$

and $x(t) = 0, t > \|x_0\|$.

In summary, the advantages of time-synchronized control are expected from the following view.

i. *Guarantee Time-Synchronized Convergence*: Naturally and straightforwardly, from (1.12), all the state elements x_i converge to the origin time-synchronously at $t = \|x_0\|$, which guarantees time-synchronized stability.
ii. *Smooth and shorten the output trajectory*: The state trajectory in (1.12) is a straight line (at least in $\mathbb{R}^2$ and $\mathbb{R}^3$ Euclidean space), optimal concerning the travel length. Note that for other time-synchronized controllers, it is not necessary for them to also generate a straight-line trajectory, especially when uncertain and disturbed systems are considered. But in general, compared with other existing control laws, time-synchronized controllers smooth and shorten the system output to a certain degree with a similar rationality as show in (1.12).

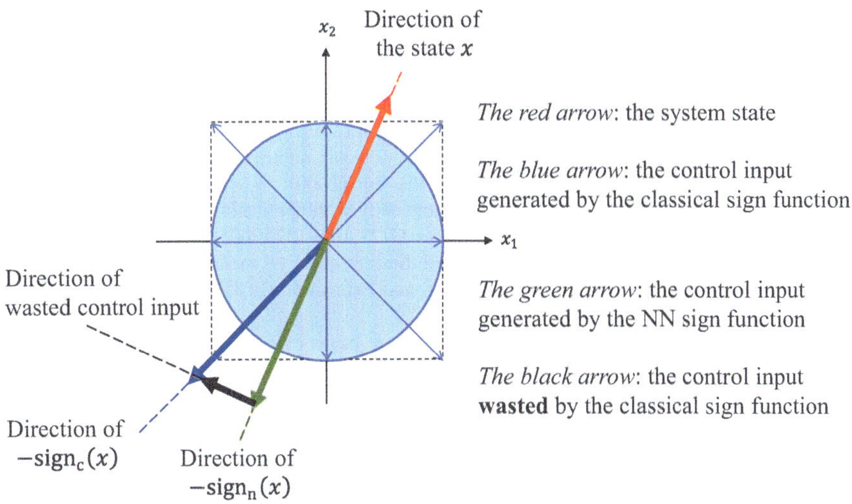

Fig. 1.1 How energy is saved by the norm-normalized (NN) sign function

iii. *Reduce the Energy Consumption*: For a time-synchronized controller designed by the norm-normalized sign function, its energy consumption is in general less than other controllers designed by the classical sign function. The rational could be found in Fig. 1.1. The control object is to drive the system state $x \in \mathbb{R}^2$ in Fig. 1.1 to the origin. A classical sign function-based controller (e.g., $u_c = -\mathrm{sign}_c(x)$) will produce an input in the direction of the blue arrow; while a norm-normalized sign function-based controller e.g., $u_n = -\mathrm{sign}_n(x)$) will produce an input in the direction of the green arrow. Clearly, the classical sign function-based controller wastes a part of the energy by additionally producing the control input in the direction of the black arrow (perpendicular to the negative direction of the system state x), while the time-synchronized controller based on the norm-normalized sign function does not.

References

1. Aldana-López, R., Gómez-Gutiérrez, D., Jiménez-Rodríguez, E., Sánchez-Torres, J.D., Defoort, M.: Enhancing the settling time estimation of a class of fixed-time stable systems. International Journal of Robust and Nonlinear Control **29**(12), 4135–4148 (2019)
2. Basin, M.: Finite- and fixed-time convergent algorithms: Design and convergence time estimation. Annual Reviews in Control **48**, 209–221 (2019). https://doi.org/10.1016/j.arcontrol.2019.05.007
3. Basin, M., Panathula, C.B., Shtessel, Y.: Multivariable continuous fixed-time second-order sliding mode control: Design and convergence time estimation. IET Control Theory and Applications **11**(8), 1104–1111 (2017)

4. Basin, M., Rodriguez-Ramirez, P., Garza-Alonso, A.: Continuous Fixed-Time Convergent Super-Twisting Algorithm in Case of Unknown State and Disturbance Initial Conditions. Asian Journal of Control **21**(1), 323–338 (2019)
5. Bejarano, F.J., Fridman, L.: High order sliding mode observer for linear systems with unbounded unknown inputs. International Journal of Control **83**(9), 1920–1929 (2010)
6. Bhat, S.P., Bernstein, D.S.: Finite-time stability of continuous autonomous systems. SIAM Journal on Control and Optimization **38**(3), 751–766 (2000)
7. Bhat, S.P., Bernstein, D.S.: Geometric homogeneity with applications to finite-time stability. Mathematics of Control, Signals and Systems **17**(2), 101–127 (2005)
8. Bidram, A., Davoudi, A., Lewis, F.L.: Finite-time frequency synchronization in microgrids. In: 2014 IEEE Energy Conversion Congress and Exposition (ECCE), pp. 2648–2654. IEEE (2014)
9. Boyd, S.P., Barratt, C.H.: Linear controller design: limits of performance. NJ: Prentice Hall, Englewood Cliffs (1991)
10. Chen, G., Lewis, F.L., Xie, L.: Finite-time distributed consensus via binary control protocols. Automatica **47**(9), 1962–1968 (2011)
11. Colombino, M., Smith, R.S., Summers, T.H.: Mutually quadratically invariant information structures in two-team stochastic dynamic games. IEEE Transactions on Automatic Control **63**(7), 2256–2263 (2018). https://doi.org/10.1109/TAC.2017.2772020
12. Di Mauro, G., Lawn, M., Bevilacqua, R.: Survey on guidance navigation and control requirements for spacecraft formation-flying missions. Journal of Guidance, Control, and Dynamics **41**(3), 1–22 (2017). https://doi.org/10.2514/1.G002868
13. Dorato, P.: Short-time stability in linear time-varying systems. Polytechnic Institute of Brooklyn (1961)
14. Du, H., Li, S., Qian, C.: Finite-time attitude tracking control of spacecraft with application to attitude synchronization. IEEE Transactions on Automatic Control **56**(11), 2711–2717 (2011)
15. Edwards, C., Spurgeon, S.: Sliding mode control: Theory and applications. Crc Press (1998)
16. Engel, R., Kreisselmeier, G.: A continuous-time observer which converges in finite time. IEEE Transactions on Automatic Control **47**(7), 1202–1204 (2002)
17. Fuller, A.T.: Relay control systems optimized for various performance criteria. IFAC Proceedings Volumes **1**(1), 520–529 (1960)
18. Galicki, M.: Finite-time control of robotic manipulators. Automatica **51**, 49–54 (2015)
19. Gutman, S.: Uncertain dynamical systems–A Lyapunov min-max approach. IEEE Transactions on Automatic Control **24**(3), 437–443 (1979)
20. Gutman, S., Leitmann, G.: Stabilizing feedback control for dynamic systems with bounded uncertainties. In: Proceedings of the IEEE Conference on Decision and Control, pp. 94–99 (1976)
21. Haimo, V.T.: Finite time controllers. SIAM Journal on Control and Optimization **24**(4), 760–770 (1986)
22. Hamayun, M.T., Edwards, C., Alwi, H.: Design and analysis of an integral sliding mode fault-tolerant control scheme. IEEE Transactions on Automatic Control **57**(7), 1783–1789 (2012)
23. Hong, Y., Wang, H.O., Bushnell, L.G.: Adaptive finite-time control of nonlinear systems. In: Proceedings of the 2001 American Control Conference.(Cat. No. 01CH37148), vol. 6, pp. 4149–4154. IEEE (2001)
24. Hong, Y., Xu, Y., Huang, J.: Finite-time control for robot manipulators. Systems & control letters **46**(4), 243–253 (2002)
25. Hong, Y., Yang, G., Bushnell, L., Wang, H.O.: Global finite-time stabilization: from state feedback to output feedback. In: Proceedings of the 39th IEEE Conference on Decision and Control (Cat. No. 00CH37187), vol. 3, pp. 2908–2913. IEEE (2000)
26. Jeon, I.S., Lee, J.I., Tahk, M.J.: Homing guidance law for cooperative attack of multiple missiles. Journal of guidance, control, and dynamics **33**(1), 275–280 (2010)
27. Jiang, B., Chen, Z., Hu, Q., Qi, J.: Fixed-time attitude control for rigid spacecraft with actuator saturation and faults. IEEE Transactions on Control Systems Technology **24**(5), 4375–4380 (2014)

28. Kamenkov, G.: On stability of motion over a finite interval of time. Journal of Applied Mathematics and Mechanics **17**(2), 529–540 (1953)
29. Khoo, S., Xie, L., Man, Z.: Robust finite-time consensus tracking algorithm for multirobot systems. IEEE/ASME Transactions on Mechatronics **14**(2), 219–228 (2009). https://doi.org/10.1109/TMECH.2009.2014057
30. Lebedev, A.: On stability of motion during a given interval of time. Journal of Applied Mathematics and Mechanics **18**, 139–148 (1954)
31. Lebedev, A.: The problem of stability in a finite interval of time. Journal of Applied Mathematics and Mechanics **18**(1), 75–94 (1954)
32. Leitmann, G.: Guaranteed asymptotic stability for some linear-systems with bounded uncertainties. Journal of Dynamic Systems Measurement and Control-Transactions of the ASME **101**(3), 212–216 (1979)
33. Levant, A., Shustin, B.: Quasi-Continuous MIMO Sliding-Mode Control. IEEE Transactions on Automatic Control **63**(9), 3068–3074 (2018). https://doi.org/10.1109/TAC.2017.2778251
34. Li, D., Ge, S.S., He, W., Ma, G., Xie, L.: Multilayer formation control of multi-agent systems. Automatica **109**, 108558 (2019). https://doi.org/10.1016/j.automatica.2019.108558
35. Li, D., Ge, S.S., Lee, T.H.: Fixed-Time-Synchronized Consensus Control of Multi-Agent Systems. IEEE Transactions on Control of Network Systems **8**(1), 89–98 (2021). https://doi.org/10.1109/TCNS.2020.3034523
36. Li, D., Ge, S.S., Lee, T.H.: Simultaneous-Arrival-to-Origin Convergence: Sliding-Mode Control through the Norm-Normalized Sign Function. IEEE Transactions on Automatic Control, in press. https://doi.org/10.1109/TAC.2021.3069816 (2021)
37. Li, D., Tee, K.P., Xie, L., Yu, H.: Time-Synchronized Control for Disturbed Systems. IEEE Transactions on Cybernetics, early access. https://doi.org/10.1109/TCYB.2021.3054589 (2021)
38. Li, D., Yu, H., Tee, K.P., Wu, Y., Ge, S.S., Lee, T.H.: On Time-Synchronized Stability and Control. IEEE Transactions on Systems, Man, and Cybernetics: Systems, in press. https://doi.org/10.1109/TSMC.2021.3050183 (2021)
39. Liu, X., Lam, J., Yu, W., Chen, G.: Finite-time consensus of multiagent systems with a switching protocol. IEEE transactions on neural networks and learning systems **27**(4), 853–862 (2015)
40. Pal, A.K., Kamal, S., Nagar, S.K., Bandyopadhyay, B., Fridman, L.: Design of controllers with arbitrary convergence time. Automatica **112**, 108710 (2020)
41. Pal, A.K., Kamal, S., Yu, X., Nagar, S.K., Xiong, X.: Free-Will Arbitrary Time Consensus for Multiagent Systems. IEEE Transactions on Cybernetics, early access, https://doi.org/10.1109/TCYB.2020.3032217 (2020)
42. Perruquetti, W., Floquet, T., Moulay, E.: Finite-time observers: application to secure communication. IEEE Transactions on Automatic Control **53**(1), 356–360 (2008)
43. Polyakov, A.: Nonlinear feedback design for fixed-time stabilization of linear control systems. IEEE Transactions on Automatic Control **57**(8), 2106–2110 (2012). https://doi.org/10.1109/TAC.2011.2179869
44. Ríos, H., Efimov, D., Moreno, J.A., Perruquetti, W., Rueda-Escobedo, J.G.: Time-Varying Parameter Identification Algorithms: Finite and Fixed-Time Convergence. IEEE Transactions on Automatic Control **62**(7), 3671–3678 (2017)
45. Sanyal, A.K., Bohn, J.: Finite-time stabilisation of simple mechanical systems using continuous feedback. International Journal of Control **88**(4), 783–791 (2015). https://doi.org/10.1080/00207179.2014.974675
46. Song, Y., Wang, Y., Holloway, J., Krstic, M.: Time-varying feedback for regulation of normal-form nonlinear systems in prescribed finite time. Automatica **83**, 243–251 (2017)
47. Song, Y.D., Lu, Y., Gan, Z.X.: Descriptor sliding mode approach for fault/noise reconstruction and fault-tolerant control of nonlinear uncertain systems. Information Sciences **367**, 194–208 (2016)
48. Sun, C., Hu, G., Xie, L., Egerstedt, M.: Robust finite-time connectivity preserving coordination of second-order multi-agent systems. Automatica **89**, 21–27 (2018)
49. Utkin, V., Guldner, J., Shi, J.: Sliding Mode Control in Electro-Mechanical Systems. CRC Press (2009)

50. Utkin, V., Poznyak, A., Orlov, Y.V., Polyakov, A.: Road map for sliding mode control design. Springer (2020)
51. Utkin, V.I.: Sliding Modes in Control and Optimization. Springer Berlin Heidelberg (1992)
52. Wang, H., Yu, W., Wen, G., Chen, G.: Finite-time bipartite consensus for multi-agent systems on directed signed networks. IEEE Transactions on Circuits and Systems I: Regular Papers **65**(12), 4336–4348 (2018)
53. Weiss, L., Infante, E.: Finite time stability under perturbing forces and on product spaces. IEEE Transactions on Automatic Control **12**(1), 54–59 (1967)
54. Xiao, B., Hu, Q., Zhang, Y.: Adaptive sliding mode fault tolerant attitude tracking control for flexible spacecraft under actuator saturation. IEEE Transactions on Control Systems Technology **20**(6), 1605–1612 (2011)
55. Xiao, F., Wang, L., Chen, J., Gao, Y.: Finite-time formation control for multi-agent systems. Automatica **45**(11), 2605–2611 (2009)
56. Yang, X., Lam, J., Ho, D.W.C., Feng, Z.: Fixed-Time Synchronization of Complex Networks With Impulsive Effects via Nonchattering Control. IEEE Transactions on Automatic Control **62**(11), 5511–5521 (2017)
57. Young, K.D., Utkin, V.I., Özgüner, Ü.: A control engineer's guide to sliding mode control. IEEE Transactions on Control Systems Technology **7**(3), 328–342 (1999). 10.1109/87.761053
58. Yu, S., Yu, X., Shirinzadeh, B., Man, Z.: Continuous finite-time control for robotic manipulators with terminal sliding mode. Automatica **41**(11), 1957–1964 (2005)
59. Zamora-Gomez, G.I., Zavala-Rio, A., Lopez-Araujo, D.J., Cruz-Zavala, E., Nuno, E.: Continuous Control for Fully Damped Mechanical Systems With Input Constraints: Finite-Time and Exponential Tracking. IEEE Transactions on Automatic Control **65**(2), 882–889 (2020)
60. Zhou, B.: Finite-time stability analysis and stabilization by bounded linear time-varying feedback. Automatica **121**, 109191 (2020)
61. Zou, A.M., Kumar, K.D., Hou, Z.G., Liu, X.: Finite-time attitude tracking control for spacecraft using terminal sliding mode and Chebyshev neural network. IEEE Transactions on Systems, Man, and Cybernetics, Part B (Cybernetics) **41**(4), 950–963 (2011)
62. Zuo, Z., Tian, B., Defoort, M., Ding, Z.: Fixed-Time Consensus Tracking for Multiagent Systems With High-Order Integrator Dynamics. IEEE Transactions on Automatic Control **63**(2), 563–570 (2018)

Chapter 2
Time-Synchronized Stability

Abstract In this chapter, time-synchronized stability is formally introduced as a unique type of finite-time stability. Its special branches include fixed-time-synchronized stability and predefined-time-synchronized stability, where stronger requirements of the synchronized settling time are imposed. Relevant Lyapunov-like theorems and respective sufficient conditions of these stability formulations are likewise suitably proposed. Among these sufficient conditions, the concept of ratio persistence is one of the key components. A system state is called to be ratio persistent when the ratio of each pair of the state elements is a constant except at the equilibrium. A similar but milder condition is called ratio restriction, requiring only time-varying but bounded ratios, rather than constant ratios. Compared with the classical sign function, we introduced a so-called norm-normalized sign function to facilitate ratio persistence (also ratio restriction) of a system state. Given a state vector x with n dimension, the norm-normalized sign function incorporates the normalization computation of the state x with its norm, while the classical sign function provides only n individual decoupled scalars. Concerning these embedded attributes, it is evidently possible to expect stringent time-synchronized performance from a system design involving the norm-normalized sign function rather than the classical one.

2.1 Properties of Sign Functions

In the literature, the classical sign function is often defined as

$$\operatorname{sign}_{\mathrm{c}}(x_i) \triangleq \begin{cases} +1, & x_i > 0, \\ 0, & x_i = 0, \\ -1, & x_i < 0, \end{cases} \tag{2.1}$$

where its vector form and exponential form are defined as

$$\operatorname{sign}_{\mathrm{c}}(x) = \left[\operatorname{sign}_{\mathrm{c}}(x_1), \ldots, \operatorname{sign}_{\mathrm{c}}(x_n)\right]^T, \tag{2.2}$$

$$\operatorname{sig}_{\mathrm{c}}^{\alpha}(x) = \left[\operatorname{sign}_{\mathrm{c}}(x_1)\,|x_1|^{\alpha}, \ldots, \operatorname{sign}_{\mathrm{c}}(x_n)\,|x_n|^{\alpha}\right]^T. \tag{2.3}$$

D. Li et al., *Time-Synchronized Control: Analysis and Design*,
https://doi.org/10.1007/978-981-16-3089-7_2

Remark 2.1 The classical sign function is not necessarily equal to zero when $x = 0$. Especially, when differential inclusions are involved and considered under the Filippov sense [4], the classical sign function can be treated as a set-valued function,

$$\mathrm{sign_c}\left(x_i\right) = \begin{cases} 1, & \text{if } x_i > 0, \\ [-1, 1], & \text{if } x_i = 0, \\ -1, & \text{if } x_i < 0. \end{cases} \tag{2.4}$$

The differential inclusion and its relevant discussions will be detailed in Chap. 3.

In this book, a modified sign function called norm-normalized sign function is defined in the following, which will play a central role in the design of (fixed/ predefined-) time-synchronized control.

Definition 2.1 *(Norm-Normalized Sign Function)* For a vector $x = [x_1, x_2, \ldots, x_n]^T \in \mathbb{R}^n$, a *norm-normalized sign function* and its exponential form are defined as

$$\mathrm{sign_n}\left(x\right) \triangleq \begin{cases} \frac{x}{\|x\|}, & x \neq 0, \\ 0, & x = 0, \end{cases} \tag{2.5}$$

$$\mathrm{sig_n^{\alpha}}\left(x\right) \triangleq \|x\|^{\alpha}\mathrm{sign_n}\left(x\right). \tag{2.6}$$

Remark 2.2 In Definition 2.1, the norm-normalized sign function is introduced. It incorporates an input vector and its norm by normalization computation, which also serves as a unit direction vector. We call it the norm-normalized sign function because $\mathrm{sign_n}$ holds many properties similar to those of a classical sign function, e.g., they both show the general direction of an input vector. In the literature, similar functions are also called as the unit vector [3] (Sect. 3.6, *The Unit Vector Approach*), the unit controller [14] (Sect. 3.5, *Unit Control*), the scaled unit vector [6], the vector-valued function [11], to name a few.

The sign functions (2.2) and (2.5) are discontinuous, which are hard to be implemented in practice and impede the control performance due to the chattering phenomenon. The exponential forms are more widely utilized in a controller, since they offer continuity and more freedom to control design due to an adjustable power α. For convenience, we refer $\mathrm{sig_c}$ and $\mathrm{sig_n}$ also as the classical sign function and the norm-normalized sign function in this book.

The function $\mathrm{sign_c}\left(x\right)$ defines 3^n vectors evenly distributed in the n-dimensional space (including $0 \in \mathbb{R}^n$ for $x = 0$), while $\mathrm{sign_n}\left(x\right)$ defines an n-dimensional unit ball. A 2-dimensional example and a 3-dimensional example have been given presented in Fig. 2.1a, b, where the blue arrows denote $\mathrm{sign_c}\left(x\right)$ and the red ones denote $\mathrm{sign_n}\left(x\right)$. Next, we give the relation between $\mathrm{sign_c}\left(x\right)$ and $\mathrm{sign_n}\left(x\right)$:

i. $\mathrm{sign_c}\left(x\right) = \mathrm{sign_n}\left(x\right)$ if and only if any one of following conditions hold: 1) x is a scalar. 2) x contains at least two elements, where one element is non-zero and the rest are all zero. 3) All elements of x are zero. For the other cases, $\mathrm{sign_c}\left(x\right) \neq \mathrm{sign_n}\left(x\right)$;

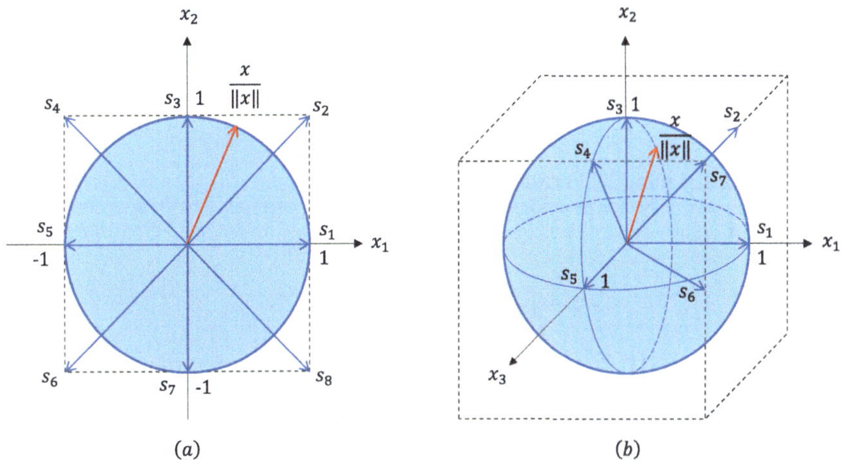

(a) 2-D example (b) 3-D example

Fig. 2.1 The functions $\mathrm{sign_c}(x)$ and $\mathrm{sign_n}(x)$ in 2- and 3-dimensions. (For $\mathrm{sign_c}(x)$ in the 3-D case, only 7 vectors $(s_1, s_2, \ldots, s_7)$ are shown for simplicity.)

ii. $\|\mathrm{sign_c}(x)\| \le \sqrt{n}$, while $\|\mathrm{sign_n}(x)\| = 1$ if $x \neq 0$;
iii. $x^T\mathrm{sign_c}(x) = \sum_{i=1}^{n} |x_i| = \|x\|_1$, while $x^T\mathrm{sign_n}(x) = \|x\|^2/\|x\| = \|x\|$;
iv. consider a positive (negative) definite matrix $A \in \mathbb{R}^{n\times n}$. $x^T A\ \mathrm{sign_n}(x)$, $x \neq 0$ is also positive (negative) definite as $x^T A\ \mathrm{sign_n}(x) = x^T A\, x/\|x\|$, while in general, no conclusion could be drawn for $x^T A\ \mathrm{sign_c}(x)$.

We next provide some useful properties of $\mathrm{sig_n^\alpha}(x)$.

Lemma 2.1 *For any real vector $x \in \mathbb{R}^n$ and constant $\alpha > 0$,*

$$\frac{\partial}{\partial x}\|x\|^{\alpha+1} = (\alpha+1)\,\|x\|^{\alpha-1}x, \tag{2.7}$$

$$\frac{\mathrm{d}}{\mathrm{d}t}\|x\|^{\alpha+1} = (\alpha+1)\,\|x\|^{\alpha-1}x^T\dot{x}, \tag{2.8}$$

$$\frac{\partial}{\partial x}\mathrm{sig_n^{\alpha+1}}(x) = \alpha\|x\|^{\alpha-2}xx^T + \|x\|^{\alpha}I_n. \tag{2.9}$$

Proof Using the fact that $\frac{\partial}{\partial x}\|x\| = \frac{x}{\|x\|}$, the proof is straightforward. □

Lemma 2.2 *For any real vector $x \in \mathbb{R}^n$, and constants $\alpha > 0$ and $\beta > 0$,*

$$\|x\|^{\alpha-\beta}xx^T\mathrm{sig_n^\alpha}(x) = \mathrm{sig_n^{2\alpha-\beta+2}}(x). \tag{2.10}$$

Proof From (2.6),

$$\|x\|^{\alpha-\beta}xx^T\mathrm{sig_n^\alpha}(x) = \|x\|^{2\alpha-\beta}xx^T\mathrm{sign_n}(x). \tag{2.11}$$

By element-wisely analyzing $xx^T \mathrm{sign_n}(x)$ in the case of $x = 0$ and the case of $x \neq 0$, we have

$$xx^T \mathrm{sign_n}(x) = \mathrm{sig}_\mathrm{n}^2(x). \tag{2.12}$$

Taking (2.12) into (2.11) results in (2.10). □

Lemma 2.3 *For*

$$z = \left[z_1^T, z_2^T, \ldots, z_N^T\right]^T \in \mathbb{R}^{Nn}, \tag{2.13}$$

where $z_i \in \mathbb{R}^n$, $i \in \{1, 2, ..., N\}$, $\alpha \in [0, 1]$, *and* N *and* n *are positive integers, we have the following inequality holds element-wisely*

$$\left[\mathrm{sig}_\mathrm{n}^\alpha(z_1)^T,\ \mathrm{sig}_\mathrm{n}^\alpha(z_2)^T, \ldots,\ \mathrm{sig}_\mathrm{n}^\alpha(z_N)^T\right]^T \geq \mathrm{sig}_\mathrm{n}^\alpha(z). \tag{2.14}$$

Proof In the case of $z_i \neq 0$, we can get

$$\mathrm{sig}_\mathrm{n}^\alpha(z_i)^T = \frac{z_i^T}{\|z_i\|^{1-\alpha}} \geq \frac{z_i^T}{\sqrt{z_1^T z_1 + z_2^T z_2 + \ldots + z_N^T z_N}^{1-\alpha}}. \tag{2.15}$$

In the case of $z_i = 0$, it reads $\mathrm{sig}_\mathrm{n}^\alpha(z_i)^T = 0$. In this case, $\mathrm{sig}_\mathrm{n}^\alpha(z_i)^T$ is equal to the corresponding element in $\mathrm{sig}_\mathrm{n}^\alpha(z)$.

Based on the above, we have

$$\left[\mathrm{sig}_\mathrm{n}^\alpha(z_1)^T,\ \mathrm{sig}_\mathrm{n}^\alpha(z_2)^T, \ldots,\ \mathrm{sig}_\mathrm{n}^\alpha(z_N)^T\right]^T \geq \mathrm{sig}_\mathrm{n}^\alpha(z). \tag{2.16}$$

Lemma 2.1 is often used in the Lyapunov stability analysis in terms of a controller built by the norm-normalized sign function, since manipulations in the time derivative of a Lyapunov function is inevitable therein. Lemma 2.2 is often applied to ease the analysis in (2.9). More specifically,

$$\begin{aligned}\frac{\partial}{\partial t}\mathrm{sig}_\mathrm{n}^{\alpha+1}(x) &= \frac{\partial}{\partial x}\mathrm{sig}_\mathrm{n}^{\alpha+1}(x)\frac{\partial x}{\partial t}\\ &= \alpha\|x\|^{\alpha-2}xx^T\dot{x} + \|x\|^\alpha\dot{x}.\end{aligned} \tag{2.17}$$

In the case of $\dot{x}$ including the norm-normalized sign function, Lemma 2.2 is useful to facilitate the investigation. Lemma 2.3 helps to study the cooperative convergence property of a networked system (each z_i can be viewed as a subsystem or an agent).

2.2 Useful Stability Theorems and Formula

We consider the following system dynamics,

$$\dot{x} = f(x), \ f(0) = 0, \ x(0) = x_0, \tag{2.18}$$

where $x = [x_1, \ldots, x_n]^T \in \mathbb{R}^n$, and with respect to an open neighborhood $\mathcal{D}_0 \subseteq \mathbb{R}^n$ of the origin, $f : \mathcal{D}_0 \to \mathbb{R}^n$ is continuous. We assume that the system (2.18) has a unique solution for all initial conditions in forward time.

Results on input-to-state stability are first recalled and introduced.

Definition 2.2 ([12]) The system (2.18) is input-to-state stable (ISS) if there exist a $\mathcal{KL}$-function γ_1 and a $\mathcal{K}$-function γ_2 such that, for each input $u \in L_\infty$ and each $x_0 \in \mathbb{R}^n$, it holds that

$$|x(t, x_0, u)| \leq \gamma_1(|x_0|, t) + \gamma_2(\|u\|), \ \forall t \geq 0. \tag{2.19}$$

Definition 2.3 ([7]) A continuous function $\alpha : [0, a) \to [0, \infty)$ belongs to class $\mathcal{K}$-functions if the following conditions hold:

i. α is strictly increasing;
ii. $\alpha(0) = 0$.

Definition 2.4 ([7]) A class $\mathcal{K}$-function $\alpha : [0, a) \to [0, \infty)$ belongs to class $\mathcal{K}_\infty$-functions if the following conditions hold:

i. $a = \infty$;
ii. $\lim_{r\to\infty} \alpha(r) = \infty$.

Definition 2.5 ([12]) A smooth function $V : \mathbb{R}^n \to \mathbb{R}_+$ is an ISS Lyapunov function if there exist functions $\alpha_1, \alpha_2 \in \mathcal{K}_\infty$ and $\alpha_3, \ell \in \mathcal{K}$, such that for $\forall x \in \mathbb{R}^n, u \in \mathbb{R}^m$, the following conditions hold

$$\alpha_1(|x|) \leqslant V(x) \leqslant \alpha_2(|x|), \tag{2.20}$$

$$|x| \geqslant \ell(|u|) \Rightarrow L_{f(x,u)}V(x) \leqslant -\alpha_3(|x|), \tag{2.21}$$

where ℓ is the Lyapunov gain.

Lemma 2.4 ([12]) *The system (2.18) is ISS iff there exists an ISS Lyapunov function for it.*

Some well-known concepts and results on finite-time stability are next introduced.

Definition 2.6 (*Finite-Time Stability*) The equilibrium of system (2.18) is finite-time stable if

i. *Finite-Time Convergence*: For an open neighborhood $\mathcal{N}_0 \subseteq \mathcal{D}_0$ of the origin, $\forall x_0 \in \mathcal{N}_0 \backslash \{0\}$, $\lim_{t \uparrow T(x_0)} x(t) = 0$, where $T(x_0)$ is the settling time.

ii. *Lyapunov stability*: Given any open neighborhood $\mathcal{U}_\varepsilon$ of 0, there exits an open subset $\mathcal{U}_\delta$ of $\mathcal{N}_0$ containing 0 such that, for $\forall x_0 \in \mathcal{U}_\delta \backslash \{0\}$, $x(t) \in \mathcal{U}_\varepsilon$, for $\forall t \in [0, T(x_0))$.

The equilibrium is globally finite-time stable if it is finite-time stable with $\mathcal{N}_0 = \mathcal{D}_0 = \mathbb{R}^n$.

Definition 2.6 is motivated by [8, 15], which differs slightly from the definition of finite-time stability in [2]. The definition in [2] requires $\forall x_0 \in \mathcal{N}_0 \backslash \{0\}$, $x(t) \in \mathcal{N}_0 \backslash \{0\}$, $\lim_{t \to T(x_0)} x(t) = 0$ while Definition 2.6 only requires $\forall x_0 \in \mathcal{N}_0 \backslash \{0\}$, $\lim_{t \uparrow T(x_0)} x(t) = 0$. This is essential when considering finite-time control for non-autonomous systems as will introduced in Sect. 2.5.

Definition 2.7 (*Fixed-Time Stability*) The equilibrium of the system (2.18) is fixed-time stable if

i. it is globally finite-time stable in the sense of Definition 2.6;
ii. the upper bound of the settling-time $T(x_0)$ is independent of any initial system state, i.e., for $\forall x_0$, we have $T(x_0) < T_m$, where T_m is a positive constant.

Since Definition 2.6, in general, involves less requirements on stabilization, the following sufficient conditions of finite-time stability and fixed-time stability in [2, 9] still hold for Definitions 2.6 and 2.7.

Lemma 2.5 ([2] Sufficient Conditions of Finite-Time Stability) *Considering a Lyapunov function $V(x)$, for any real numbers $c > 0$ and $\alpha \in (0, 1)$, the origin of the system (2.18) is finite-time stable if*

$$\dot{V}(x) + cV^{\alpha}(x) \leq 0, \tag{2.22}$$

where the settling time is estimated by

$$T(x_0) \leq \frac{V^{1-\alpha}(x_0)}{c(1-\alpha)}. \tag{2.23}$$

Lemma 2.6 ([9]) *The equilibrium of the system (2.18) is fixed-time stable, if there exists a Lyapunov function $V(x)$ such that, along the state trajectory of (2.18), we have*

$$\dot{V}(x) \leq -\left(\alpha V^{p}(x) + \beta V^{g}(x)\right)^{\chi}, \tag{2.24}$$

where α, β, p, g and χ are positive constants, with $p\chi < 1$ and $g\chi > 1$. The settling time $T(x_0)$ is bounded as

$$T(x_0) \leq \frac{1}{\alpha^{\chi}(1-p\chi)} + \frac{1}{\beta^{\chi}(g\chi - 1)}. \tag{2.25}$$

In the literature, when encountering disturbances and uncertainties, the system state is usually driven to a small bounded neighborhood around the origin. Here, the concepts of semi-globally uniform ultimate boundedness and practical finite-time stability, and their sufficient Lyapunov conditions are introduced together in the following lemmas. Stability of semi-globally uniform ultimate boundedness is first proposed in [5], and then practical finite-time stability is introduced in [16]. As shown below, they essentially hold the same definitions.

Definition 2.8 (*Semi-Global Uniform Ultimate Boundedness* [5]) The equilibrium of the system (2.18) is semi-globally uniformly ultimately bounded, if for any open neighborhood $\mathcal{N}_0 \subseteq \mathcal{D}_0$ of the origin, a compact subset of $\mathbb{R}^n$ and all $x_0 \in \mathcal{N}_0$, there exist a constant $\mu > 0$ and a settling time $T(\mu, x_0)$, s.t., $\forall t > t_0 + T$, $\|x(t)\| < \mu$.

Lemma 2.7 (Sufficient Conditions of Semi-Global Uniform Ultimate Boundedness [5]) *The system (2.18) is semi-globally uniformly ultimately bounded, if there exists a Lyapunov function $V(x)$ such that,*

$$\dot{V} \leq -\rho V + c, \tag{2.26}$$

where ρ and c are positive constants.

Proof Multiplying (2.26) by $e^{-\rho t}$, we have the following by integrating both sides,

$$\begin{aligned} V(t) \leq & V(x_0)e^{-\rho_i t} + \frac{c}{\rho}\left(1 - e^{-\rho_i t}\right) \\ \leq & V(x_0) + \frac{c}{\rho}. \end{aligned} \tag{2.27}$$

Definition 2.9 (*Practical Finite-Time Stability* [16]) The equilibrium of the system (2.18) is practical finite-time stable, if for any open neighborhood $\mathcal{N}_0 \subseteq \mathcal{D}_0$ of the origin, a compact subset of $\mathbb{R}^n$ and all $x_0 \in \mathcal{N}_0$, there exist a constant $\mu > 0$ and a settling time $T(\mu, x_0)$, s.t., $\forall t > t_0 + T$, $\|x(t)\| < \mu$.

Lemma 2.8 (Sufficient Conditions of Practical Finite-Time Stability [16]) *For the system (2.18), if there exists a continuous function $V(x)$, s.t.,*

$$\dot{V}(x) \leq -\lambda V^{\ell}(x) + \eta, \tag{2.28}$$

where λ, ℓ, and η are scalars satisfying that $\lambda > 0$, $0 < \ell < 1$, and $0 < \eta < \infty$, the equilibrium of the system (2.18) is practical finite-time stable, i.e., the trajectories are bounded in finite time T_m as

$$\lim_{t \to T_m} x \in \left\{ x \in \mathbb{R}^d | V^{\ell}(x) \leq \frac{\eta}{(1-\theta)\lambda} \right\}, \tag{2.29}$$

$$T_m \leq \frac{V^{1-\ell}(x_0)}{\lambda\theta(1-\ell)}, \tag{2.30}$$

where θ is a scalar $\in (0, 1)$.

From Definitions 2.8 and 2.9, it is clear that they refer to the same characterization of system stability. It is interesting that they have different Lyapunov-like conditions in Lemmas 2.7 and 2.8. In the following lemma, we prove that, actually, they are likewise the same formulation.

Lemma 2.9 (Relationship between Semi-Globally Uniform Ultimate Boundedness and Practical Finite-Time Stability) *If the system (2.18) meets the condition in Lemma 2.8, it is also semi-globally uniformly ultimately bounded. If the system (2.18) meets the condition in Lemma 2.7, it is also practical finite-time stable.*

Proof First, if the system (2.18) meets the condition in Lemma 2.8, it is clearly practical finite-time stable according to Lemma 2.8. Accordingly, we have

$$\lim_{t\to T_m} x \in \left\{ x \in \mathbb{R}^d | V^{\ell}(x) \le \frac{\eta}{(1-\theta)\lambda} \right\}, \tag{2.31}$$

$$T_m \le \frac{V^{1-\ell}(x_0)}{\lambda\theta(1-\ell)}. \tag{2.32}$$

The above formula satisfies the definition of semi-global uniform ultimate boundedness in Definition 2.8.

Next, if the system (2.18) meets the condition in Lemma 2.7, we have $\dot{V} \le -\rho V + c$. Further, after some manipulations, it reads

$$\begin{aligned}
\dot{V} &\le -\rho V + c \\
&\le -\rho\left(\frac{1}{2}V + \frac{1}{2}\left(V^{\frac{1}{2}} - V^{\frac{1}{4}}\right)^2 - \frac{1}{2}V^{\frac{1}{2}} + V^{\frac{3}{4}}\right) + c \\
&\le -\rho\left(\frac{1}{2}V - \frac{1}{2}V^{\frac{1}{2}} + V^{\frac{3}{4}}\right) + c \\
&\le -\rho\left(\frac{1}{2}V - \frac{1}{2}\left(\frac{1}{2}V + \frac{1}{2}\right) + V^{\frac{3}{4}}\right) + c \\
&\le -\rho V^{\frac{3}{4}} + c + \frac{\rho}{4}.
\end{aligned} \tag{2.33}$$

The inequality (2.33) satisfies the conditions in Lemma 2.8. Thus, the system (2.18) is also practical finite-time stable. □

Remark 2.3 Stability, in general, refers to a specific equilibrium with regard to the target system. In this book, we mainly focus on the equilibrium at the origin of a given system. Therefore, we may also refer to the stability of a system under certain cases without ambiguity.

2.3 Time-Synchronized Stability

In the sequel, time-synchronized stability based on Lyapunov analysis is proposed. Then, a time-synchronized controller is designed for a class of affine systems under disturbances to achieve time-synchronized stability.

We formally define time-synchronized stability as follows.

Definition 2.10 (*Time-Synchronized Stability*) The equilibrium of the system (2.18) is *time-synchronized stable* if

i. it is finite-time stable in the sense of Definition 2.6;
ii. for an open neighborhood $\mathcal{N}_0 \subseteq \mathcal{D}_0$ of the origin, there exists a function $T : \mathcal{N}_0 \backslash \{0\} \to (0, \infty)$, called the *synchronized settling-time function*, such that for $\forall x_0 \in \mathcal{N}_0 \backslash \{0\}$ and $i \in \{1, 2, \ldots, n\}$, we have $x(t) \in \mathcal{N}_0$, $\forall t \in [0, \infty)$, and for $x_i(0) \neq 0$,

$$x_i(t) \neq 0, \lim_{t \uparrow T(x_0)} x_i(t) = 0, \forall t \in [0, T(x_0)), \tag{2.34}$$

where $T(x_0)$ is the *synchronized settling time*.

The equilibrium of the system (2.18) is *globally time-synchronized stable* if it is time-synchronized stable with $\mathcal{N}_0 = \mathcal{D}_0 = \mathbb{R}^n$.

Remark 2.4 In Definition 2.10, the solution to system (2.18) is only considered for $t \in [0, T(x_0))$ which is not considered for $t \geq T(x_0)$. According to Definition 2.10, time-synchronized stability already requires the system state to converge to the equilibrium in finite time. Therefore, time-synchronized stability can be regarded as a stricter, special form of finite-time stability, as it meets the finite-time requirements in Definition 2.6. Comparing Definitions 2.6 and 2.10, one can observe that the main difference is that finite-time stability allows some of the state elements to arrive at the origin before $T(x_0)$ as long as the remaining state elements converge to zero as $t \to T(x_0)$, while time-synchronized stability requires all state elements x_i ($x_i(0) \neq 0$) to arrive at the origin at the same time $T(x_0)$. Clearly, the requirement of time-synchronized stability is stricter than that of finite-time stability.

We introduce an *indirect* method to achieve time-synchronized stability, which involves proving that system (2.18) is finite-time stable, and during convergence, the ratio of each pair of state elements is always constant. Since all the state elements arrive at the origin at the same time, the above two conditions sufficiently lead to time-synchronized stability. This intuitive insight motivates us to develop the following notion of *ratio persistence*.

Definition 2.11 (*Ratio Persistence*) The closed-loop state x of the system (2.18) is *ratio persistent* if for $x \neq 0$, we have

$$\frac{x}{\|x\|} = \zeta \frac{f(x)}{\|f(x)\|}, \tag{2.35}$$

where ζ is the *direction of ratio persistence*, which can be either 1 or -1.

According to Definition 2.11, the condition of ratio persistence seems to be a little strict, when in fact, many known control methods in the literature (e.g. [6, 11, 13]) can allow a given system to meet this condition. In addition, the classical proportional control (know as the P control) also enables a system state to be ratio persistent, which is the simplest version of a ratio persistent system. Next, we give some exercises.

Example 2.1 Which of the following systems is/are ratio persistent?

1. $\dot{x} = -\mathrm{sign}_{\mathrm{n}}(x)$, where $x \in \mathbb{R}^n$.
2. $\dot{x} = \mathrm{sign}_{\mathrm{n}}(x) + \mathrm{sig}_{\mathrm{n}}^{\alpha}(x)$, where $x \in \mathbb{R}^n$ and α is a positive constant.
3. $\dot{x} = -\mathrm{sign}_{\mathrm{c}}(x) - \mathrm{sig}_{\mathrm{c}}^{\alpha}(x)$, where $x \in \mathbb{R}^n$ and α is a positive constant.
4. $\dot{x} = kx + \mathrm{sign}_{\mathrm{n}}(x) + \mathrm{sig}_{\mathrm{n}}^{\alpha}(x)$, where $x \in \mathbb{R}^n$, and α and k are positive constants.
5. $\dot{x} = -\mathrm{sig}_{\mathrm{n}}^{\alpha}(x) - \mathrm{sig}_{\mathrm{n}}^{\beta}(x) + \delta(t)$, where $x \in \mathbb{R}^n$, α and β are positive constants, and $\delta(t)$ is a vector function bounded by a small positive constant, i.e.,

$$\|\delta(t)\| \leq \delta_M. \tag{2.36}$$

In the following, several useful properties of ratio persistence are introduced.

Lemma 2.10 *If the system state x is ratio persistent, for $\forall x_i(t), x_j(t) \neq 0$, $i, j \in \{1, 2, \ldots, n\}$, $i \neq j$, we have*

$$\frac{x_i(t)}{x_j(t)} = c_{ij}, \tag{2.37}$$

where $c_{ij} = x_i(0)/x_j(0)$ is a non-zero constant.

Proof According to Definition 2.11, a ratio persistent state x satisfies

$$\dot{x} = \kappa(x)x, \tag{2.38}$$

where the function $\kappa : \mathcal{N}_0 \backslash \{0\} \to \mathbb{R}$ is defined as

$$\kappa(x) = \frac{\|f(x)\|}{\zeta\|x\|}. \tag{2.39}$$

From (2.35), we know $\kappa(x) \neq 0$ when $x \neq 0$. Next, for $\forall x_i(t), x_j(t) \neq 0$, $i, j \in \{1, 2, \ldots, n\}$, $i \neq j$, we have

$$\frac{\mathrm{d}}{\mathrm{d}t}\left(\frac{x_i(t)}{x_j(t)}\right) = \frac{\dot{x}_i(t)x_j(t) - x_i(t)\dot{x}_j(t)}{x_j^2(t)}. \tag{2.40}$$

Taking $\dot{x}_i = \kappa(x)x_i$ and $\dot{x}_j = \kappa(x)x_j$ into (2.40), we can get

$$\frac{\mathrm{d}}{\mathrm{d}t}\left(\frac{x_i(t)}{x_j(t)}\right) = 0, \tag{2.41}$$

which further lead to (2.37) and completes the proof. □

Lemma 2.11 *If the system state x is ratio persistent, the unique solution to the system takes the form*

$$x(t) = \exp\left(\int_0^t \kappa\left(x\left(s\right)\right) \mathrm{d}s\right) x_0, \tag{2.42}$$

where $\kappa(x)$ is defined in (2.39).

Proof The proof is trivial by integrating (2.38). This solution (2.42) also directly proves Lemma 2.10. □

Straightforwardly and intuitively, when the direction of ratio persistence $\zeta = -1$, the state x of the system (2.18) is bounded.

Lemma 2.12 *For a ratio persistent system (2.18) with the negative direction of ratio persistence ($\zeta = -1$), the system state x is bounded.*

Proof Consider a Lyapunov function $V = \frac{1}{2}x^T x$. Based on Definition 2.11, the time derivative of V has the following form

$$\dot{V} = \zeta^{-1} \frac{x^T x}{\|x\|} \left\|f\left(x\right)\right\|. \tag{2.43}$$

Integrating both sides of (2.43), we have

$$V\left(t\right) - V\left(0\right) = \int_0^t \zeta^{-1} \frac{x^T x}{\|x\|} \left\|f\left(x\right)\right\| \mathrm{d}t. \tag{2.44}$$

After some manipulations, it reads

$$V\left(t\right) = V\left(0\right) - \underbrace{\int_0^t \|x\| \left\|f\left(x\right)\right\| \mathrm{d}t}_{\geq 0}, \tag{2.45}$$

which ends the proof. □

Remark 2.5 A ratio persistent trajectory of the state x is the shortest path between any initial state and the equilibrium if x denotes the system position in $\mathbb{R}^2$ or $\mathbb{R}^3$ Euclidean space. In this case, if x is ratio persistent, its trajectory will be a straight line, taking the form $x_1 = c_{12}x_2$ for $x \in \mathbb{R}^2$ and $x_1 = c_{12}x_2 = c_{12}c_{23}x_3$ for $x \in \mathbb{R}^3$. This suggests that in certain cases, the output trajectory generated by a ratio persistent system can be optimal in terms of its travel length.

According to the above discussion, a sufficient condition of time-synchronized stability is that system (2.18) is finite-time stable and the closed-loop state x is ratio persistent, i.e.,

> *"Finite-Time Stable"* + *"Ratio Persistent"* $\Rightarrow$ *"Time-Synchronized Stable"*.

For example, we consider an autonomous system in $\mathbb{R}^n$,

$$\dot{x} = -\mathrm{sig}_{\mathrm{n}}^{\alpha}(x), \tag{2.46}$$

where $\alpha \in (0, 1)$.

By using Lemma 2.5, considering a Lyapunov function $V = x^T x$, the solution of the system (2.46) can be proved to be finite-time stable with settling time

$$T(x_0) = \frac{\|x_0\|^{1-\alpha}}{1-\alpha}. \tag{2.47}$$

Next, one can easily verify the ratio persistent property of the system (2.46) according to Definition 2.11, by showing that for $x \neq 0$,

$$\frac{x}{\|x\|} = \zeta \frac{f(x)}{\|f(x)\|}, \tag{2.48}$$

where $\zeta = -1$ and $f(x) = -x\|x\|^{\alpha-1}$. As discussed in Lemma 2.10, for $\forall x_i(t)$, $x_j(t) \neq 0$, $i, j \in \{1, 2, \ldots, n\}$, $i \neq j$, we have

$$\frac{\mathrm{d}}{\mathrm{d}t}\left(\frac{x_i(t)}{x_j(t)}\right) = \frac{\frac{x_i(t)}{\|x(t)\|} x_j(t) - x_i(t) \frac{x_j(t)}{\|x(t)\|}}{x_j^2(t)} = 0. \tag{2.49}$$

Evidently, time-synchronized stability is achieved since the system state converges in finite time and the ratio of each pair of the state element is constant (the formal proof will be provided in Theorem 2.2).

Note that the system (2.46) is continuous and locally Lipschitz except at the origin. We can directly get the analytical solution of the system for any initial condition x_0 in $\mathbb{R}^n \backslash \{0\}$, which explicitly shows how and when the system state converges. The solution takes the following form

$$x = \left(\|x_0\|^{1-\alpha} - (1-\alpha)t\right)^{\frac{1}{1-\alpha}} \frac{x_0}{\|x_0\|}, \tag{2.50}$$

when $t < \|x_0\|^{1-\alpha}/(1-\alpha)$, and $x = 0$ when $t \geq \|x_0\|^{1-\alpha}/(1-\alpha)$. This solution is clearly ratio persistent. Moreover it infers that all the non-zero elements of $x(t)$

converge to the origin at the same time $t = \|x_0\|^{1-\alpha}/(1-\alpha)$. The synchronized settling time is in line with (2.47).

Remark 2.6 In general, for a MIMO system, it is difficult to ensure time-synchronized convergence by designing individual finite-time controllers for every sub-system. As presented in Lemma 2.5, only the upper bound of the settling time can be estimated. Therefore, even if the estimated upper bounds of the settling time of the sub-systems could be the same, the actual settling times of different sub-systems may vary greatly, especially under external disturbances.

Based on the above discussion, we are now ready to present the fundamental Lyapunov result of time-synchronized stability based on ratio persistence and finite-time stability, which will be applied throughout this book. Under some additional constraints, the analytical solution of system (2.18) and its synchronized settling time are also derived.

Theorem 2.1 (General Sufficient Conditions of Time-Synchronized Stability) *The equilibrium of the system (2.18) is time-synchronized stable if*

i. It is finite-time stable in the sense of Definition 2.6;
ii. The state x is ratio persistent.

Proof The proof is given by contradiction. Assume that at least two state elements x_i and x_j reach the equilibrium at different time instants T_i and T_j. Without loss of generality, let x_j be the latter one to arrive, i.e., $T_j > T_i$. According to the property of ratio persistence, we have

$$\lim_{t \to T_i^-} \frac{\mathrm{d}}{\mathrm{d}t}\left(\frac{x_i(t)}{x_j(t)}\right) = 0, \tag{2.51}$$

$$\lim_{t \to T_i^-} x_i(t) = c_{ij} \lim_{t \to T_i^-} x_j(t), \tag{2.52}$$

where $c_{ij} = x_i(0)/x_j(0)$. Further, since the element x_i arrives at the origin at T_i, it yields

$$\lim_{t \to T_i^-} x_i(t) = 0 \Rightarrow c_{ij} \lim_{t \to T_i^-} x_j(t) = 0, \tag{2.53}$$

which contradicts the premise that x_j reaches zero at T_j. Therefore, all the state elements reach the equilibrium at the same time, and the closed-loop system is time-synchronized stable, which concludes the proof in the case of an arbitrary $V(x)$ with $V(x) + cV^{\alpha}(x) \leq 0$. □

A more practical condition to prove time-synchronized stability is then given based on Theorem 2.1.

Theorem 2.2 (Sufficient Conditions of Time-Synchronized Stability) *Considering a Lyapunov function $V(x)$, the origin of the system (2.18) is time-synchronized stable, with the synchronized settling time*

$$T(x_0) \leq \frac{V^{1-\alpha}(x_0)}{c(1-\alpha)}, \tag{2.54}$$

if the following conditions hold:

i. *there exist constants $c > 0$ and $\alpha \in (0, 1)$ such that*

$$\dot{V}(x) + cV^{\alpha}(x) \leq 0; \tag{2.55}$$

ii. *the state x is ratio persistent.*

As a special case, if for $k > 0$,

$$V(x) = kx^T x, \quad \dot{V}(x) + cV^{\alpha}(x) = 0, \tag{2.56}$$

the system (2.18) has an explicit synchronized settling time

$$T(x_0) = \frac{V^{1-\alpha}(x_0)}{c(1-\alpha)}, \tag{2.57}$$

and its analytical solution takes the following form,

$$x = \begin{cases} \left(\|x_0\|^{2-2\alpha} - c(1-\alpha)k^{\alpha-1}t\right)^{\frac{1}{2-2\alpha}} \frac{x_0}{\|x_0\|}, & t < T(x_0), \ x_0 \neq 0, \\ 0, & t \geq T(x_0), \ x_0 \neq 0, \\ 0, & t \geq 0, \quad x_0 = 0. \end{cases} \tag{2.58}$$

Proof We first consider the case of an arbitrary $V(x)$ with $V(x) + cV^{\alpha}(x) \leq 0$.

Noticing (2.55) and using Lemma 2.5, we can prove that the system (2.18) is finite-time stable with the settling time

$$T(x_0) \leq \frac{V^{1-\alpha}(x_0)}{c(1-\alpha)}. \tag{2.59}$$

Based on Theorem 2.1, the closed-loop system is time-synchronized stable.

Next, consider the case of $V(x) = kx^T x$ with $\dot{V}(x) + cV^{\alpha}(x) = 0$.

By (2.56) and Lemma 2.5, the system (2.18) is finite-time stable with the settling time

$$T(x_0) = \frac{V^{1-\alpha}(x_0)}{c(1-\alpha)}. \tag{2.60}$$

From Definition 2.6 and the system (2.18), when $f(0) = 0$, it infers

$$x(t) = 0, \ t \geq T(x_0). \tag{2.61}$$

In the following, by analyzing the case of $x(t) \neq 0,\ t < T(x_0)$, we will prove that all the state elements reach the origin synchronously, and derive the analytical solution of the closed-loop system (2.18).

Since $V(x) = kx^T x$, we have

$$\dot{V}(x) = 2kx^T \dot{x}. \tag{2.62}$$

Note that according to (2.56), we can obtain

$$\begin{aligned}\dot{V}(x) &= -cV^{\alpha}(x) = -ck^{\alpha}\|x\|^{2\alpha} \\ &= -ck^{\alpha}x^T \operatorname{sig}_{\mathrm{n}}^{2\alpha-1}(x),\end{aligned} \tag{2.63}$$

Substituting (2.62) into (2.63), we can further get

$$2kx^T \dot{x} = -ck^{\alpha}x^T \operatorname{sig}_{\mathrm{n}}^{2\alpha-1}(x). \tag{2.64}$$

Since the system state x is ratio persistent, for $x \neq 0$, we have

$$\dot{x} = \kappa(x)\, x, \tag{2.65}$$

where $\kappa(x) \in \mathbb{R}$ is defined as in (2.38).

Taking (2.65) into (2.64), for $x \neq 0$, it yields

$$\kappa(x) = -\frac{ck^{\alpha-1}}{2\|x\|^{2-2\alpha}}. \tag{2.66}$$

Therefore, we can get that for $x \neq 0$,

$$\dot{x} = \kappa(x)\, x = -\frac{c}{2}k^{\alpha-1}\frac{x}{\|x\|^{2-2\alpha}}. \tag{2.67}$$

Again, as the system state x is ratio persistent, for $x \neq 0$, we know $x = \lambda e$, where $\lambda(t) \in \mathbb{R}$ is an unknown scalar variable to be solved and $e \in \mathbb{R}^n$ is a unit vector defined as $e = x_0/\|x_0\|$.

Then, (2.67) leads to

$$\dot{\lambda} = -\frac{c}{2}k^{\alpha-1}\frac{\lambda}{|\lambda|^{2-2\alpha}}. \tag{2.68}$$

Define $\lambda_0 = \lambda(0)$. The closed form solution is given by

$$\lambda = \left(|\lambda_0|^{2-2\alpha} - c(1-\alpha)k^{\alpha-1}t\right)^{\frac{1}{2-2\alpha}}. \tag{2.69}$$

Based on (2.69), we obtain an expression for x as follows:

$$x = \left(\|x_0\|^{2-2\alpha} - c\,(1-\alpha)\,k^{\alpha-1}t\right)^{\frac{1}{2-2\alpha}} \frac{x_0}{\|x_0\|}. \tag{2.70}$$

where $x_0 \neq 0$.

Due to the continuity of the system (2.18), the solution (2.70) shows that the all the state elements $x_i\,(t)$ will arrive at the origin when

$$t \to \frac{\|x_0\|^{2-2\alpha}}{c\,(1-\alpha)\,k^{\alpha-1}}. \tag{2.71}$$

One can verify from (2.60) that

$$T\,(x_0) = \frac{\|x_0\|^{2-2\alpha}}{c\,(1-\alpha)\,k^{\alpha-1}}, \tag{2.72}$$

which is in line with (2.71).

According to (2.61) and (2.70), the analytical solution of the closed-loop system state x takes the form as in (2.58), which infers that all the state elements $x_i\,(t)$ converge to the origin at the same time $T\,(x_0)$. The solution (2.58) meets the requirements in Definition 2.10 of time-synchronized stability, which completes the proof here. □

Remark 2.7 Actually, in the analytical solution (2.58) from Theorem 2.2, we have

$$x\,(t) = 0,\ t \geq T\,(x_0)\,, \tag{2.73}$$

although it is required by neither finite-time stability and time-synchronized stability. It can be verified by differentiating (2.69) and taking $T(\lambda_0)$ into it. We have,

$$\dot{\lambda} = \frac{1}{2-2\alpha}\left(|\lambda_0|^{2-2\alpha} - c(1-\alpha)k^{\alpha-1}t\right)^{\frac{1}{2-2\alpha}-1} \cdot \left(-c(1-\alpha)k^{\alpha-1}\right). \tag{2.74}$$

Taking $T(\lambda_0) = \frac{|\lambda_0|^{2-2\alpha}}{c(1-\alpha)k^{\alpha-1}}$ into (2.74), it reads when $t = T(x_0)$ $(t = T(\lambda_0))$, $\dot{\lambda} = 0$ $(\dot{x} = 0)$. Since $x(T(x_0)) = 0$, we further have

$$x(t) = 0,\ \forall t \geq T\,(x_0)\,. \tag{2.75}$$

as in (2.61). In this case, α should satisfy $\alpha \in (0, 1/2)$ to avoid the singularity in (2.74). Therefore, this remark is only for a further interpretation of Theorem 2.2, and $\alpha \in (0, 1/2)$ is not necessary for Theorem 2.2.

Remark 2.8 Theorem 2.2 provides not only sufficient conditions to prove time-synchronized stability but also a simple guide to time-synchronized control design, which will be detailed in the subsequent chapters. Note that the Lyapunov function of the special case in Theorem 2.2 is restricted to a quadratic form $V\,(x) = kx^T x$. However, it still represents a class of commonly used Lyapunov function candidates. More importantly, this special case allows us to obtain the analytical solution of the closed-loop system as well as the explicit synchronized settling time, which are important for quantifying system performance and tuning/optimizing design parameters.

Remark 2.9 Ratio persistence is a vital condition of the special case in Theorem 2.2, because if we only have the equality (2.56) without ratio persistence of state x, the conclusion of time-synchronized stability no longer holds. Specifically, one cannot get (2.67) from (2.64) due to the lack of ratio persistence, since

$$x^T\left(2k\dot{x} + ck^\alpha \mathrm{sig}_\mathrm{n}^{2\alpha-1}(x)\right) = 0 \tag{2.76}$$

does not necessarily mean that

$$\dot{x} = -\frac{c}{2}k^{\alpha-1}\mathrm{sig}_\mathrm{n}^{2\alpha-1}(x). \tag{2.77}$$

As mentioned earlier, despite looking like a restrictive condition, many systems studied in the literature actually already satisfy or have the potential to satisfy Definition 2.11 (see e.g. [6, 11, 13]).

In fact, to achieve time-synchronized stability, it is not necessary to have ratio "persistence". If the ratio of each pair of the system state elements is not constant but a bounded non-zero variable, we can get a more relaxed sufficient condition for time-synchronized stability.

Definition 2.12 (*Ratio Restriction*) The closed-loop state x of the system (2.18) is *ratio restricted* if for $x \neq 0$, we have that the ratio of each pair of the state elements $r_{ij}(t) = x_i(t)/x_j(t)$ exits. The function $r_{ij}(t)$ is constrained as

$$R_m \leq \left\|r_{ij}(t)\right\| \leq R_M, \tag{2.78}$$

where R_m and R_M can be any positive constants.

Lemma 2.13 *The equilibrium of the system (2.18) is time-synchronized stable, if the following conditions hold:*

i. the system (2.18) is finite-time stable;
ii. the state x is ratio restricted.

Proof Similar to the proof in Theorem 2.1, the proof herein is likewise shown by contradiction. Assume that at least two state elements x_i and x_j reach the equilibrium at different time instants T_i and T_j. Without loss of generality, let x_j be the latter one to arrive ($T_j > T_i$). Based on ratio restriction, we have

$$R_m \lim_{t\to T_i^-} x_i(t) \leq \lim_{t\to T_i^-} x_j(t) \leq R_M \lim_{t\to T_i^-} x_i(t), \tag{2.79}$$

Further, since the element x_i arrives at the origin at T_i, it yields

$$\lim_{t \to T_i^-} x_i(t) = 0 \Rightarrow \lim_{t \to T_i^-} x_j(t) = 0. \tag{2.80}$$

Due to the continuity of system (2.18), we have $x_j(T_i) = 0$, which contradicts the premise that x_j reaches zero at T_j. Therefore, all the state elements reach the equilibrium at the same time (the synchronized settling time is bounded by (2.54)), which finishes the proof. □

Remark 2.10 Despite the more generalized conditions of time-synchronized stability in Lemma 2.13, in most cases, we are fond of building a time-synchronized controller with the ratio persistence property (rather than the ratio restriction property). The main reason is that a ratio persistent system trajectory is the shortest path in 2- and 3-dimensional Euclidean space. In addition, if the ratio restriction property is properly utilized to achieve time-synchronized convergence, it is still complicated to analyze the system performance before the convergence since the ratio constraints R_m and R_M could be arbitrary small/large. In this book, we thus naturally focus on the control design based on ratio persistence.

2.4 Fixed-Time-Synchronized Stability

In this section, we further explore the concept of fixed-time-synchronized stability, where the estimation of the synchronized settling time is invariant with any initial state.

Definition 2.13 (*Fixed-Time-Synchronized Stability*) The equilibrium of the system (2.18) is *fixed-time-synchronized stable* if

i. it is globally time-synchronized stable;
ii. the upper bound of the synchronized settling time $T(x_0)$ is independent of any initial system state, i.e., for $\forall x_0$, we have $T(x_0) \le T_m$, where T_m is a positive constant.

Remark 2.11 According to Definition 2.13, fixed-time-synchronized stability already requires the system state to converge to the equilibrium in fixed time. Therefore, fixed-time-synchronized stability can be regarded as a special kind of fixed-time stability. Further, comparing Definition 2.7 and Definition 2.13, one can find that the main difference is that fixed-time stability allows some of the state elements to arrive at the origin before $T(x_0)$ as long as the remaining state elements converge to zero as $t \to T(x_0)$, while fixed-time-synchronized stability does not.

In the sequel, we introduce an *indirect* method to achieve fixed time-synchronized stability, which is to guarantee that: (i) the system (2.18) is fixed-time stable; and (ii) during the convergence, the ratio of each pair of the state elements is constant. Since all the state elements arrive at the origin at the same time, the above two conditions sufficiently lead to fixed-time-synchronized stability, which further motivates the following notion.

According to the above discussion, one sufficient condition of fixed-time-synchronized stability becomes that the system (2.18) is fixed-time stable and the closed-loop state x is ratio persistent, i.e.,

"Fixed-Time Stable" + *"Ratio Persistent"* $\Rightarrow$ *"Fixed-Time-Synchronized Stable"*.

For example, we consider an autonomous system $\in \mathbb{R}^n$,

$$\dot{x} = -\alpha \mathrm{sig}_{\mathrm{n}}^{p}(x) - \beta \mathrm{sig}_{\mathrm{n}}^{g}(x). \tag{2.81}$$

By using Lemma 2.6, considering a Lyapunov function $V = x^T x$, the solution of the system (2.81) can be proved to be fixed-time stable with settling time bounded by

$$T(x_0) \leq \frac{1}{\alpha(1-p)} + \frac{1}{\beta(g-1)}. \tag{2.82}$$

Next, from (2.81), we have

$$\frac{\dot{x}}{\|\dot{x}\|} = \zeta \frac{\alpha \frac{x}{\|x\|}\|x\|^p + \beta \frac{x}{\|x\|}\|x\|^g}{\alpha\|x\|^p + \beta\|x\|^g} = \zeta \frac{x}{\|x\|}, \tag{2.83}$$

where ζ is defined as in Definition 2.11.

Based on Definition 2.11, the state x of the system (2.81) is ratio persistent. Following Lemma 2.10, for $\forall x_i(t), x_j(t) \neq 0, i, j \in \{1, 2, \ldots, n\}, i \neq j$, we get

$$\frac{x_i(t)}{x_j(t)} = c_{ij}, \tag{2.84}$$

where in addition, we can also verify that

$$\begin{aligned} \frac{\mathrm{d}}{\mathrm{d}t}\left(\frac{x_i}{x_j}\right) =& \frac{-\alpha x_i x_j \|x\|^{p-1} - \beta x_i x_j \|x\|^{g-1}}{x_j^2} \\ &+ \frac{\alpha x_j x_i \|x\|^{p-1} + \beta x_j x_i \|x\|^{g-1}}{x_j^2} = 0. \end{aligned} \tag{2.85}$$

Intuitively, the system (2.81) is fixed-time-synchronized stable since the system state x converges to the origin in fixed time and during the convergence, the ratio of each pair of the state elements is constant.

Theorem 2.3 (Sufficient Conditions of Fixed-Time-Synchronized Stability) *Considering a Lyapunov function $V(x)$, the origin of the system (2.18) is fixed-time-synchronized stable, with the synchronized settling time*

$$T(x_0) \leq \frac{1}{\alpha^{\chi}(1 - p\chi)} + \frac{1}{\beta^{k}(g\chi - 1)}, \tag{2.86}$$

if the following conditions hold:

i. there exist positive constants α, β, p, g, and χ satisfying $p\chi < 1$ and $g\chi > 1$, such that

$$\dot{V}(x) \leq -(\alpha V^{p}(x) + \beta V^{g}(x))^{\chi}. \tag{2.87}$$

ii. the state x is ratio persistent.

Proof From Lemma 2.6, it is clear that the system (2.18) is fixed-time stable, with the synchronized settling time

$$T(x_0) \leq \frac{1}{\alpha^{\chi}(1 - p\chi)} + \frac{1}{\beta^{k}(g\chi - 1)}. \tag{2.88}$$

Next, following the contradiction in the proof of Theorem 2.1, fixed-time-synchronized stability is proved. □

Remark 2.12 Note that Theorem 2.3 has no special case providing an analytical solution of the closed-loop system as Theorem 2.2 does, since given a Lyapunov function $V(x) = kx^T x$, where k is a positive constant, if there exist positive constants α, β, p, and g satisfying $p < 1$ and $g > 1$, such that

$$\dot{V}(x) = -\alpha V^{p}(x) - \beta V^{g}(x), \tag{2.89}$$

the settling time will still be estimated by

$$T(x_0) \leq \frac{1}{\alpha(1 - p)} + \frac{1}{\beta(g - 1)}, \tag{2.90}$$

rather than a determinate value

$$T(x_0) = \frac{1}{\alpha(1 - p)} + \frac{1}{\beta(g - 1)}. \tag{2.91}$$

Based on (2.90), we cannot derive the analytical solution of the closed-loop system (2.18) as Theorem 2.2 does. It thus hinders the full extension to the case of fixed-time stability, as the contribution to explicitly show how and when x converges no longer exists.

Therefore, it is generally not easy to obtain the analytical solution of a fixed-time (-synchronized) stable system. However, despite the discussion in Remark 2.12, we can still explore the analytical solution of a fixed-time synchronized stable system based on a modified condition of fixed-time stability. First, Lemma 2.6 is rewritten as follows.

Lemma 2.14 *The equilibrium of the system (2.18) is fixed-time stable, if there exists a Lyapunov function $V(x)$ such that along the state trajectory of (2.18),*

$$\dot{V}(x) \leq \begin{cases} -cV^{g}(x), \text{ if } V(x) > 1, \\ -cV^{p}(x), \text{ if } V(x) \leq 1, \end{cases} \tag{2.92}$$

where c, p and g are positive constants, satisfying $p < 1$ and $g > 1$. The settling time $T(x_0)$ is bounded as

$$T(x_0) \leq \frac{1}{c(1-p)} + \frac{1}{c(g-1)}. \tag{2.93}$$

Proof The proof can be given by following that of Lemma 1 in [9]. □

Next, readily follows from Lemma 2.14, the analytical solution of a fixed-time-synchronized system is inspected.

Theorem 2.4 *Considering a Lyapunov function $V(x) = kx^T x$ where $k > 0$, if the state x is ratio persistent and the following equality holds*

$$\dot{V}(x) = \begin{cases} -cV^{g}(x), \text{ if } V(x) > 1, \\ -cV^{p}(x), \text{ if } V(x) \leq 1, \end{cases} \tag{2.94}$$

where constants $c > 0$, $g > 1$ and $p \in (0, 1)$, the origin of the system (2.18) is fixed-time-synchronized stable. The system (2.18) has an explicit synchronized settling time

$$T(x_0) = \begin{cases} \frac{1-V^{1-g}(x_0)}{c(g-1)} + \frac{1}{c(1-p)}, & \text{if } V(x_0) > 1, \\ \frac{V^{1-p}(x_0)}{c(1-p)}, & \text{if } V(x_0) \leq 1. \end{cases} \tag{2.95}$$

The analytical solution of the closed-loop system (2.18) takes the form: (1) in the case of $V(x_0) > 1$,

$$x = \begin{cases} \left(\|x_0\|^{2-2g} - \vartheta_g t\right)^{\frac{1}{2-2g}} \frac{x_0}{\|x_0\|}, & \text{if } t < T_1, \\ \left(k^{p-1} - \vartheta_p (t - T_1)\right)^{\frac{1}{2-2p}} \frac{x_0}{\|x_0\|}, & \text{if } t \geq T_1; \end{cases} \tag{2.96}$$

and (2) in the case of $V(x_0) \leq 1$,

$$x = \left(\|x_0\|^{2-2p} - \vartheta_p t\right)^{\frac{1}{2-2p}} \frac{x_0}{\|x_0\|}, \tag{2.97}$$

where $T_1 = \left(1 - V^{1-g}(x_0)\right) / (c(g-1))$, $\vartheta_p = c(1-p)k^{p-1}$, *and* $\vartheta_g = c(1-g)k^{g-1}$. *In addition, for* $\forall x_0 \in \mathbb{R}^n$, *the synchronized settling time is uniformly bounded by*

$$T(x_0) \leq \frac{1}{c(1-p)} + \frac{1}{c(g-1)}. \tag{2.98}$$

Proof The proof is given in two cases.

Case 1: $V(x_0) > 1$. Considering $V(x) > 1$, we have $\|x\| > \sqrt{k}/k$. Since the state x is ratio persistent and $\dot{V}(x) = -\alpha V^g(x)$, according to Theorem 2.2, the analytical solution takes the form

$$x = \left(\|x_0\|^{2-2g} - \vartheta_g t\right)^{\frac{1}{2-2g}} \frac{x_0}{\|x_0\|}. \tag{2.99}$$

Following the proof of Theorem 2.2, the derivative of the state x is derived

$$\dot{x} = -\frac{c}{2}k^{g-1}\frac{x}{\|x\|^{2-2g}}. \tag{2.100}$$

Based on the solution (2.99), the settling time for $\|x\| \to \sqrt{k}/k$ can be calculated

$$T_1 = \frac{1}{c(g-1)}\left(1 - \frac{1}{\|x_0\|^{2g-2}k^{g-1}}\right). \tag{2.101}$$

Considering $V(x) \leq 1$, we have $\|x\| \leq \sqrt{k}/k$. Form Theorem 2.2, it reads

$$x = \left(k^{p-1} - \vartheta_p(t + T_1)\right)^{\frac{1}{2-2p}} \frac{x_0}{\|x_0\|}. \tag{2.102}$$

Similarly, we can get

$$\dot{x} = -\frac{c}{2}k^{p-1}\frac{x}{\|x\|^{2-2p}}. \tag{2.103}$$

It is worth noting that, when $\|x\| = \sqrt{k}/k$, (2.100) and (2.103) are the equal, which guarantees the continuity of the derivative of the state x.

The corresponding settling time is derived from (2.102),

$$T_2 = \frac{1}{c(1-p)}. \tag{2.104}$$

The total synchronized settling time is obtained

$$T = T_1 + T_2 = \frac{1 - V^{1-g}(x_0)}{c(g-1)} + \frac{1}{c(1-p)}. \tag{2.105}$$

In addition, based on the above analysis, the synchronized settling time T (2.105) is ultimately bounded by

$$T = T_1 + T_2 \le \frac{1}{c\,(g-1)} + \frac{1}{c(1-p)}. \tag{2.106}$$

Case 2: $V(x_0) < 1$. It is clear that the analytical solution takes the form

$$x = \left(\|x_0\|^{2-2p} - \vartheta_p t\right)^{\frac{1}{2-2p}} \frac{x_0}{\|x_0\|}. \tag{2.107}$$

The synchronized settling time in this case is

$$T = \frac{V^{1-p}(x_0)}{c(1-p)}, \tag{2.108}$$

which is also bounded by

$$T \le \frac{1}{c\,(g-1)} + \frac{1}{c(1-p)}, \tag{2.109}$$

and this completes the proof. □

Similar to Lemma 2.13, we can get a more relaxed condition of fixed-time-synchronized stability.

Lemma 2.15 *Considering a Lyapunov function $V(x)$, the origin of the system (2.18) is fixed-time-synchronized stable, with the synchronized settling time*

$$T(x_0) \le \frac{1}{\alpha^{\chi}(1-p\chi)} + \frac{1}{\beta^{k}(g\chi - 1)}, \tag{2.110}$$

if the following conditions hold:

i. there exist positive constants α, β, p, g, and χ satisfying $p\chi < 1$ and $g\chi > 1$, such that

$$\dot{V}(x) \le -(\alpha V^{p}(x) + \beta V^{g}(x))^{\chi}. \tag{2.111}$$

ii. the state x is ratio restricted.

Proof The proof can be given following those in Lemma 2.13 and Theorem 2.3. □

2.5 Predefined-Time-Synchronized Stability

In this section, we move forward to the concept of "predefined-time-synchronized stability", under which, a control system is fixed-time-synchronized stable and the upper bound of the synchronized settling time is selected as a system parameter in advance. The definition of predefined-time-synchronized stability is, in fact, inspired by the concept of predefined-time stability in [1, 10].

We first rewrite the system dynamics (2.18) as a non-autonomous one,

$$\dot{x} = f(t, x; \varphi), \ f(t, 0; \varphi) = 0, \ x(0) = x_0, \tag{2.112}$$

where $x = [x_1, \ldots, x_n]^T \in \mathbb{R}^n$ is the system state, φ is a vector with proper dimensions consisting of constant system parameters, f is continuous with respect to x and piecewise continuous with respect to t. We assume that the system (2.18) has a unique solution for all initial conditions in forward time.

Definition 2.14 (*Predefined-Time-Synchronized Stability*) For the parameter vector φ of system (2.112) and an arbitrarily selected positive constant $T_p := T_p(\varphi)$ called *predefined synchronized time*, the system (2.112) is

i. weak predefined-time-synchronized stable if it is fixed-time-synchronized stable and the synchronized settling time $T(x_0) \leq T_p, \ \forall x_0 \in \mathbb{R}^n$;
ii. strong predefined-time-synchronized stable if it is fixed-time-synchronized stable and the synchronized settling time $\sup_{x_0 \in \mathbb{R}^n} T(x_0) = T_p$;
iii. true predefined-time-synchronized stable if it is fixed-time-synchronized stable and the synchronized settling time $T(x_0) = T_p, \ \forall x_0 \in \mathbb{R}^n$.

Largely motivated by the finite-time analysis in [15], by using the comparison principle in [7], Chap. 3, the following theorem is proposed.

Theorem 2.5 (*Sufficient Conditions of Time-Synchronized Stability for Non-Autonomous Systems*) *Considering system (2.112), assume that there exists a constant $T = T(x_0) > 0$, which may depend on the initial condition x_0, two functions $v_i \in \mathcal{K}$, $i = 1, 2$, a continuous function $\mu(t) : [0, T) \to \mathbb{R}$, a constant $\kappa_0 > 0$, a continuous function $\kappa(t) : [0, T) \to [\kappa_0, +\infty)$, and a function $V(t, x) : [0, T) \times \mathbb{R}^n \to \mathbb{R}_{\geq 0}$, such that*

$$\kappa(t) v_1(\|x\|) \leq V(t, x), \forall t \in [0, T), \tag{2.113}$$

$$\mathbf{D}^+ V(t, x)\big|_{(}2.112) \leq \mu(t) V(t, x), \forall t \in [0, T), \tag{2.114}$$

$$V(0, x) \leq v_2(\|x\|). \tag{2.115}$$

Define $\rho(t) : [0, T) \to \mathbb{R}$ as

$$\rho(t) = \ln(\kappa(t)) - \int_0^t \mu(s) \mathrm{d}s. \tag{2.116}$$

The non-autonomous system (2.112) is globally time-synchronized stable if the state x is ratio persistent for $t \in [0, T)$, and

$$\lim_{t \uparrow T} \rho(t) = +\infty. \tag{2.117}$$

The system state is bounded as

$$\nu_1(\|x(t)\|) \leq \frac{\exp\left(\int_0^t \mu(s)\mathrm{d}s\right)}{\kappa(t)} \nu_2(\|x_0\|), \ \forall t \in [0, T). \tag{2.118}$$

Proof Note that for $\mathbf{D}^+ V(t, x) = \mu(t)V(t, x)$, we have

$$V(t, x) = \exp\left(\int_0^t \mu(s)\mathrm{d}s\right) V(0, x_0). \tag{2.119}$$

According to the comparison principle [7], it follows from (2.114) and (2.115) that,

$$\begin{aligned} V(t, x) \leq & \exp\left(\int_0^t \mu(s)\mathrm{d}s\right) V(0, x_0) \\ \leq & \exp\left(\int_0^t \mu(s)\mathrm{d}s\right) \nu_2(\|x_0\|), \ \forall t \in [0, T). \end{aligned} \tag{2.120}$$

From $\rho(t) : [0, T) \to \mathbb{R}$ and (2.117), these exists a constant c_0 satisfying $\rho(t) \geq c_0$. We thus have

$$\mathrm{e}^{-\rho(t)} = \frac{\exp\left(\int_0^t \mu(s)\mathrm{d}s\right)}{\kappa(t)} \leq \mathrm{e}^{-c_0}, \ \forall t \in [0, T). \tag{2.121}$$

Based on (2.113), (2.120) and (2.121), we have

$$\begin{aligned} \nu_1(\|x(t)\|) \leq & \frac{V(t, x(t))}{\kappa(t)} \\ \leq & \frac{\exp\left(\int_0^t \mu(s)\mathrm{d}s\right)}{\kappa(t)} \nu_2(\|x_0\|) \\ \leq & \mathrm{e}^{-c_0} \nu_2(\|x_0\|), \ \forall t \in [0, T), \end{aligned} \tag{2.122}$$

which leads to (2.118). For $\forall \varepsilon > 0$, let $\delta = \delta(\varepsilon) > 0$ satisfying

$$\mathrm{e}^{-c_0} \nu_2(\delta) \leq \nu_1(\varepsilon). \tag{2.123}$$

For $\forall \|x_0\| \leq \delta$, $\forall t \in [0, T)$, it readily follows from (2.122) that

$$v_1(\|x(t)\|) \le e^{-c_0} v_2(\|x_0\|) \le e^{-c_0} v_2(\delta) \le v_1(\varepsilon). \tag{2.124}$$

The above further yields

$$\|x(t)\| \le \varepsilon, \ \forall t \in [0, T). \tag{2.125}$$

Next, from (2.117) and (2.118), we have

$$\lim_{t \uparrow T} v_1(\|x(t)\|) \le e^{-\rho(t)} v_2(\|x_0\|) = 0, \tag{2.126}$$

which implies

$$\lim_{t \uparrow T} \|x(t)\| = 0. \tag{2.127}$$

From (2.125) and (2.127), according to Definition 2.6, the system is finite-time stable. Since the state x is ratio persistent for $t \in [0, T)$, following the contradiction in proof of Theorem 2.1, globally time-synchronized stability is proved. □

Based on Theorem 2.5, the following result is drawn.

Lemma 2.16 *Considering system (2.112), assume that there exists a constant $T = T(x_0) > 0$, which may depend on the initial condition x_0, two functions $w_i \in \mathcal{K}$, $i = 1, 2$, a continuous function $\mu(t) : [0, T) \to \mathbb{R}$, and a function $U(t, x) : [0, T) \times \mathbb{R}^n \to \mathbb{R}_{\ge 0}$, such that*

$$w_1(\|x\|) \le U(t, x) \le w_2(\|x\|), \ \forall t \in [0, T), \tag{2.128}$$

$$\mathbf{D}^+ U(t, x)\big|_(2.112) \le \mu(t) \frac{e^{U(t,x)} - 1}{e^{U(t,x)}}, \ \forall t \in [0, T), \tag{2.129}$$

Define $\rho(t) : [0, T) \to \mathbb{R}$ as

$$\rho(t) = -\int_0^t \mu(s) \mathrm{d}s. \tag{2.130}$$

The non-autonomous system (2.112) is globally time-synchronized stable if the state x is ratio persistent for $t \in [0, T)$, and

$$\lim_{t \uparrow T} \rho(t) = +\infty. \tag{2.131}$$

The system state is bounded as

$$w_1(\|x(t)\|) \le \ln\left(1 + e^{\int_0^t \mu(s)\mathrm{d}s} \left(e^{w_2(\|x_0\|)} - 1\right)\right). \tag{2.132}$$

Proof Define

$$V(t,x) = \mathrm{e}^{U(t,x)} - 1, \tag{2.133}$$

and from (2.128), we have

$$v_1(\|x\|) \le V(t,x) \le v_2(\|x\|), \tag{2.134}$$

where two $\mathcal{K}$-functions are defined as

$$v_1(\|x\|) \triangleq \mathrm{e}^{w_1(\|x\|)} - 1, \tag{2.135}$$

$$v_2(\|x\|) \triangleq \mathrm{e}^{w_2(\|x\|)} - 1, \tag{2.136}$$

Note that the inequality in (2.134) satisfies the requirements in (2.113) and (2.115) (where $\kappa(t) = 1$). From (2.129) and (2.133), we have $\forall t \in [0, T)$,

$$\begin{aligned} \mathbf{D}^+ V(t,x)\big|_{(2.112)} &= \mathrm{e}^{U(t,x)} \mathbf{D}^+ U(t,x)\big|_{(2.112)} \\ &\le \mathrm{e}^{U(t,x)} \mu(t) \frac{\mathrm{e}^{U(t,x)} - 1}{\mathrm{e}^{U(t,x)}} \\ &= \mu(t) V(t,x), \end{aligned} \tag{2.137}$$

which further satisfies the requirement in (2.114). Then, the proof is concluded by the direct application of Theorem 2.5. □

From Lemma 2.16, we provide the following sufficient conditions of predefined-time-synchronized stability.

Lemma 2.17 *Considering system (2.112), assume that there exists a positive constant T (independent of x_0 and adjustable by changing φ), two functions $w_i \in \mathcal{K}$, $i = 1, 2$, a continuous function $\mu(t) = -a/(T - t)$, $t \in [0, T)$, $a > 0$, and a function $U(t,x) : [0, T) \times \mathbb{R}^n \to \mathbb{R}_{\ge 0}$, such that*

$$w_1(\|x\|) \le U(t,x) \le w_2(\|x\|), \forall t \in [0, T), \tag{2.138}$$

$$\mathbf{D}^+ U(t,x)\big|_(2.112) \le \mu(t) \frac{\mathrm{e}^{U(t,x)} - 1}{\mathrm{e}^{U(t,x)}}, \quad \forall t \in [0, T), \tag{2.139}$$

Define $\rho(t) : [0, T) \to \mathbb{R}$ as

$$\rho(t) = -\int_0^t \mu(s) \mathrm{d}s. \tag{2.140}$$

The non-autonomous system (2.112) is predefined-time-synchronized stable if the state x is ratio persistent for $t \in [0, T)$. The system state is bounded as

$$w_1(\|x(t)\|) \le \ln\left(1 + \mathrm{e}^{\int_0^t \mu(s)\mathrm{d}s} \left(\mathrm{e}^{w_2(\|x_0\|)} - 1\right)\right). \tag{2.141}$$

If the equality is achieved in (2.139), system (2.112) is true predefined-time-synchronized stable. In this case, if $U = kx^T x$ is chosen, the analytical solution

$$x = \frac{\sqrt{k}}{k}\left(\ln\left(\left(\frac{T-t}{T}\right)^a \left(e^{k\|x_0\|^2} - 1\right) + 1\right)\right)^{\frac{1}{2}} \frac{x_0}{\|x_0\|}. \tag{2.142}$$

Proof From (2.140) and $\mu(t) = -a/(T-t)$, we have

$$\int_0^t \mu(s)\mathrm{d}s = -\int_0^t \frac{a\,\mathrm{d}s}{T-s} = a\ln\left(\frac{T-t}{T}\right), \tag{2.143}$$

which satisfies (2.131). Based on Lemma 2.16 and noticing T is a constant independent of x_0, it is evident that system (2.112) is predefined-time-synchronized stable according to Definition 2.14.

Define $V(t,x) = e^{U(t,x)} - 1$. From (2.120), we obtain $\forall t \in [0, T)$,

$$\begin{aligned} V(t, x(t)) &\leq \exp\left(\int_0^t \mu(s)\mathrm{d}s\right) V(0, x(0)) \\ &= \left(\frac{T-t}{T}\right)^a V(0, x(0)). \end{aligned} \tag{2.144}$$

Substituting $V(t,x) = e^{U(t,x)} - 1$ into (2.144), yields

$$e^{U(t,x)} - 1 \leq \left(\frac{T-t}{T}\right)^a \left(e^{U(0,x_0)} - 1\right), \tag{2.145}$$

from which we have

$$U(t,x) \leq \ln\left(\left(\frac{T-t}{T}\right)^a \left(e^{U(0,x_0)} - 1\right) + 1\right). \tag{2.146}$$

Based on (2.146), we obtain

$$U(t,x) = \ln\left(\left(\frac{T-t}{T}\right)^a \left(e^{U(0,x_0)} - 1\right) + 1\right). \tag{2.147}$$

provided that the equality is achieved in (2.139). In this case, true predefined-time-synchronized stability is proved according to Definition 2.14, since the time-synchronized convergence is achieved exactly at the time instant T independent of x_0 that can be adjusted by φ. In the case of $U = kx^T x$, from Lemma 2.11, we define

$$x = \lambda(t,x)\frac{x_0}{\|x_0\|}, \tag{2.148}$$

where $\lambda(t,x) : [0, T) \times \mathbb{R}^n \to \mathbb{R}_{\geq 0}$ is an unknown function. Thus, we have

$$\lambda = \sqrt{\frac{U}{k}} = \frac{\sqrt{k}}{k}\left(\ln\left(\left(\frac{T-t}{T}\right)^{a}\left(e^{k\|x_0\|^2}-1\right)+1\right)\right)^{\frac{1}{2}}, \tag{2.149}$$

which further yields the analytical solution

$$x = \frac{\sqrt{k}}{k}\left(\ln\left(\left(\frac{T-t}{T}\right)^{a}\left(e^{k\|x_0\|^2}-1\right)+1\right)\right)^{\frac{1}{2}}\frac{x_0}{\|x_0\|}. \tag{2.150}$$

The proof ends here. □

It is clear that Lemma 2.17 is a special case of Lemma 2.16, providing one solution to achieve true predefined-time-synchronized stability.

2.6 Summary

This chapter has revisited the properties of sign functions. The norm-normalized sign function serves as an essential role in the achievement of time-synchronized convergence. Note that this norm-normalized sign function has wide applicability not only in control system design but also in fields of engineering and various other sciences, which possibly endows this approach an extended contribution. Next, we have articulated the formulations of ratio persistence, ratio restriction, time-synchronized stability, fixed-time-synchronized stability and predefined-time-synchronized stability. It has been revealed that ratio persistence and finite/fixed/predefined-time stability together contribute to (fixed-/predefined-) time-synchronized stability. This fundamental formula plays an essential role in the construction of (fixed-/predefined-) time-synchronized controllers in the subsequent chapters. In addition, under certain conditions, the analytical trajectories of a time-synchronized stable system have been explicitly provided. This merit grants a control engineer more flexibly to predefine the control parameters, to preview the control performance, and to predict the system output.

A relationship of different types of time-synchronized stability is provided in Fig. 2.2 to clearly clarify their connections.

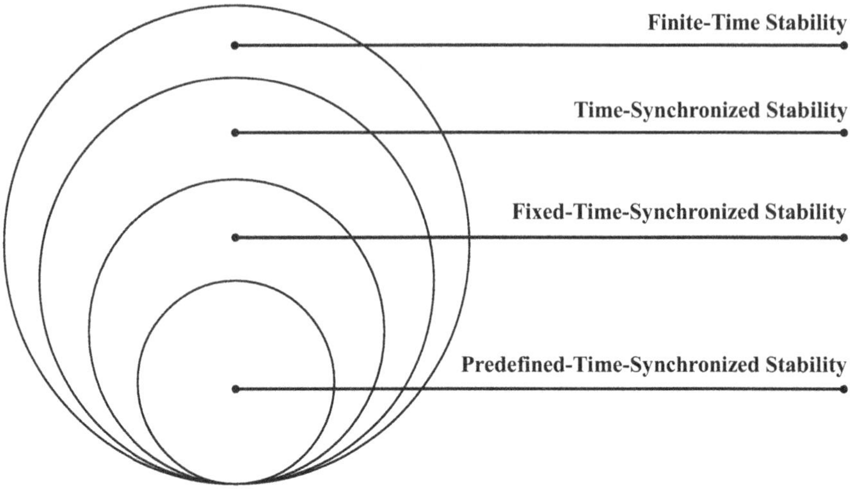

Fig. 2.2 Relationship of different types of time-synchronized stability

References

1. Aldana-López, R., Gómez-Gutiérrez, D., Jiménez-Rodríguez, E., Sánchez-Torres, J.D., Defoort, M.: Enhancing the settling time estimation of a class of fixed-time stable systems. International Journal of Robust and Nonlinear Control **29**(12), 4135–4148 (2019)
2. Bhat, S.P., Bernstein, D.S.: Finite-time stability of continuous autonomous systems. SIAM Journal on Control and Optimization **38**(3), 751–766 (2000)
3. Edwards, C., Spurgeon, S.: Sliding mode control: Theory and applications. CRC Press (1998)
4. Filippov, A.F.: Differential Equations with Discontinuous Right-Hand Side. Dordrecht, The Netherlands: Kluwer (1988)
5. Ge, S.S., Hang, C.C., Lee, T.H., Zhang, T.: Stable adaptive neural network control. Springer, Boston (2002)
6. Hamayun, M.T., Edwards, C., Alwi, H.: Design and analysis of an integral sliding mode fault-tolerant control scheme. IEEE Transactions on Automatic Control **57**(7), 1783–1789 (2012)
7. Khalil, H.K.: Nonlinear systems. Prentice hall, New Jersey (1996)
8. Pal, A.K., Kamal, S., Nagar, S.K., Bandyopadhyay, B., Fridman, L.: Design of controllers with arbitrary convergence time. Automatica **112**, 108710 (2020)
9. Polyakov, A.: Nonlinear feedback design for fixed-time stabilization of linear control systems. IEEE Transactions on Automatic Control **57**(8), 2106–2110 (2012). https://doi.org/10.1109/TAC.2011.2179869
10. Sánchez-Torres, J.D., Gómez-Gutiérrez, D., López, E., Loukianov, A.G.: A class of predefined-time stable dynamical systems. IMA Journal of Mathematical Control and Information **35**(Supplement_1), i1—-i29 (2018)
11. Sanyal, A.K., Bohn, J.: Finite-time stabilisation of simple mechanical systems using continuous feedback. International Journal of Control **88**(4), 783–791 (2015). https://doi.org/10.1080/00207179.2014.974675
12. Sontag, E.D., Wang, Y.: On characterizations of the input-to-state stability property. Systems & Control Letters **24**(5), 351–359 (1995)
13. Sun, C., Hu, G., Xie, L., Egerstedt, M.: Robust finite-time connectivity preserving coordination of second-order multi-agent systems. Automatica **89**, 21–27 (2018)

14. Utkin, V., Guldner, J., Shi, J.: Sliding Mode Control in Electro-Mechanical Systems. CRC Press (2009)
15. Zhou, B.: Finite-time stability analysis and stabilization by bounded linear time-varying feedback. Automatica **121**, 109191 (2020)
16. Zhu, Z., Xia, Y., Fu, M.: Attitude stabilization of rigid spacecraft with finite-time convergence. International Journal of Robust and Nonlinear Control **21**(6), 686–702 (2011)

Chapter 3
Time-Synchronized Sliding Modes

Abstract Sliding-mode techniques have wide applications in control system design and are robust to matched disturbances when properly invoked. In this chapter, we integrate the essentials of time-synchronized convergence with the structure of sliding modes, where various time-synchronized sliding modes are introduced to force all the system state elements to converge time-synchronously on the sliding surface. It is observed and discussed that when the norm-normalized sign function is properly involved in the design of sliding modes, the outcome of time-synchronized convergence is naturally expected. In addition, in light of the convergence rate, continuity, and singularity, the time-synchronized sliding modes are classified into specific types. Control engineers or researchers are able to select proper time-synchronized sliding modes based on the requirements of their individual tasks.

3.1 Quasi-Continuous Time-Synchronized Control Design

In the sequel, we introduce the concept of differential inclusions and some relevant preliminaries. Then, a quasi-continuous time-synchronized controller is proposed to showcase the characteristics of time-synchronized convergence. First, the following definitions are given.

Definition 3.1 (*Open Cube*) A product of open intervals

$$U = (a_1, b_1) \times (a_2, b_2) \times \cdots \times (a_n, b_n), \tag{3.1}$$

is called an *open cube*, and its *volume* is defined as

$$\mathrm{vol}(U) = (b_1 - a_1)(b_2 - a_2) \cdots (b_n - a_n). \tag{3.2}$$

Definition 3.2 (*Measure Zero*) [7] A set $X \in R^n$ has *measure zero* if $\forall \epsilon > 0$, $\exists U_1, U_2, \cdots$ (open cubes), s.t. $X \subseteq \bigcup_{i=1}^{\infty} U_i$ and $\sum_{i=1}^{\infty} \mathrm{vol}(U_i) < \epsilon$.

Definition 3.3 (*Quasi-Continuity*) [10] Let X be a topological space. A real-valued function $f : X \to \mathbb{R}$ is quasi-continuous at a point $x \in X$ if for any $\epsilon > 0$ and

D. Li et al., *Time-Synchronized Control: Analysis and Design*,
https://doi.org/10.1007/978-981-16-3089-7_3

any open neighborhood U of x there is a non-empty open set $G \subset U$ such that $|f(x) - f(y)| < \epsilon, \ \forall y \in G$.

Note that in the above definition, it is not necessary that $x \in G$.

Consider system dynamics with discontinuous right-hand sides

$$\dot{x} = f(t, x), \ t \in \mathbb{R}, \ x \in \mathbb{R}^n, \tag{3.3}$$

where the function $f : \mathbb{R} \times \mathbb{R}^n \to \mathbb{R}^n$ is quasi-continuous and the set of discontinuous points has *measure zero*.

The concept of *differential inclusions* is then introduced, where the quasi-continuous system (3.3) is handled and transformed into dynamics (called differential inclusion) with *multi-value right-hand sides*,

$$\dot{x} \in F(t, x), \tag{3.4}$$

where $t \in \mathcal{D}_t \subset \mathbb{R}$, $x \in \mathcal{D}_x \subset \mathbb{R}^n$, and $F(t, x)$ is a multi-value function, mapping points (t, x) of the region $\mathcal{D} = \mathcal{D}_t \times \mathcal{D}_x$ to the set $F(t, x)$ of vectors $\in \mathbb{R}^n$. Clearly, the essential behind differential inclusions involves a set-valued map, mapping a point to a set of points.

Definition 3.4 (*Absolutely Continuous Function*) A mapping $\varphi : [a, b] \to \mathbb{R}$ is an *absolutely continuous function*, if $\forall \varepsilon \in (0, \infty)$, $\exists \delta \in (0, \infty)$, s.t., given an arbitrary finite disjoint open intervals $\{(a_1, b_1), \ldots, (a_n, b_n)\} \in [a, b]$ with

$$\sum_{i=1}^{n} (b_i - a_i) < \delta, \tag{3.5}$$

the following inequality holds

$$\sum_{i=1}^{n} |\gamma(b_i) - \gamma(a_i)| < \varepsilon. \tag{3.6}$$

Definition 3.5 (*Solution of Differential Inclusion*) [1] A absolutely continuous function $x(t) \in \mathbb{R}^n$ is *a solution of differential inclusion* (3.4) if when $\dot{x}(t)$ exists,

$$\dot{x}(t) \in f(t, x(t)). \tag{3.7}$$

If the solution of differential inclusions is said to be understood in the *Filippov sense* [3], it meets the following *Filippov set-valued map*. The core of *Filippov solution* lies in observing the vector field within the neighborhood of each point.

Definition 3.6 (*Filippov Solution*) [4] For a function $f : \mathbb{R}^n \to \mathbb{R}^n$, the *Filippov Set-Valued Map* is defined as $F[f] : \mathbb{R}^n \to \mathfrak{B}(\mathbb{R}^n)$ by

$$F[f](x) \triangleq \bigcap_{\delta>0} \bigcap_{\mu(S)=0} \overline{\text{co}}\{f(\mathfrak{B}(x,\delta)\backslash S)\}, \quad x \in \mathbb{R}^n, \tag{3.8}$$

where $\overline{\text{CO}}$ represents convex closure, μ Lebesgue measure, and

$$\mathfrak{B}(x,\delta) = \left\{y \in R^n | \|y-x\| < \delta\right\}. \tag{3.9}$$

If $x(t)$ is a solution of the following differential inclusion

$$\dot{x} \in F[f](x), \tag{3.10}$$

then $x(t)$ is a *Filippov solution* of the system (3.3).

From Definition 3.6, in terms of a point x, we know that $F[f](x)$ is independent of the actual value of $f(x)$. Except uniqueness, the Filippov solution involves most of the well-known standard properties.

Simple first-order affine dynamics in $\mathbb{R}^n$ is considered,

$$\dot{x} = f(x) + b(x)u, \tag{3.11}$$

where $x = [x_1, x_2, \ldots, x_n]^T \in \mathbb{R}^n$ denotes the state vector, $u \in \mathbb{R}^n$ the control input, and $f(x) \in \mathbb{R}^n$ and $b(x) \in \mathbb{R}^{n\times n}$ $(\det(b(x)) \neq 0)$ the known parts.

We now propose a time-synchronized control law u_2, and introduce a finite-time control law u_1 for comparison

$$u_c = -b^{-1}(x)\left(k\text{sign}_\text{c}(x) + f(x)\right), \tag{3.12}$$

$$u_n = -b^{-1}(x)\left(k\text{sign}_\text{n}(x) + f(x)\right). \tag{3.13}$$

Taking the controllers (3.12) and (3.13) respectively into the system (3.11), we have

$$\dot{x} = -k\text{sign}_\text{c}(x), \tag{3.14}$$

$$\dot{x} = -k\text{sign}_\text{n}(x). \tag{3.15}$$

Note that the closed-loop systems (3.14) and (3.15) have discontinuous right-hand sides. They should be understood in the Filippov sense. Then, the closed-loop dynamics (3.14) and (3.15) become

$$\dot{x} \in F\left[-k\text{sign}_\text{c}(x)\right](x), \tag{3.16}$$

$$\dot{x} \in F\left[-k\text{sign}_\text{n}(x)\right](x). \tag{3.17}$$

Since the Filippov set-valued maps

$$F\left[-k\text{sign}_\text{n}(x)\right](x) \text{ or } F\left[-k\text{sign}_\text{c}(x)\right](x), \tag{3.18}$$

at the point x_0 ($x_0 = 0$ or any element of x_0 is equal to zero) are independent of the actual value of $\mathrm{sign_n}(x_0)$ or $\mathrm{sign_c}(x_0)$, one can thus redefine/generalize the multi-variable classical sign function $\mathrm{sign_c}$ and the norm-normalized sign function $\mathrm{sign_n}$ as follows,

$$\mathrm{sign_c}(x) = \left[\mathrm{sign_c}(x_1), ..., \mathrm{sign_c}(x_n)\right]^T, \tag{3.19}$$

$$\mathrm{sign_n}(x) = \frac{x}{\|x\|}, \tag{3.20}$$

where

$$\mathrm{sign_c}(x_i) = \begin{cases} 1, & \text{if } x_i > 0, \\ [-1, 1], & \text{if } x_i = 0, \\ -1, & \text{if } x_i < 0. \end{cases} \tag{3.21}$$

We next give the main result on the performance of control laws (3.12) and (3.13).

Theorem 3.1 *For the system dynamics (3.11), under the finite-time control law (3.12), the closed-loop system is finite-time stable; and under the time-synchronized control law (3.13), the system state x is ratio persistent, and the closed-loop system is time-synchronized stable.*

Proof We will first prove the finite-time convergence of system dynamics (3.14). Consider a Lyapunov function $V_1 = \frac{1}{2}x^T x$. Taking the derivative of V_1, we have $\dot{V}_1 = -x^T \mathrm{sgn}_1(x) = -\|x\|_1$. According to [6], to any $x \subseteq \mathbb{R}^n$, it holds

$$\|x\|_1 \geq \|x\|, \tag{3.22}$$

where $\|x\| = \sqrt{x^T x}$, $\|x\|_1 = \sum_{i=1}^{n} |x_i|$, x_i is ith element of vector x. According to (3.22), we can get $\dot{V}_1 \leq -\|x\| = -\sqrt{2}V_1^{\frac{1}{2}}$. According to Lemma 2.5 (Sufficient Conditions of Finite-Time Stability), the system (3.14) is finite-time stable with the settling time

$$T_1(x_0) \leq \sqrt{2}V_1^{\frac{1}{2}}(x_0). \tag{3.23}$$

We next move to the system dynamics (3.15). Following Definition 2.11, it is clear that the system state$x$ is ratio persistent, since for $x \neq 0$, one can get

$$\frac{x}{\|x\|} = \zeta \frac{-k\mathrm{sign_n}(x)}{\|-k\mathrm{sign_n}(x)\|}, \tag{3.24}$$

where the direction of ratio persistence $\zeta = -1$.

Based on Theorem 2.2 (Sufficient Conditions of Time-Synchronized Stability), by analyzing the same Lyapunov function $V_2 = \frac{1}{2}x^T x$, we know that the system (3.15) is time-synchronized stable with the synchronized settling time

$$T_2(x_0) = \sqrt{2}V_2^{\frac{1}{2}}(x_0). \tag{3.25}$$

More specifically, the time-synchronized solution takes the form

$$x = \begin{cases} \left(\|x_0\| - \frac{\sqrt{2}}{2}k^{\frac{1}{2}}t\right)\frac{x_0}{\|x_0\|}, & t < T(x_0),\ x_0 \neq 0, \\ 0, & t \geq T(x_0),\ x_0 \neq 0, \\ 0, & t \geq 0,\quad x_0 = 0. \end{cases} \tag{3.26}$$

□

Remark 3.1 Please note that a Lyapunov function $V(t)$ is not necessarily differentiable along the absolutely continuous Filippov solutions of (3.14) and (3.15). In [2], Proposition 14.1, the generalized Lyapunov theorem actually requires only continuity of $V(t)$, which can be applied to system dynamics involving differential inclusions. In contrast, the classical Lyapunov theorem requires the continuous right-hand side of the differential equation and/or a differentiable (or at least locally Lipschitz continuous) Lyapunov function candidate. Obviously, the Lyapunov functions utilized in the proof of Theorem 3.1 meet none of the above conditions. This issue has also been discussed in [5, 9].

Remark 3.2 In terms of the controller (3.12), since the control input u_c of the system (3.11) is in $\mathbb{R}^n$, the controller (3.12) generates n decoupled scalar systems, where time-synchronized convergence cannot be expected.

Simulation examples for control laws (3.12) and (3.13) are shown in Figs. 3.1, 3.2 and 3.3. The initial state is given as $x_0 = [4, -6, 8]^T$. As presented in Figs. 3.1, 3.2 and 3.2, both control laws (3.12) and (3.13) guarantee that the system state x converges to the origin in finite time. In Fig. 3.2, the state elements x_1, x_2, and x_3

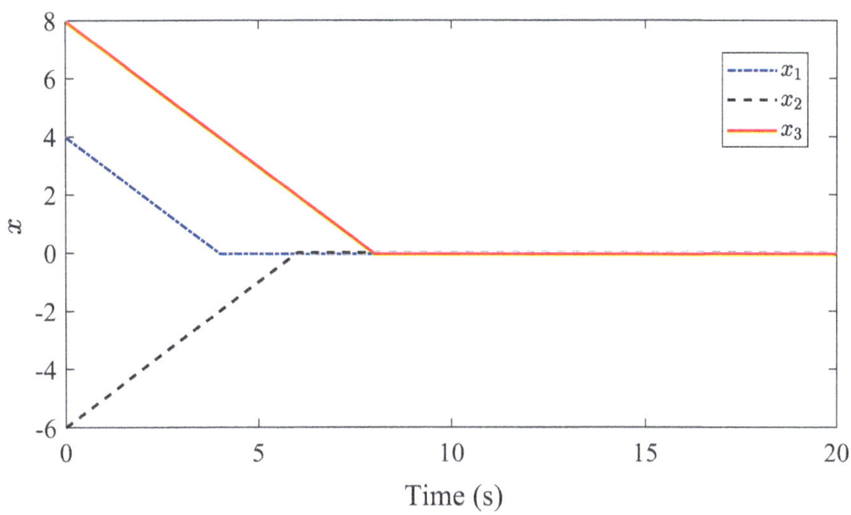

Fig. 3.1 Performance of the control law (3.12)

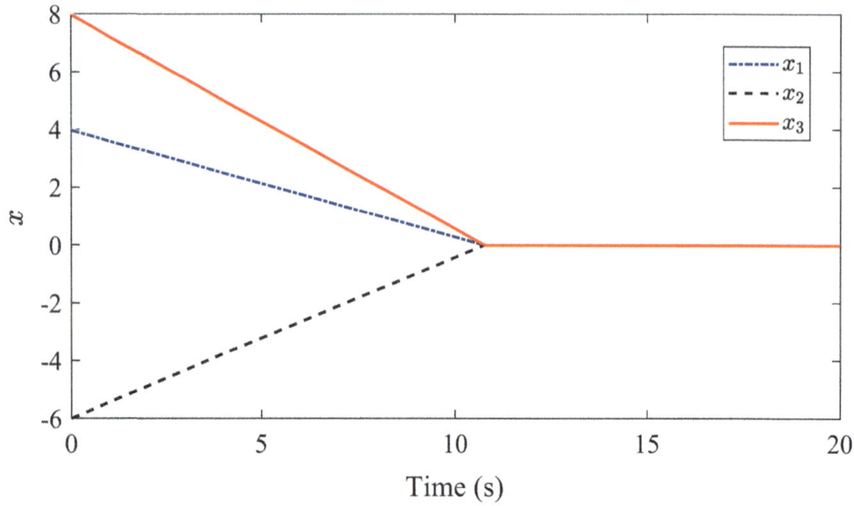

Fig. 3.2 Performance of the control law (3.13)

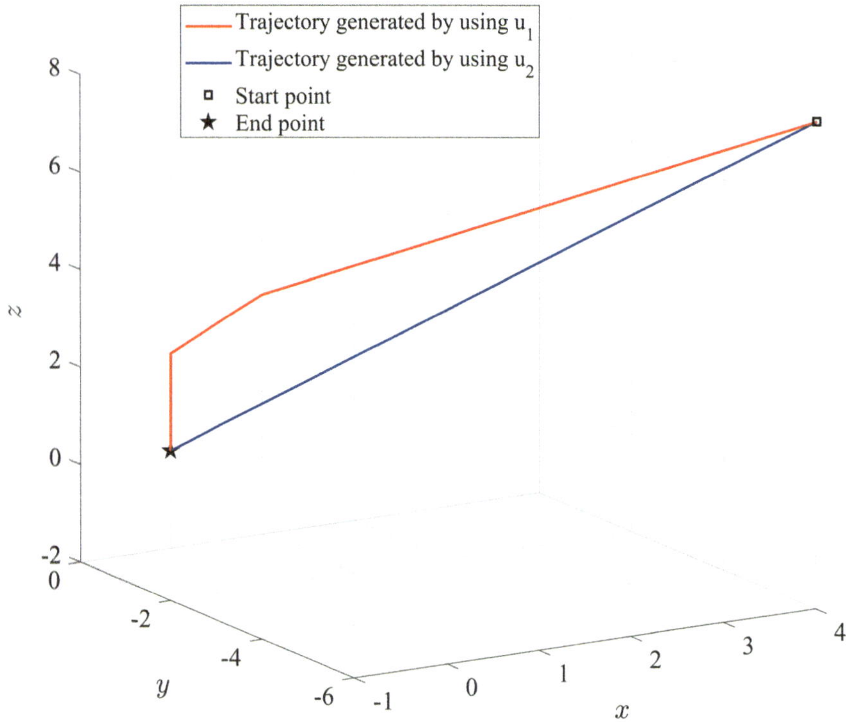

Fig. 3.3 Comparison of the trajectories governed by control laws (3.12) and (3.13)

arrive at the origin at the same time, which demonstrates that under the control law (3.13), the closed-loop system achieves SATO convergence. According to Fig. 3.3, it is shown that under the controller (3.13), the output trajectory is smoother and shorter compared with the system governed by the control law (3.12).

3.2 Continuous Time-Synchronized Control Design

In this section, for a comprehensive comparison of time-synchronized stability and finite-time stability, based on Theorem 2.2, the norm-normalized sign function (2.6) and the classical sign function (2.3) are utilized to design a continuous time-synchronized control law and a continuous finite-time control law, respectively.

For the system dynamics (3.11), we now propose a time-synchronized control law u_2, and introduce a finite-time control law u_1 for comparison

$$u_1 = -b^{-1}(x)\left(k\mathrm{sig}_{\mathrm{c}}^{\alpha}(x) + f(x)\right), \tag{3.27}$$

$$u_2 = -b^{-1}(x)\left(k\mathrm{sig}_{\mathrm{n}}^{\alpha}(x) + f(x)\right), \tag{3.28}$$

where k is a positive control gain, and α is defined as $0 < \alpha = \alpha_1/\alpha_2 < 1$ with positive odd integers α_1 and α_2.

Theorem 3.2 *Under control laws (3.27) and (3.28), i.e., $u = u_1$ and $u = u_2$, the system system (3.11) is finite-time stable and time-synchronized stable, respectively. In addition, under the controller (3.28), the closed-loop state x is ratio persistent.*

Proof Consider a Lyapunov function $V_1 = x^T x$. Taking the derivative of V_1, based on the finite-time control law (3.27), we have

$$\dot{V}_1 = -2kx^T \mathrm{sig}_{\mathrm{c}}^{\alpha}(x) \le -2kV_1^{\frac{\alpha+1}{2}}. \tag{3.29}$$

According to Lemma 2.5, we can derive the settling time of the finite-time convergence

$$T_1(x_0) \le \frac{V_1^{\frac{1-\alpha}{2}}(x_0)}{k(1-\alpha)}. \tag{3.30}$$

Next, for the system under the time-synchronized control law (3.28), we consider a second Lyapunov function $V_2 = x^T x$. Accordingly, its time derivative reads

$$\dot{V}_2 = -2kx^T \mathrm{sig}_{\mathrm{n}}^{\alpha}(x) = -2kV_2^{\frac{\alpha+1}{2}}. \tag{3.31}$$

On the other hand, clearly, x is ratio persistent. According to Theorem 2.2, it proves that the closed-loop system is time-synchronized stable, and the synchronized settling time is

$$T_2(x_0) = \frac{V_2^{\frac{1-\alpha}{2}}(x_0)}{k(1-\alpha)}. \tag{3.32}$$

The proof ends here. □

Remark 3.3 As presented above, the performances of the time-synchronized controller (3.28) and the finite-time controller (3.27) differ substantially. Observe from the structures of (3.28) and (3.27) that the only difference between the two controllers lies in the utilization of different sign functions (the norm-normalized sign function and the classical sign function, respectively). More specifically, the outputs of $\mathrm{sig}_{\mathrm{c}}^{\alpha}(x)$ are n decoupled individual scalar, while each output element of $\mathrm{sig}_{\mathrm{n}}^{\alpha}(x)$ is coupled with the norm of the whole input vector. Thus from a qualitative aspect, the time-synchronized convergence property cannot be expected from a controller integrating only the classical sign function $\mathrm{sig}_{\mathrm{c}}^{\alpha}(x)$.

For the introduced affine system (3.11), we here evaluate the performance of the time-synchronized control law (3.28) in comparison with the finite-time controller (3.27). We consider an extremely simple case for the system (3.11), where $f(x) = 0$ and $b(x) = I_3$, leading to a single integrator in $\mathbb{R}^3$, $\dot{x} = u$.

Relative simulation examples are given in Figs. 3.4 and 3.5. The simulation parameters are chosen as $x_0 = [1, -3, 6]^T$ and $\alpha = 1/3$. As shown in Figs. 3.4a, b, both control laws (3.27) and (3.28) can guarantee that the system state x converges to the origin in finite time. In Fig. 3.4a, three state elements converge to the equilibrium separately. It takes x_1 a much longer time to converge as its initial value is far from zero. On the contrary, under the control law (3.28), time-synchronized stability is revealed in Fig. 3.4b, where x_1, x_2 and x_3 arrive at the origin synchronously.

The 3-dimensional trajectories under the finite-time controller (3.27) and the time-synchronized control law (3.28) are shown in Fig. 3.5. It suggests that time-synchronized stability helps the system generate a smoother and shorter output, as according to Theorem 3.2, the trajectory of the closed-loop system under the time-synchronized control law (3.28) is ratio persistent, being a (shortest) straight line in the 3-dimensional space.

3.3 Time-Synchronized Sliding-Mode Design

In this section, we propose a series of different time-synchronized sliding manifolds. They can force the system dynamics to move according to some predefined desirable patterns.

The time-synchronized sliding mode is governed by the following first-order nonlinear differential equations,

$$s = \dot{x} + k\mathrm{sign}_{\mathrm{n}}(x), \tag{3.33}$$

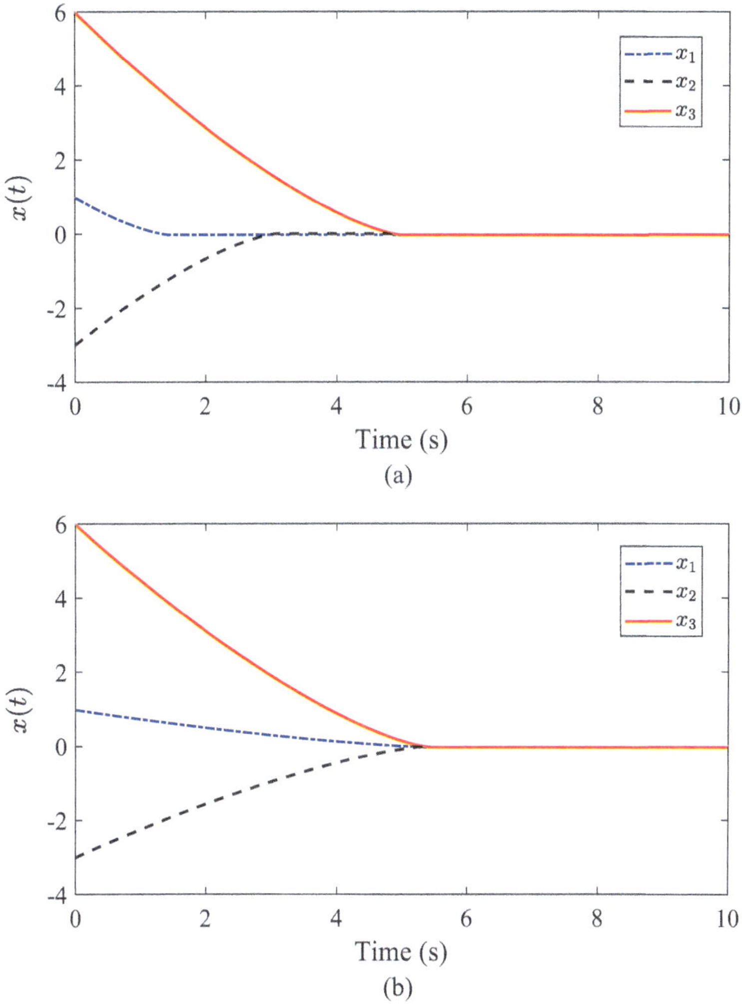

Fig. 3.4 System state x: **a** performance of the finite-time control law (3.27); **b** performance of the time-synchronized control law (3.28)

where s is the sliding manifold, x the system state, and k a positive constant. We have the following result for the time-synchronized sliding mode (3.33).

Lemma 3.1 *On the time-synchronized sliding manifold (3.33) ($s = 0$), the system state x is ratio persistent, and the closed-loop system is time-synchronized stable, with synchronized settling time*

$$T = \frac{V^{\frac{1-\alpha}{2}}(x_0)}{k(1-\alpha)}, \tag{3.34}$$

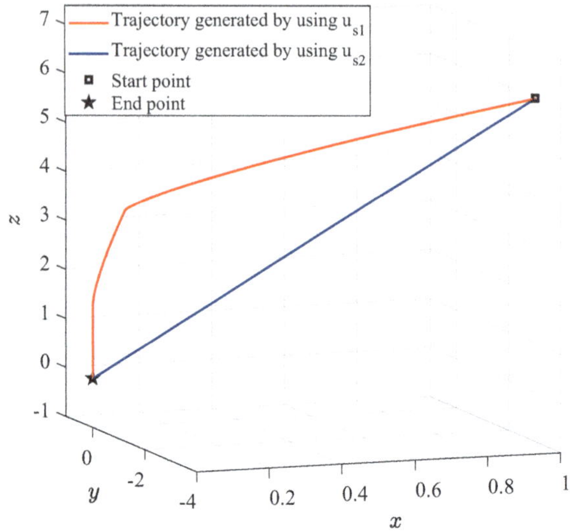

Fig. 3.5 Comparison of the trajectories under control laws (3.27) and (3.28)

where $V = x^T x$.

Proof On the sliding manifold $s = 0$, (3.33) becomes

$$\dot{x} = -k\text{sign}_\text{n}(x). \tag{3.35}$$

Then, the proof is given by following that of Theorem 3.1. □

Remark 3.4 For $x \in \mathbb{R}^n$, let us consider two sliding modes,

$$s_n = \dot{x} + k\text{sign}_\text{n}(x), \tag{3.36}$$
$$s_c = \dot{x} + k\text{sign}_\text{c}(x). \tag{3.37}$$

The sliding mode s_c transforms a multi-input control problem (with n control inputs) into a decoupled problem involving n individual subsystems associated with n single inputs. Sliding mode structures similar to s_c undergo discontinuities should any element of x changes its sign. On the contrary, due to the inherent properties of the norm-normalized sign function sign_n, the time-synchronized sliding mode s_n is absolutely continuous until $x = 0$ is reached.

As discussed in Sect. 3.1, the solution trajectories of system dynamics (3.33) should be understood in the Filippov sense. We next introduced another time-synchronized sliding mode with a continuous right-hand side and an additional adjustable power constant, i.e.,

$$s = \dot{x} + k\text{sig}_\text{n}^\alpha(x), \tag{3.38}$$

where s is the sliding manifold, x the system state, k a positive constant, and α a positive constant satisfying $1/2 < \alpha = \alpha_1/\alpha_2 < 1$ with positive odd integers α_1 and α_2.

In terms of the sliding mode (3.38), we have the following result.

Lemma 3.2 *On the time-synchronized sliding manifold (3.38) ($s = 0$), the system state x is ratio persistent, and the closed-loop system is time-synchronized stable, with time-synchronized settling time*

$$T(x_0) = \frac{\|x_0\|^{1-\alpha}}{k(1-\alpha)}. \tag{3.39}$$

Proof On the sliding manifold $s = 0$, (3.38) becomes

$$\dot{x} = -k\mathrm{sig}_{\mathrm{n}}^{\alpha}(x). \tag{3.40}$$

Then, the proof is given by following that of Theorem 3.2. □

When design a time-synchronized control law using the sliding mode (3.38), possible singularities could occur. One the sliding mode $s = 0$, we have

$$\dot{x} = -k\mathrm{sig}_{\mathrm{n}}^{\alpha}(x), \tag{3.41}$$

Differentiating $\dot{x}$ with respect to time, we have

$$\ddot{x} = -k(\alpha - 1)\|x\|^{\alpha-3}xx^T\dot{x} - k\|x\|^{\alpha-1}\dot{x}. \tag{3.42}$$

Taking (3.41) into (3.42), and using Lemma 2.2, we can get

$$\begin{aligned}\ddot{x} &= k(\alpha - 1)\|x\|^{\alpha-3}xx^T\left(k\mathrm{sig}_{\mathrm{n}}^{\alpha}(x)\right) + k\|x\|^{\alpha-1}\left(k\mathrm{sig}_{\mathrm{n}}^{\alpha}(x)\right)\\ &= k^2(\alpha - 1)\mathrm{sig}_{\mathrm{n}}^{2\alpha-1}(x) + k^2\mathrm{sig}_{\mathrm{n}}^{2\alpha-1}(x)\\ &= \alpha k^2\mathrm{sig}_{\mathrm{n}}^{2\alpha-1}(x),\end{aligned} \tag{3.43}$$

Here, by choosing $\alpha > \frac{1}{2}$, the possible singularity involved in the term $\mathrm{sig}_{\mathrm{n}}^{2\alpha-1}(x)$ could be avoided on the time-synchronized sliding mode $s = 0$. Unfortunately, in the case of $s \neq 0$ while $x = 0$, the possible singularity issue still exists.

For convenience, given a vector $z \in \mathbb{R}^n$ and a constant α, define an operation $\lceil z \rfloor^{\alpha} = \left[z_1^{\alpha}, z_2^{\alpha}, ..., z_2^{\alpha}\right]^T$.

Next, we propose a non-singular time-synchronized sliding mode.

$$s = \lceil \dot{x} \rfloor^{\frac{1}{\alpha}} + k^{\frac{1}{\alpha}}\left\lceil \mathrm{sig}_{\mathrm{n}}^{\alpha}(x)\right\rfloor^{\frac{1}{\alpha}}, \tag{3.44}$$

where s is the sliding manifold, x the system state, k a positive constant, and α a positive constant satisfying $1/2 < \alpha = \alpha_1/\alpha_2 < 1$ with positive odd integers α_1 and α_2.

The below result follows from the sliding mode (3.44).

Lemma 3.3 *On the non-singular time-synchronized sliding manifold (3.44) ($s = 0$), the system state x is ratio persistent, and the closed-loop system is time-synchronized stable, while no singularity occurs before or after the system reaches the sliding manifold.*

Proof In the case of $s = 0$, from (3.44),

$$\lceil \dot{x} \rfloor^{\frac{1}{\alpha}} = -k^{\frac{1}{\alpha}} \left\lceil \mathrm{sig}_{\mathrm{n}}^{\alpha}(x) \right\rfloor^{\frac{1}{\alpha}}, \tag{3.45}$$

$$= -k^{\frac{\alpha_2}{\alpha_1}} \left\lceil \mathrm{sig}_{\mathrm{n}}^{\frac{\alpha_1}{\alpha_2}}(x) \right\rfloor^{\frac{\alpha_2}{\alpha_1}}. \tag{3.46}$$

Taking $\lceil \bullet \rfloor^{\frac{\alpha 1}{\alpha 2}}$ from both sides of (3.46),

$$\dot{x} = -k\mathrm{sig}_{\mathrm{n}}^{\alpha}(x). \tag{3.47}$$

The transformed sliding manifold (3.47) has the same form as (3.41). Following the proof of Theorem 3.2, on the sliding manifold $s = 0$, we know the system state x is ratio persistent, and the closed-loop system is time-synchronized stable. Further, following (3.41)-(3.43), it can be verified that there exists no singularity in this case.

In the case of $x \neq 0$, taking the time derivative of s, it yields

$$\begin{aligned} \dot{s} =& \frac{1}{\alpha}\mathrm{diag}\left(\dot{x}^{\frac{1}{\alpha}-1}\right)\ddot{x} + \frac{1}{\alpha}k^{\frac{1}{\alpha}}\|x\|^{1-\frac{1}{\alpha}}\mathrm{diag}\left(x^{\frac{1}{\alpha}-1}\right)\dot{x} \\ &+ \left(1-\frac{1}{\alpha}\right)k^{\frac{1}{\alpha}}\|x\|^{-\frac{1}{\alpha}-1}\lceil x \rfloor^{\frac{1}{\alpha}}x^T\dot{x}. \end{aligned} \tag{3.48}$$

Since $\alpha \in (1/2, 1)$, by analyzing (3.48) element-wisely, we find that (3.48) is quasi-continuous and thus non-singular. This completes the proof. □

Remark 3.5 The sliding modes (3.38) and (3.44) have the same form on the manifold $s = 0$, i.e., $\dot{x} = -k\mathrm{sig}_{\mathrm{n}}^{\alpha}(x)$. These two sliding mode converge slowly when $\|x\| > 1$. This is because $\|x\|^{\alpha} \leq \|x\|$ when $\alpha < 1$ and $\|x\| > 1$. To be more specifically, when the system state x is far from the origin, even a linear sliding mode

$$s = \dot{x} + kx, \tag{3.49}$$

converges faster than the time-synchronized sliding modes (3.38) and (3.44).

Please note that on the manifold ($s = 0$) of the linear sliding mode (3.49) (which is, in fact, proportional control), the system state is ratio persistent, while no conclusion of time-synchronized convergence could be drawn since it is not finite-time stable (or we can say that time-synchronized convergence is achieved when $t \to \infty$).

To increase the convergence rate of the time-synchronized sliding modes, we propose the following fast time-synchronized sliding mode,

$$s = \dot{x} + k_1 x + k_2 \mathrm{sig}_{\mathrm{n}}^{\alpha}(x), \tag{3.50}$$

where s is the sliding manifold, x the system state, k_1 and k_2 a positive constant, and α a positive constant satisfying $1/2 < \alpha = \alpha_1/\alpha_2 < 1$ with positive odd integers α_1 and α_2.

The following result is then given.

Lemma 3.4 *On the fast time-synchronized sliding manifold (3.50) ($s = 0$), the system state x is ratio persistent, and the closed-loop system is time-synchronized stable.*

Proof On the sliding manifold $s = 0$, (3.50) becomes

$$\dot{x} = -k_1 x - k_2 \mathrm{sig}_{\mathrm{n}}^{\alpha}(x). \tag{3.51}$$

From Definition 2.11, system dynamics (3.51) is ratio persistent. Consider a Lyapunov function $V = x^T x$. Its time derivative takes the form

$$\begin{aligned}\dot{V} &= -2k_1 x^T x - 2k_2 x^T \mathrm{sig}_{\mathrm{n}}^{\alpha}(x) \\ &= -2k_1 V - 2k_2 V^{\frac{1+\alpha}{2}} \\ &\leq -2k_2 V^{\frac{1+\alpha}{2}}.\end{aligned} \tag{3.52}$$

Based on Theorem 2.2, the system state achieves time-synchronized convergence on the manifold $s = 0$, with the synchronized settling time

$$T \leq \frac{V^{\frac{1-\alpha}{2}}(x_0)}{k_2(1-\alpha)}, \tag{3.53}$$

where x_0 denotes the initial state. □

Clearly, the fast time-synchronized sliding mode (3.50) has a faster convergence rate compared with the sliding modes (3.38) and (3.44) since (3.50) involves an extra linear term when $\|x\| > 1$.

Based on the discussion in the proof of Lemma 3.3, the fast time-synchronized sliding mode (3.50) is capable of avoiding singularities in the case of $s = 0$ by choosing $\alpha \in (1/2, 1)$. However, in the case of $s \neq 0$ and $x = 0$, possible singularity issues still exist.

In the sequel, we provide another solution to avoid singularity from the perspective of building a switching manifold. The following time-synchronized sliding mode is proposed,

$$s = \dot{x} + ks_s, \tag{3.54}$$

where k is a positive constant. The switching law s_s in (3.54) takes the form,

$$s_s = \begin{cases} \mathrm{sig}_\mathrm{n}^\alpha(x), & \text{if } s^* = 0 \text{ or } s^* \neq 0,\ \|x\| > \varepsilon, \\ l_1 x + l_2 \mathrm{sig_n}^4(x), & \text{if } s^* \neq 0,\ \|x\| \leq \varepsilon, \end{cases} \tag{3.55}$$

where l_1 and l_2 are two constants satisfying

$$l_1 = \left(\frac{4}{3} - \frac{\alpha}{3}\right) \|\varepsilon\|^{\alpha-1}, \tag{3.56}$$

$$l_2 = \left(\frac{\alpha}{3} - \frac{1}{3}\right) \|\varepsilon\|^{\alpha-4}, \tag{3.57}$$

ε denotes a small constant, and the variable of the triggering condition is designed as

$$s^* = \dot{x} + k\mathrm{sig}_\mathrm{n}^\alpha(x). \tag{3.58}$$

Differentiating the switching law s_s with respect to time yields,

$$\dot{s}_s = \begin{cases} (\alpha - 1)\|x\|^{\alpha-3} x x^T \dot{x} + \|x\|^{\alpha-1}\dot{x}, & \text{if } s^* = 0 \text{ or } s^* \neq 0,\ \|x\| > \varepsilon, \\ l_1\dot{x} + 3l_2 \|x\| x x^T \dot{x} + l_2\|x\|^3\dot{x}, & \text{if } s^* \neq 0,\ \|x\| \leq \varepsilon, \end{cases} \tag{3.59}$$

By checking the switching law (3.55) and its derivative (3.59), it could be verified that they are both continuous. The key idea behind this design lies in that, in the case of $x \neq 0$ and $x \to 0$, the sliding mode is switched to a non-singular one, i.e.,

$$s = \dot{x} + kl_1 x + kl_2 \mathrm{sig}_\mathrm{n}^4(x). \tag{3.60}$$

One interesting finding is that the system state x is also ratio persistent on the general sliding manifold $s = \dot{x} + kl_1 x + kl_2 \mathrm{sig}_\mathrm{n}^4(x)$ when $s = 0$.

In accordance with the switching-based time-synchronized sliding-mode design, the following result holds.

Lemma 3.5 *On the switching-based time-synchronized sliding manifold (3.54) ($s = 0$), the system state x is ratio persistent, and the closed-loop system is time-synchronized stable, while no singularity occurs before or after the system reaches the sliding manifold.*

Proof On the sliding manifold $s = 0$, from (3.54),

$$\dot{x} = -k\mathrm{sig}_\mathrm{n}^\alpha(x). \tag{3.61}$$

Time-synchronized convergence thus readily follows from Theorem 2.2 as system dynamics (3.61) is ratio persistent and finite-time stable.

Next, we discuss the singularity problem. In the case of $s = 0$, from (3.54)

$$\begin{aligned}\ddot{x} =& -k\left(\mathrm{sig}_{\mathrm{n}}^{\alpha}(x)\right)' \\ =& -k\|x\|^{\alpha-1}\dot{x} - k(\alpha-1)\|x\|^{\alpha-3}xx^T\dot{x} \\ =& -k\|x\|^{\alpha-1}\left(-k\mathrm{sig}_{\mathrm{n}}^{\alpha}(x)\right) - k(\alpha-1)\|x\|^{\alpha-3}xx^T\left(-k\mathrm{sig}_{\mathrm{n}}^{\alpha}(x)\right) \\ =& \alpha k^2\|x\|^{2\alpha-2}x.\end{aligned} \tag{3.62}$$

The term $\|x\|^{2\alpha-2}x$ meets the following inequality,

$$\left\| \|x\|^{2\alpha-2}x \right\| \le \|x\|^{2\alpha-1}. \tag{3.63}$$

Here, the singularity is avoided by simply selecting $\alpha > \frac{1}{2}$.

In the case of $x \neq 0$ and $\|x\| \le \varepsilon$, the sliding mode is switched to the non-singular form, i.e.,

$$s = \dot{x} + l_1 x + l_2 \mathrm{sig}_{\mathrm{n}}^{4}(x). \tag{3.64}$$

Taking the time derivative of s, it yields

$$\dot{s} = \ddot{x} + l_1\dot{x} + l_2\left(3\|x\|xx^T\dot{x} + \|x\|^3\dot{x}\right), \tag{3.65}$$

which is clearly non-singular. The proof ends here. □

To achieve a faster convergence rate, one can also apply the idea of the fast time-synchronize sliding mode (3.50) to the switching-based sliding mode (3.54), based on which, we can rewrite the sliding mode (3.54) as follows,

$$s = \dot{x} + k_1 x + k_2 s_s, \tag{3.66}$$

where the switching law s_s takes the same form as (3.55). And the following result for the sliding mode (3.66) follows.

Lemma 3.6 *On the fast switching-based time-synchronized sliding manifold (3.66) ($s = 0$), the system state x is ratio persistent, and the closed-loop system is time-synchronized stable, while no singularity occurs before or after the system reaches the sliding manifold.*

Proof The proof readily follows from those of Lemmas 3.4 and 3.5. □

Tables 3.1 and 3.2 summarize the abbreviations and features of time-synchronized sliding modes. Their types of singularity, continuity and convergence rate are provided.

Remark 3.6 The time-synchronized sliding mode introduced in this subsection can also be extended and applied to design the fixed/predefined-time-synchronized control law, based on the sufficient conditions of fixed-time-synchronized stability and predefined-time-synchronized stability. And this will be explored in Chaps. 5 and 6.

Table 3.1 Abbreviations for time-synchronized sliding modes

Abbreviations	Time-synchronized sliding modes
TSM	Time-synchronized sliding mode
FTSM	Fast time-synchronized sliding mode
NTSM	Non-singular time-synchronized sliding mode
FNTSM	Fast non-singular time-synchronized sliding mode

Table 3.2 Features of time-synchronized sliding modes

Structures	Classes	Singularity	Continuity	Convergence Rate
$s = \dot{x} + k\,\mathrm{sign}_{\mathrm{n}}(x)$	TSM	Possible	Quasi-continuous	Constant rate
$s = \dot{x} + k\mathrm{sig}_{\mathrm{n}}^{\alpha}(x)$	TSM	Possible	Continuous	Slow when $\|x\| \geq 1$
$s = \lceil \dot{x} \rfloor^{\frac{1}{\alpha}} + k^{\frac{1}{\alpha}} \left(\mathrm{sig}_{\mathrm{n}}^{\alpha}(x)\right)^{\frac{1}{\alpha}}$	NTSM	Non-singular	Quasi-continuous[a]	Slow when $\|x\| \geq 1$
$s = \dot{x} + k_1 x + k_2 \mathrm{sig}_{\mathrm{n}}^{\alpha}(x)$	FTSM	Possible	Continuous	Improved rate
$s = \dot{x} + k s_s$[b]	NTSM	Non-singular	Continuous	Slow when $\|x\| \geq 1$
$s = \dot{x} + k_1 x + k_2 s_s$[b]	FNTSM	Non-singular	Continuous	Improved rate

[a]The sliding mode s itself is continuous while its time derivative $\dot{s}$ is quasi-continuous
[b]The switching law s_s takes the form as in (3.55)

Remark 3.7 Actually, the proposed time-synchronized sliding modes in this subsection are closely related to the concept of the "terminal sliding mode". In general, a terminal sliding mode guarantees finite-time convergence of the closed-loop system state on the sliding manifold. It can be transformed into a time-synchronized sliding mode by properly integrating additional property of "ratio persistence". In this way, we are able to extend most results on terminal sliding modes to time-synchronized ones based on Theorem 2.2. On the other hands, despite a series of introduced time-synchronized sliding modes, there are far more elegant sliding mode techniques in the literature, which do not necessarily hold finite-time convergence, e.g., the abundant results in the survey papers (and the references therein) on sliding model control [8, 11–13]. Here, a relationship of different types of time-synchronized sliding modes is provided in Fig. 3.6 to clearly clarify their connections.

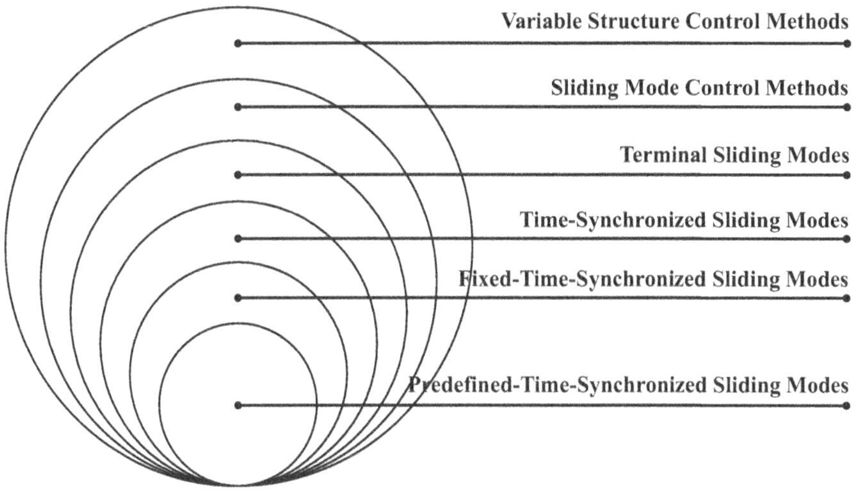

Fig. 3.6 Relationship of different types of time-synchronized sliding modes

3.4 Summary

A series of time-synchronized sliding modes have been well introduced in this chapter. The quasi-continuous time-synchronized sliding modes have been articulated by differential inclusions. Next, the time-synchronized sliding mode, fast time-synchronized sliding mode, non-singular time-synchronized sliding mode, etc., have been constructed. It is noteworthy that all these time-synchronized sliding modes are capable of achieving also fixed-/predefined-time-synchronized convergence after proper modifications. The sliding mode method is in general robust to matched disturbances. This property will be utilized in the design of time-synchronized control laws for disturbed systems in the following chapters.

References

1. Clempner, J.B., Yu, W.: New perspectives and applications of modern control theory: In honor of Alexander S. Poznyak. Springer (2017)
2. Deimling, K.: Multivalued differential equations. Walter de Gruyter (2011)
3. Filippov, A.F.: Differential Equations with Discontinuous Right-Hand Side. Dordrecht, The Netherlands: Kluwer (1988)
4. Filippov, A.F.: Differential equations with discontinuous righthand sides: control systems, vol. 18. Springer Science & Business Media (2013)
5. Gonzalez, T., Moreno, J.A., Fridman, L.: Variable gain super-twisting sliding mode control. IEEE Transactions on Automatic Control **57**(8), 2100–2105 (2011)
6. Horn, R., Johnson, C.: Matrix analysis. Cambridge university press, Cambridge UK (1990)
7. Jeffreys, H., Jeffreys, B., Swirles, B.: Methods of mathematical physics. Cambridge university press (1999)

8. Latosiński, P.: Sliding mode control based on the reaching law approach - A brief survey. 2017 22nd International Conference on Methods and Models in Automation and Robotics, MMAR 2017 pp. 519–524 (2017). https://doi.org/10.1109/MMAR.2017.8046882
9. Nagesh, I., Edwards, C.: A multivariable super-twisting sliding mode approach. Automatica **50**(3), 984–988 (2014). https://doi.org/10.1016/j.automatica.2013.12.032
10. Neubrunn, T.: Quasi-continuity. Real Analysis Exchange **14**(2), 259–306 (1988)
11. Panchade, V.M., Chile, R.H., Patre, B.M.: A survey on sliding mode control strategies for induction motors. Annual Reviews in Control **37**(2), 289–307 (2013). https://doi.org/10.1016/j.arcontrol.2013.09.008
12. Pisano, A., Usai, E.: Sliding mode control: A survey with applications in math. Mathematics and Computers in Simulation **81**(5), 954–979 (2011). https://doi.org/10.1016/j.matcom.2010.10.003
13. Šabanovic, A.: Variable structure systems with sliding modes in motion control - A survey. IEEE Transactions on Industrial Informatics **7**(2), 212–223 (2011). https://doi.org/10.1109/TII.2011.2123907

Chapter 4
Time-Synchronized Control for Disturbed Systems

Abstract In this chapter, we first propose time-synchronized control design for a class of affine systems and the Euler-Lagrange systems under matched disturbances. Next, a more general MIMO system is investigated to achieve also time-synchronized stability, under matched and mismatched disturbances. To expedite the control design, the general MIMO system is classified into two types based on the dimension of the system output and input: (1) the input-dimension-dominant MIMO system; and (2) the state-dimension-dominant MIMO system. According to the attributes of the two types of systems, robust time-synchronized controllers are structured respectively. In addition, for a state-dimension-dominant MIMO system, the time-synchronized controller has a more complex form since the right inverse of a crucial system matrix no longer exists. As a consequence, only semi-time-synchronized convergence is achieved herein, where only a subset of, rather than all of the state elements converge simultaneously. Actually, as a pertinent aside, there already exist elegant control schemes to stabilize different classes of MIMO systems with practical considerations of disturbances and model uncertainties. The methods to be introduced in this chapter are non-trivial because the state of the art existing works cannot be directly applied due to the inherent but distinctive requirements of time-synchronized stability.

4.1 Time-Synchronized Control Design for Disturbed Systems

In this section, based on the formulations of time-synchronized stability and it relevant sufficient conditions, we will explore the time-synchronized control design for system dynamics under disturbances.

D. Li et al., *Time-Synchronized Control: Analysis and Design*,
https://doi.org/10.1007/978-981-16-3089-7_4

4.1.1 Problem Formulation

We consider the following affine system in $\mathbb{R}^n$,

$$\dot{x} = h_0(x) + g_0(x)u + \Delta, \tag{4.1}$$

where $x = [x_1, x_2, \ldots, x_n]^T \in \mathbb{R}^n$ is the state vector, $h_0(x) \in \mathbb{R}^n$ and $g_0(x) \in \mathbb{R}^{n\times n}$ are the known parts, and $\Delta \in \mathbb{R}^n$ represents the unknown disturbance and uncertainties that can be lumped into the disturbance term. Usually, the disturbance and uncertainties are inevitable in practical control systems [14]. The objective here is to design a time-synchronized control law for system dynamics (4.1) such that, in spite of the unknown disturbance and uncertainties Δ, the closed-loop system state x is time-synchronized stable in accordance with Definition 2.10. In the sequel, we assume that the following assumption on the unknown system dynamics is in force.

Assumption 4.1 $\dot{\Delta}$ is bounded by a known scalar δ, i.e., $||\dot{\Delta}|| \leq \delta$.

Remark 4.1 Assumption 4.1 is mild with regard to the considered disturbance, which are common in the literature, e.g., in [2, 12]. In real-world applications, when the upper bound of the unknown disturbances is hard to be acquired in prior, the control engineers can first choose a relative large bound and then adjust its value by the trial and error method.

4.1.2 Control Design

In what follows, we introduce a disturbance observer to accurately estimate the unknown term Δ in finite time. Based on the disturbance observer, a time-synchronized control scheme is proposed. Then, according to the proposed Lyapunov conditions, the equilibrium of the closed-loop system (4.1) is proved to be time-synchronized stable.

Inspired by the multivariable super-twisting sliding mode approach [12], we construct the following observer,

$$\dot{z}_0 = -k_1 \frac{z_0 - x}{\|z_0 - x\|^{1/2}} - k_2(z_0 - x) + h_0(x) + g_0(x)u + z_1, \tag{4.2}$$

$$\dot{z}_1 = -k_3 \frac{z_0 - x}{\|z_0 - x\|} - k_4(z_0 - x), \tag{4.3}$$

where z_0 and z_1 are used to approximate the state vector x and the unknown term Δ, respectively, and k_1, k_2, k_3 and k_4 are positive observer gains.

Defining $\tilde{z}_0 = z_0 - x$ and $\tilde{z}_1 = z_1 - \Delta$, we can further get the error dynamics

$$\dot{\tilde{z}}_0 = -k_1 \frac{\tilde{z}_0}{\|\tilde{z}_0\|^{1/2}} - k_2 \tilde{z}_0 + \tilde{z}_1, \tag{4.4}$$

$$\dot{\tilde{z}}_1 = -k_3 \frac{\tilde{z}_0}{\|\tilde{z}_0\|} - k_4 \tilde{z}_0 - \dot{\Delta}. \tag{4.5}$$

Readily follow from the result in [12], we make the following assumption.

Assumption 4.2 $g_0(x)$ is an $n \times n$ non-singular mapping $(\det(g_0(x)) \neq 0)$.

Assumption 4.2 is common in sliding-mode control results for affine systems [5]. We recall the following result for the system (4.4)–(4.5).

Lemma 4.1 *[12] Considering the system* (4.4) *and* (4.5), *under Assumptions 4.1–4.2, a range of values for* k_1, k_2, k_3 *and* k_4 *can be chosen to drive* $\tilde{z}_0$ *and* $\tilde{z}_1$ *to the origin in finite time bounded by*

$$T_1 \leq \frac{2}{c} V_0^{\frac{1}{2}}, \quad V_0 = X_0^T P X_0, \tag{4.6}$$

where

$$X_0 = \left(\frac{\tilde{z}_0^T(0)}{\|\tilde{z}_0(0)\|^{1/2}}, \tilde{z}_0^T(0), \tilde{z}_1^T(0) \right)^T, \tag{4.7}$$

$$c = \frac{2\lambda_{\max}(P)}{\lambda_{\min}(\Omega)\sqrt{\lambda_{\min}(P)}}, \tag{4.8}$$

and $P \in \mathbb{R}^{3n \times 3n}$ *and* $\Omega \in \mathbb{R}^{3 \times 3}$ *are two positive definite matrices.*

Proof For details, refer to Proposition 1 in [12]. □

Remark 4.2 The observer (4.2)–(4.3) has discontinuous right hand sides, whose solution should be understood in the Filippov sense [7].

Remark 4.3 The disturbance observer (4.2)–(4.3) introduced from [12] has a multivariable super-twisting structure. In the literature, many disturbance observers decouples a multi-input control problem into a couple of single input control problems (e.g., [9] and [4]). This may result in discontinuities when any of the state elements changes its sign. On the contrary, the state function (4.2)–(4.3) is continuous until $\tilde{z}_0 = 0$ is reached.

Note that h_0 and g_0 are considered to be known in (4.1). If h_0 is unknown and $\dot{h}_0$ is bounded by a known scalar, the result still holds as it can be tackled by the super-twisting observer. The problem with unknown g_0 is more challenging which is usually tackled by incorporating additional adaptive techniques into the control design.

Based on the disturbance observer (4.2)–(4.3), we propose a time-sychronized control law for the system (4.1),

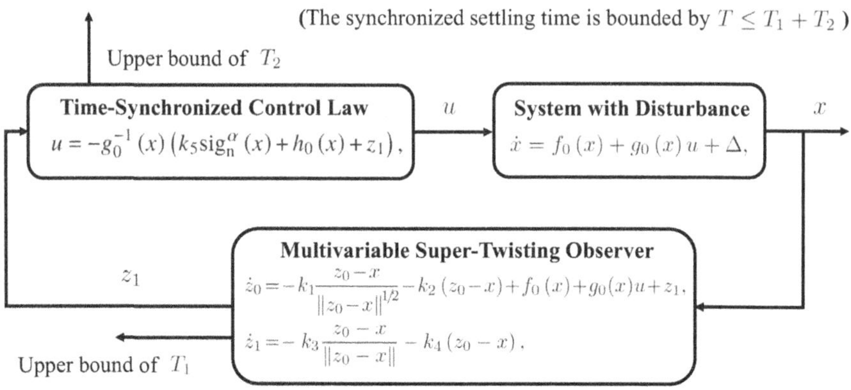

Fig. 4.1 Block diagram of the control scheme

$$u = -g_0^{-1}(x)\left(k_5 \mathrm{sig}_{\mathrm{n}}^{\alpha}(x) + h_0(x) + z_1\right), \tag{4.9}$$

where k_5 is a positive control gain, and α is defined as $0 < \alpha = \alpha_1/\alpha_2 < 1$ with positive odd integers α_1 and α_2.

Remark 4.4 For clarity, a block diagram of the control scheme (4.9) is shown in Fig. 4.1. The control architecture is simple, and the proposed algorithm contains only standard mathematical operations, using only 6 time-invariant parameters (k_1, k_2, ..., k_5 and α). Based on many numerical simulations, we found that the control is feasible over a wide range of sampling frequencies. Also, the computational complexity of the proposed control is comparable with the state of the art finite-time control, despite the additional time-synchronized performance that our proposed control guarantees.

The stability result of the time-synchronized controller (4.9) is provided here.

Theorem 4.1 *Considering the system* (4.1), *if Assumptions 4.1–4.2 hold, under the time-synchronized control law* (4.9) *and the disturbance observer* (4.2)–(4.3), *the closed-loop system is time-synchronized stable, and the system state x is ratio persistent after the observer error* $\tilde{z}_1$ *converges to zero. The synchronized settling-time is bounded by*

$$T(x_0) \le T_1 + \frac{V^{\frac{1-\alpha}{2}}(x_0)}{k_5(1-\alpha)}, \tag{4.10}$$

where T_1 *is defined in Lemma 4.1. When* $T_1 \le t \le T(x_0)$, *the analytical solution of the closed-loop system is*

$$x = \left(\|x(T_1)\|^{1-\alpha} - k_5(1-\alpha)t\right)^{\frac{1}{1-\alpha}} \frac{x(T_1)}{\|x(T_1)\|}. \tag{4.11}$$

Proof For the system (4.1) under the time-synchronized control law (4.9), we consider a simple Lyapunov function

$$V = x^T x. \tag{4.12}$$

According to the time-synchronized control law (4.9), the time derivative of V reads

$$\begin{aligned}\dot{V} &= 2x^T \left(h_0(x) + g_0(x) u + \Delta\right) \\ &= 2x^T \left(-k_5 \mathrm{sig}_\mathrm{n}^\alpha(x) - z_1 + \Delta\right).\end{aligned} \tag{4.13}$$

In the case when $t \in (0, T_1)$, based on Lemma 4.1, it is clearly that the estimation error $\tilde{z}_1$ is bounded. Then, by showing that

$$\dot{V} \leq -2k_5 V^{\frac{\alpha+1}{2}} + b, \tag{4.14}$$

where b is a positive constant subject to $\max\limits_{t\in(0,T_1)} (\tilde{z}_1(t))$, the state of the closed-loop system (4.1) is also bounded.

Next, for $\forall t \geq T_1$, from Lemma 4.1, we have $z_1 = \Delta$. Substituting the controller (4.9) into the system (4.1), we have

$$\dot{x} = -k_5 \mathrm{sig}_\mathrm{n}^\alpha(x). \tag{4.15}$$

Thus, we know x is ratio persistent when $t \geq T_1$.

Further, taking (4.15) into (4.13), it reads

$$\dot{V} = -2k_5 V^{\frac{\alpha+1}{2}}. \tag{4.16}$$

On the other hand, when $t \geq T_1$, under the time-synchronized control law (4.9), the closed-loop system becomes $\dot{x} = -k_5 \|x\|^\alpha \upsilon(x)$, which is ratio persistent as also discussed in (2.48).

Based on Theorem 2.2, the closed-loop system is time-synchronized stable, with the synchronized settling time bounded by $T \leq T_1 + T_2$, where

$$T_2(x_0) = \frac{V^{\frac{1-\alpha}{2}}(x_0)}{k_5(1-\alpha)}. \tag{4.17}$$

In addition, according to Theorem 2.2, when $T_1 < t \leq T_1 + T_2$, we have the following analytical solution of the closed-loop system

$$x = \left(\|x(T_1)\|^{1-\alpha} - k_5(1-\alpha)t\right)^{\frac{1}{1-\alpha}} \frac{x(T_1)}{\|x(T_1)\|}. \tag{4.18}$$

This concludes the proof. □

4.1.3 An Extension to Euler-Lagrange Systems

Time-synchronized control is not limited to the disturbed affine systems (4.1). In terms of the control design for more practical systems, a time-synchronized controller can be constructed on the basis of back-stepping and sliding-mode techniques.

We consider an Euler-Lagrange system

$$M(q)\ddot{q} + C(q,\dot{q})\dot{q} + g(q) = \tau(t) + \Delta, \tag{4.19}$$

where $q \in \mathbb{R}^n$ and $\dot{q} \in \mathbb{R}^n$, respectively, denote the general position vector and velocity vector, $M(q) \in \mathbb{R}^{n\times n}$ the inertia matrix, $C(q,\dot{q}) \in \mathbb{R}^{n\times n}$ the vector of Coriolis and centrifugal matrix, $g(q) \in \mathbb{R}^n$ the gravitational force, and Δ the unknown disturbance and the system uncertainties that can be lumped into the disturbance term. The objective here is to design a time-synchronized control law for system dynamics (4.19) such that, in spite of the unknown disturbance and uncertainties Δ, the closed-loop system state q is time-synchronized stable in accordance with Definition 2.10. Note that Assumption 4.1 still holds in the subsequent subsection.

We design a sliding-mode variable

$$s = \dot{q} + \gamma_1 \mathrm{sig}_{\mathrm{n}}^{p_1}(q), \tag{4.20}$$

where γ_1 is a positive control gain, and p_1 is a constant satisfying $0 < p_1 = p_1^*/p_2^* < 1$, with positive odd integers p_1^* and p_2^*.

For this sliding-mode design, we have the following result.

Lemma 4.2 *The state vector q converges to the origin time-synchronously on the surface $s = 0$ of the sliding manifold* (4.20).

Proof On the sliding surface $s = 0$, (4.20) becomes

$$\dot{q} = -\gamma_1 \mathrm{sig}_{\mathrm{n}}^{p_1}(q), \tag{4.21}$$

and the proof can be given following that of Theorem 4.1. □

For the Euler-Lagrange system (4.19), we design a time-synchronized sliding-mode control law

$$u = -M\left(\gamma_2 \mathrm{sig}_{\mathrm{n}}^{p_2}(s) + \rho(q,\dot{q})\right) + C(q,\dot{q})\dot{q} + g(q) - z_1, \tag{4.22}$$

$$\rho = \gamma_1 (p_1 - 1)\left\|q\right\|^{p_1-3} q q^T \dot{q} + \gamma_1 \left\|q\right\|^{p_2-1} \dot{q}, \tag{4.23}$$

where γ_2 is a positive control gain, and p_2 is defined satisfying $0 < p_2 = p_3^*/p_4^* > 1$, with positive odd integers p_3^* and p_4^*. With such control design, we obtain the following result for the closed-loop system.

Theorem 4.2 *Considering the Euler-Lagrange system* (4.19) *under Assumption 4.1, the time-synchronized control law* (4.22), *and the disturbance observer* (4.2)–(4.3), *the closed-loop system is time-synchronized stable. After the observer error $\tilde{z}_1$ converges to zero, the sliding-mode variable s is ratio persistent, while the closed-loop system state q is ratio persistent on the sliding surface $s = 0$.*

Proof The proof can be given based on Lemma 4.2 and Theorem 4.1. Note that one should replace x by $\dot{q}$ in (4.2)–(4.3) to make the disturbance observer compatible with the Euler-Lagrange system (4.19). Details are omitted here. □

Remark 4.5 In Theorem 4.2, as a suitable extension, the time-synchronized control law (4.22) of the Euler-Lagrange system (4.19) is proposed. The similar idea can be applied to an nth-order system by incorporating the control design (4.22) into a back-stepping control framework.

Remark 4.6 From an application perspective, if the Euler-Lagrange dynamics (4.19) models a dual-arm robot, where $q \in \mathbb{R}^n$ contains joint information of the dual arms. According to Theorem 4.2, under the controller (4.22), q is ratio persistent on the sliding surface $s = 0$, indicating that the arm joints will reach their desired values time-synchronously. Further, ratio persistence helps the two arms reach an object at the same time to achieve the goal of catching or grasping. Specifically, for a dual-arm system setup (each arm with 2 joints), the overall system can be modeled by (4.19) after some mathematical manipulations,

$$M(q)\ddot{q} + C(q, \dot{q})\dot{q} + g(q) = \tau(t) + \Delta,$$

where $q \in \mathbb{R}^4$ contains information of the 4 joint angles. We can properly build a time-synchronized controller based on the control design (4.22), which guarantees "Ratio Persistence" of q. Therefore, at the application level of this dual-arm system, "Ratio Persistence" of q indicates that the 4 joints will reach their desired values time-synchronously. Further, in this example, "Ratio Persistence" will help the two arms reach an object at the same time to achieve the goal of catching or grasping (the position of the target could be localized by computer vision).

It it noteworthy that the time-synchronized controller (4.22) has a possible singularity problem. More specifically, the term (4.23) in the controller (4.22) is like to involve negative power terms. In the case of $s = 0$ and $q \to 0$, taking (4.21) into the controller (4.23) and using Lemma 2.2, it yields

$$\begin{aligned}
\rho &= \gamma_1 (p_1 - 1) \|q\|^{p_1 - 3} q q^T \dot{q} + \gamma_1 \|q\|^{p_2 - 1} \dot{q} \\
&= -\gamma_1^2 (p_1 - 1) \|q\|^{p_1 - 3} q q^T \mathrm{sig}_{\mathrm{n}}^{p_1}(q) - \gamma_1^2 \|q\|^{p_2 - 1} \mathrm{sig}_{\mathrm{n}}^{p_1}(q) \\
&= -\gamma_1^2 (p_1 - 1) \mathrm{sig}_{\mathrm{n}}^{2p_1 - 1}(q) - \gamma_1^2 \mathrm{sig}_{\mathrm{n}}^{2p_1 - 1}(q) \\
&= -p_1 \gamma_1^2 \mathrm{sig}_{\mathrm{n}}^{2p_1 - 1}(q).
\end{aligned} \tag{4.24}$$

By choosing $p_1 > 1/2$, the singularity is avoided. However, in the case of $s \neq 0$ and $q \to 0$, the singularity still exists.

To suitably solve this problem, a non-singular time-synchronized controller is proposed by applying the following switching time-synchronized sliding mode (3.54) in Chap. 3, Sect. 3.3,

$$s = \dot{q} + \gamma_1 s_s, \tag{4.25}$$

$$s_s = \begin{cases} \mathrm{sig}_{\mathrm{n}}^{p_1}(q), & \text{if } s^* = 0 \text{ or } s^* \neq 0, \ \|q\| > \varepsilon, \\ l_1 q + l_2 \mathrm{sig_n}^4(q), & \text{if } s^* \neq 0, \ \|q\| \leq \varepsilon, \end{cases} \tag{4.26}$$

where l_1 and l_2 are two constants satisfying

$$l_1 = \left(\frac{4}{3} - \frac{p_2}{3}\right) \|\varepsilon\|^{p_2 - 1}, \tag{4.27}$$

$$l_2 = \left(\frac{p_2}{3} - \frac{1}{3}\right) \|\varepsilon\|^{p_2 - 4}, \tag{4.28}$$

ε denotes a small constant, and the variable of the triggering condition is designed as

$$s^* = \dot{q} + \gamma_1 \mathrm{sig}_{\mathrm{n}}^{p_1}(q). \tag{4.29}$$

Remark 4.7 As discussed in Chap. 3, Sect. 3.3, the switching law s and its derivative $\dot{s}$ are both continuous. In the case of $x \neq 0$ and $x \to 0$, the sliding mode is switched to a non-singular one, which directly avoids the singularity.

Applying the above switching sliding mode (4.25), the previous time-synchronized controller (4.22) becomes a non-singular one,

$$u = -M\left(\gamma_2 \mathrm{sig}_{\mathrm{n}}^{p_2}(s) + \rho\,(q, \dot{q})\right) + C(q, \dot{q})\dot{q} + g(q) - z_1. \tag{4.30}$$

The following result holds.

Lemma 4.3 *Considering the Euler-Lagrange system* (4.19) *under Assumption 4.1, the time-synchronized control law* (4.30)*, and the disturbance observer* (4.2)–(4.3)*, the closed-loop system is time-synchronized stable. After the observer error $\tilde{z}_1$ converges to zero, the sliding-mode variable s is ratio persistent, while the closed-loop system state q is ratio persistent on the sliding surface $s = 0$. No singularity occurs before or after the system reaches the sliding manifold.*

Proof The proof can be given based on Lemma 3.5, Lemma 4.2, and Theorem 4.1. □

4.1.4 Numerical Results

Case 1: Comparison with Finite-Time Control:

We examine the time-synchronized controller (4.9) to showcase the properties and merits of time-synchronized stability. Consider the plant (4.1) with functions $h_0 = 3x$, $g_0 = 2I_3$, disturbance $\Delta = [2, 3\sin(0.02\pi t), 4\sin(0.05\pi t + \pi/2)]^T$, and initial state $x_0 = [2, -7, 10]^T$. In what follows, to yield a comprehensive comparison, we introduce a typical finite-time controller.

Lemma 4.4 *For the system* (4.1)*, under the following finite-time control law*

$$\overline{u} = -g_0^{-1}(x)\left(k_5\mathrm{sig}_c^{\alpha}(x) + h_0(x) + z_1\right), \tag{4.31}$$

and the disturbance observer (4.2)–(4.3)*, the closed-loop system is finite-time stable.*

Proof The proof is clear by using Lemmas 2.5 and 4.1. □

Simulation parameters are shown in Table 4.1, where the time-synchronized controller and the finite-time controller share the same control and observer parameters to produce a fair comparison.

As presented in Figs. 4.2 and 4.3, under the time-synchronized control law (4.9), all the system state elements reach the equilibrium at the same time, while under

Table 4.1 Parameters for simulation

Parameters	Values	Parameters	Values
k_1	10	k_2	50
k_3	10	k_4	20
k_5	0.5	α	7/11

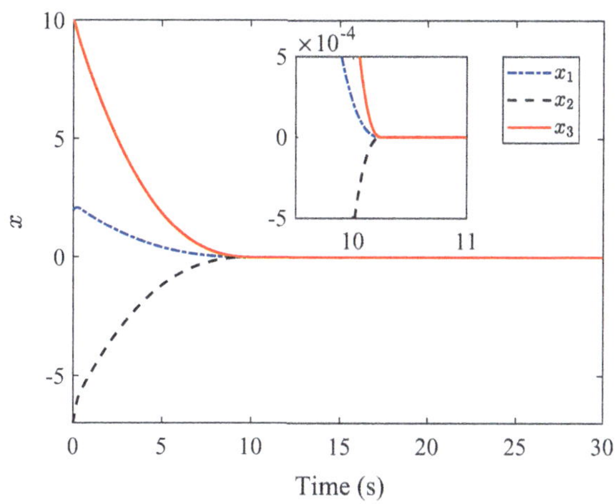

Fig. 4.2 System state x under the time-synchronized control law (4.9)

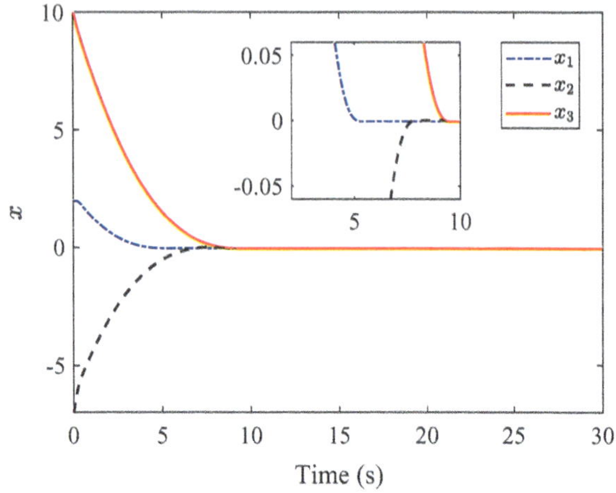

Fig. 4.3 System state x under the finite-time control law (4.31)

the finite-time law (4.31), three state elements arrive at the equilibrium at different times. Moreover, in Fig. 4.2, the state elements converge time-synchronously even when we zoom in at the level of $\times 10^{-4}$.

The property of ratio persistence is shown in Fig. 4.4. It evidently approves declared time-synchronized stability under the controller (4.9), where after T_1, the ratio of each pair of the state elements is time-invariant. Specific subfigures are also included to showcase how the ratio converges to a constant.

The 3-dimensional state trajectories under the time-synchronized control law (4.9) and the finite-time law (4.31) are presented in Fig. 4.5. The trajectory under the time-synchronized control law (4.9) (the blue curve) is divided into two parts: the trajectory s.t. $t < T_1$ and the trajectory s.t. $t \geq T_1$. It is noted that the trajectory s.t. $t < T_1$ is a curve since the estimation from the disturbance observer (4.2)–(4.3) is not

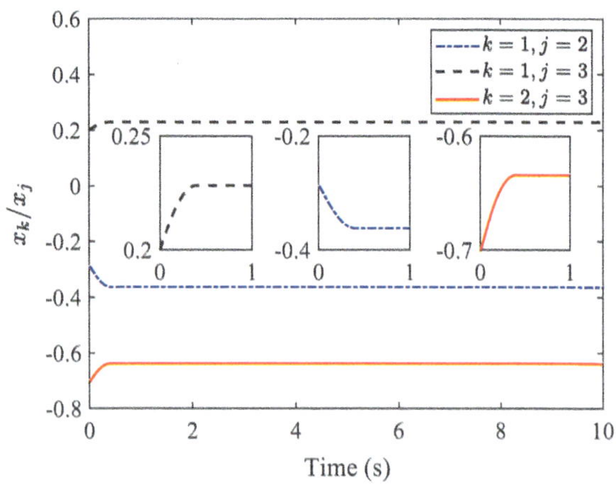

Fig. 4.4 Ratio of each pair of the state elements

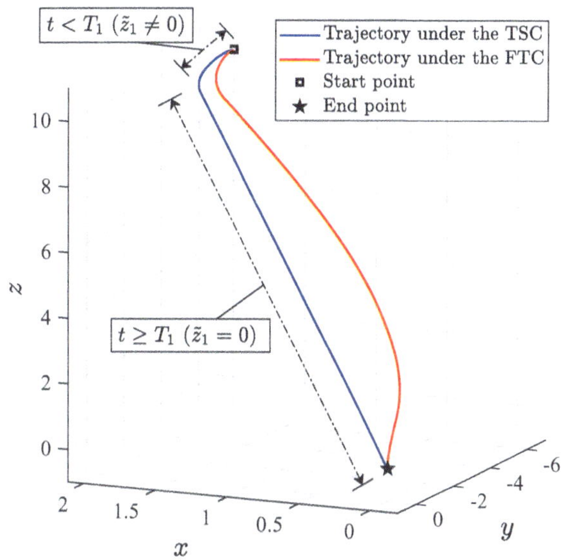

Fig. 4.5 Trajectories of x under the time-synchronized controller (TSC) and under the finite-time controller (FTC). (It can be observed that the trajectory of x governed by the TSC is the shortest path in the three-dimensional space when $t \geq T_1$.)

accurate, i.e., $\tilde{z}_1 \neq 0$. Given a precise disturbance estimation, it suggests that time-synchronized stability helps the system generate a smoother and shorter path. As according to Theorem 4.1, when $\tilde{z}_1 = 0, t \geq T_1$, the output trajectory of the closed-loop system under the time-synchronized control law (4.9) is ratio persistent, being a straight line in the 3-dimensional space. On the contrary, the trajectory governed by the finite-time law (4.31) (the red curve) is much longer.

From Fig. 4.6, the disturbance observer (4.2)–(4.3) yields an accurate estimation for both controllers (4.9) and (4.31) within 0.5 s ($T_1 < 0.5$ s). Note that the elements of the estimation error in Fig. 4.6 seem to arrive at the origin at the same time. That

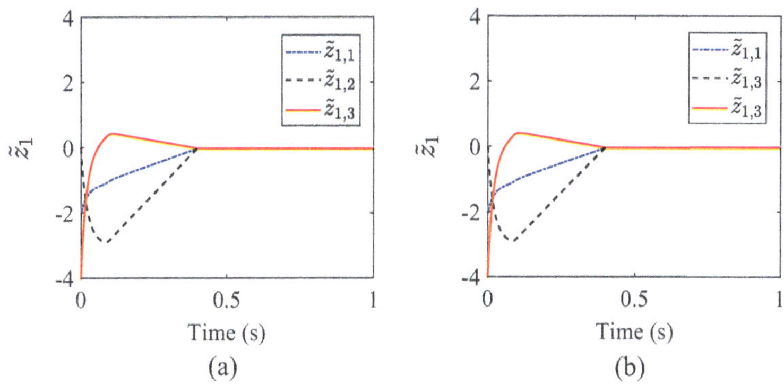

Fig. 4.6 Estimation errors $\tilde{z}_1$: **a** under the time-synchronized control law (4.9); and **b** under the finite-time control law (4.31)

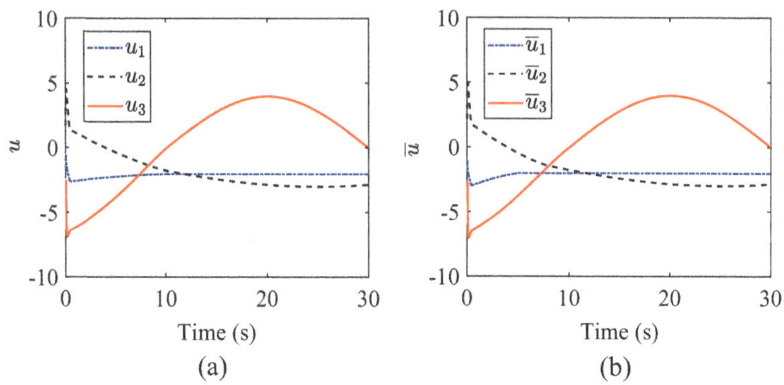

Fig. 4.7 Control inputs: **a** of the time-synchronized control law u; and **b** of the finite-time control law $\overline{u}$

is because, the error dynamics (4.4) is ratio persistent given $\tilde{z}_1 = 0$ (which is certainly achieved in finite time), while the rigorous proof of its "time-synchronized-like behavior" is beyond the scope of this subsection. As mentioned, the ratio persistence widely exists in the literature of control system design, which grants the proposed result wide applicability.

The relevant control inputs are shown in Fig. 4.7, while it seems that no significant difference can be observed between the two controllers. Further, the detailed initial values of the control inputs are provided in Fig. 4.8. It is clear that the finite-time controller requires a larger control input compared with the time-synchronized controller. This explains why compared with the time-synchronized controller, the finite-time controller seems to achieve a faster convergence rate as in Figs. 4.2 and 4.3. The control inputs in Fig. 4.7 still have huge fluctuation after T_1 since the controllers need to neutralize the external disturbance Δ. This validates the effectiveness of the controllers and the disturbance observers.

Further, we carefully test several other cases with different parameters. It shows that when $k_5 = 0.61$ is chosen for the time-synchronized controller ($k_5 = 0.5$ for the finite-time controller), under the finite-time controller, the convergence times of x_1, x_2 and x_3 are about 5.0 s, 8.0 s and 9.0 s (which is in line with the settling time in Fig. 4.3), and under the time-synchronized controller, the convergence times of x_1, x_2 and x_3 are about 8.3 s. The convergence of the system state is shown in Figs. 4.9 and 4.10. In this case, as shown in Figs. 4.11 and 4.12, the initial control input of the time-synchronized controller is still in general smaller than that of the finite-time controller.

Case 2: Comparison with Predefined-Time Control:

Predefined-time control forces the system state to converge to the origin before or at a given prespecified time instant. One may wonder whether time-synchronized convergence can be achieved by designing predefined-time controllers for each subsystem, and how time-synchronized control is different from predefined-time control.

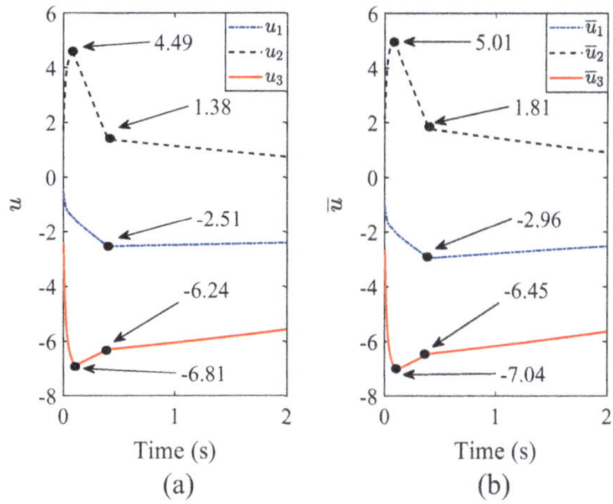

Fig. 4.8 Control inputs with detailed values: **a** of the time-synchronized control law u; and **b** of the finite-time control law $\overline{u}$. (The values with arrows denote $|u_i|$ or $|\overline{u}_i|$, $i = 1, 2, 3$ at black solid points.)

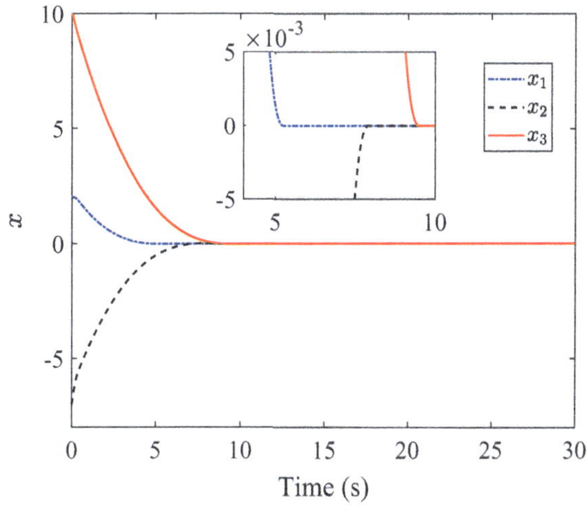

Fig. 4.9 System state x under the finite-time controller

To fully showcase the differences compared with predefined-time control, considering the plant (4.1) (the system parameters f_0, g_0, and the disturbance Δ are the same as those in *Case 1* for fairness of comparison), we would like to introduce the following predefined-time controller that is proposed in the very recent literature [13] and is also analyzed in the state of the art work [20],

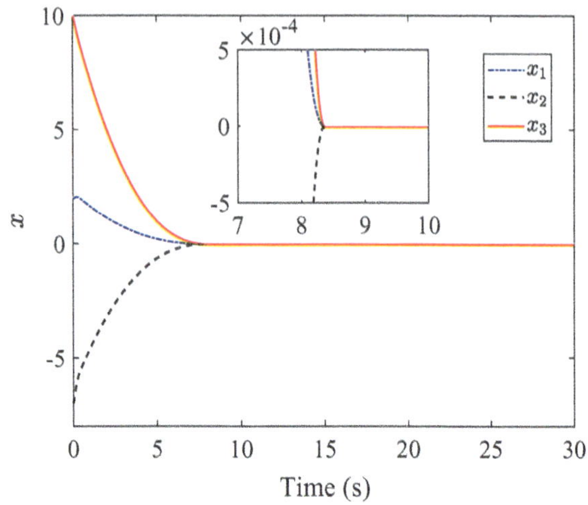

Fig. 4.10 System state x under the time-synchronized controller

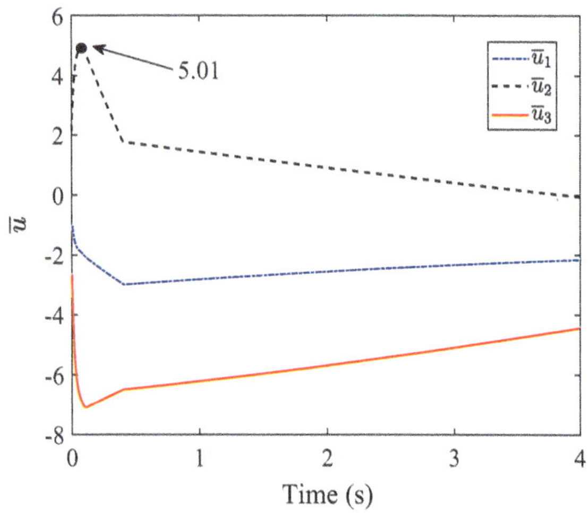

Fig. 4.11 Control inputs of the finite-time controller. (The values with arrows denote $|u_i|$ or $|\overline{u}_i|$, $i = 1, 2, 3$ at black solid points.)

$$\bar{u}_i = u_{ni} - \kappa_2 \text{sign}_c(s_i) - 3x_i, \tag{4.32}$$

$$u_{ni} = \begin{cases} \frac{-\kappa_1 (e^{x_i} - 1)}{e^{x_i}(t_f - t)}, & \text{if } 0 \le t < t_f, \\ 0, & \text{otherwise}, \end{cases} \tag{4.33}$$

$$s_i = x_i - x_i(0) - \int_0^t u_{ni} d\tau, \tag{4.34}$$

where $i = 1, 2, 3$, κ_1, κ_2 are positive control gains, and t_f is the predefined settling time. The basic idea behinds this design lies in that $u_{ni} \to \infty$ as $t \to t_f$, which guarantees the convergence at the predefined time instant t_f. The detailed proof follows from the results in [13].

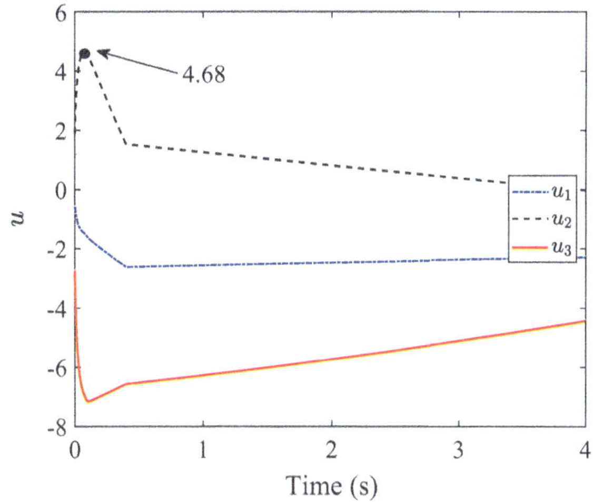

Fig. 4.12 Control inputs of the time-synchronized controller. (The values with arrows denote $|u_i|$ or $|\bar{u}_i|$, $i = 1, 2, 3$ at black solid points.)

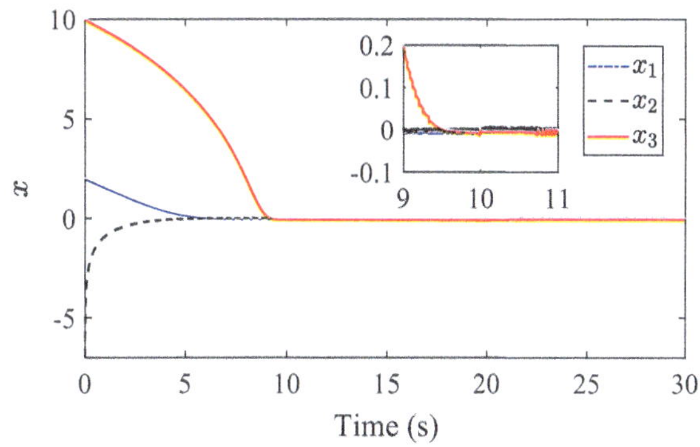

Fig. 4.13 The system state x under the predefined-time control law (4.32)

The simulation parameters for the controller (4.32) are selected as $\kappa_1 = 5$, $\kappa_2 = 4$, and $t_f = 10$. Figure 4.13 shows that all the system state elements converge to the origin before $t = 10$ s. In Fig. 4.14, the control input $\bar{u}$ is provided. The state trajectories under the time-synchronized control law (4.9) and the predefined-time law (4.32) are given in Fig. 4.15. Given a precise disturbance estimation, it suggests that the time-synchronized controller helps the system yield a smoother and shorter path. In accordance with Theorem 4.1, when $\tilde{z}_1 = 0, t \geq T_1$, the system output under the time-synchronized control law (4.9) is ratio persistent, which is a straight line. In contrast, the trajectory governed by the predefined-time control law (4.32) is longer. In summary, we note the following facts (note that the comparison is conducted under the same initial state and disturbance):

i. For the predefined-time controller (4.32), despite the same predefined settling time t_f is chosen for each subsystem, some state elements may still converge to the origin faster or slower due to the variance of the parameter selection and the

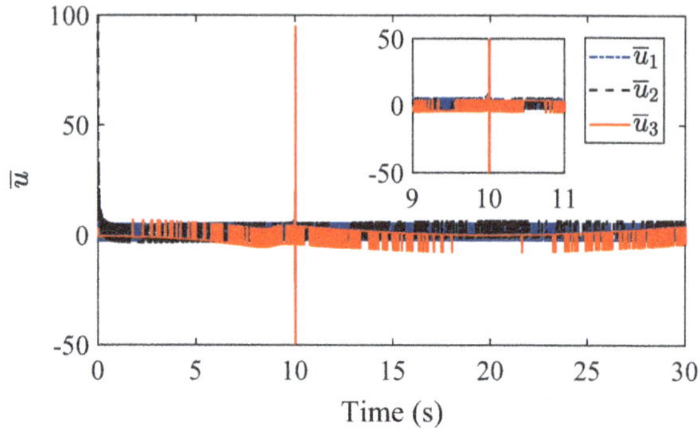

Fig. 4.14 The control input $\bar{u}$ of the predefined-time control law (4.32)

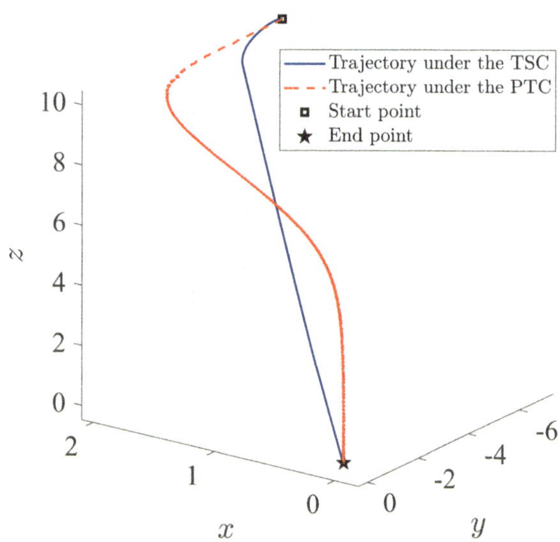

Fig. 4.15 Trajectories of x under the time-synchronized controller (TSC) and under the predefined-time controller (PTC). (It can be observed that the trajectory of x governed by the TSC is the shortest path in the three-dimensional space after an accurate disturbance estimation is acquired by the disturbance observer.)

initial state. In this case, during the convergence, some state elements may be significantly close to the origin while the others are still far from the origin (see Fig. 4.13). On the contrary, the time-synchronized controller (4.9) guarantees the same convergence time while preserving the ratio of the state elements (see Fig. 4.2).

ii. To cancel the disturbances, the chattering behavior is inevitable in the predefined-time controller (4.32) (see Fig. 4.14).
iii. When $t \to t_f$, the control input of the predefined-time (4.32) may be relevant large (see the peak in Fig. 4.14).
iv. The time-synchronized controller (4.9) yields a shorter and smoother trajectory due to the property of ratio persistence (see Fig. 4.15).
v. The predefined-time controller (4.32) strictly ensures the convergence before a predefined settling time t_f while the predefined-time controller (4.32) does not.

Table 4.2 Parameters for simulation: case 3

Parameters	Values	Parameters	Values
γ_1	0.1	γ_2	0.1
p_2	7/11	p_2	7/11
k_1	10	k_2	50
k_3	10	k_4	20
$\sigma(0)$	$[0.08, -0.04, 0.02]^T$		
Δ	$[0.05, 0.03\sin(0.02\pi t), 0.04\sin(0.05\pi t + \pi/2)]^T$		

The above observation and discussion are also in line with the analysis and simulation in [13]. Actually, the predefined-time controllers in [13, 20] hold the merit which guarantees the convergence before/at a predefined settling time t_f that explicitly integrated into the controller. In conclusion, predefined-time control and time-synchronized control have different focuses and their respective advantages.

Therefore, interesting possible future directions may include time-synchronized control with predefined settling time as will introduced in Chap. 6.

Case 3: Time-Synchronized Control For Euler-Lagrange Systems:

In this subsection, simulation results for practical Euler-Lagrange systems are provided to further verify the effectiveness.

We consider the spacecraft attitude dynamics as follows,

$$M(\sigma)\ddot{\sigma} + C(\sigma, \dot{\sigma})\dot{\sigma} = u(t) + \Delta, \tag{4.35}$$

where σ the modified Rodrigues parameters [1]. The system model (4.35) and its parameters are the same as those in [10], where the inertia matrix is

$$[11, 0.3, 0.2; 0.4, 9, 0.5; 0.2, 0.4, 8] \ \left(\text{kg m}^2\right). \tag{4.36}$$

Other simulation parameters for the controller (4.22) are given in Table 4.2.

The following simulation results are provided to fully verify the performance. In Figs. 4.16 and 4.17, both the system state and its norm are presented for clarity, where all the state elements arrive at the origin synchronously. This approves the claimed time-synchronized convergence property.

4.2 Time-Synchronized Control for General MIMO Systems under Matched and Unmatched Disturbances

In the sequel, we consider a MIMO system with matched and unmatched disturbances,

$$\dot{x} = Ax + Bu + d(t, x, u), \tag{4.37}$$

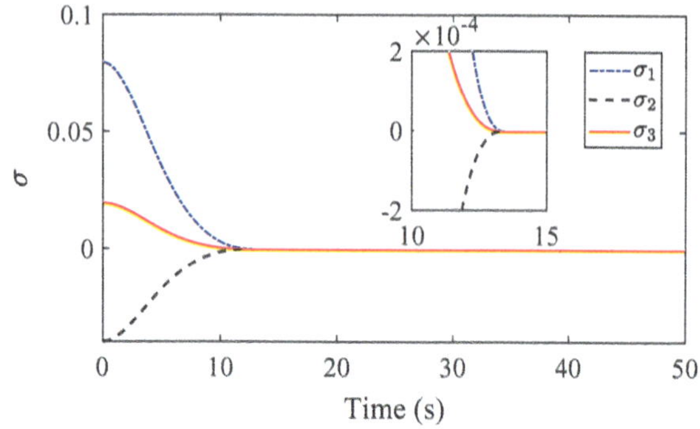

Fig. 4.16 The spacecraft attitude σ under the time-synchronized controller (4.22)

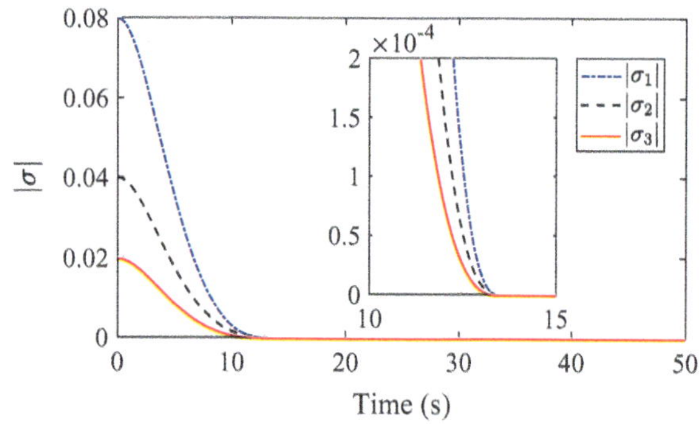

Fig. 4.17 Norm of the spacecraft attitude σ under the time-synchronized control law (4.22)

where $x \in \mathbb{R}^n$ denotes the system state, $u \in \mathbb{R}^m$ $(1 \le m \le n)$ the control input, $d(t, x, u)$ the general disturbances involving model uncertainties (e.g., parameter variations) and external disturbances, and $A \in \mathbb{R}^{n \times n}$ and $B \in \mathbb{R}^{n \times m}$ system matrices (with proper dimensions).

The following assumption on the system matrices is made.

Assumption 4.3 The matrices A and B lead to a controllable pair (A, B). The matrix B is of full rank, and by reordering the components, it is partitioned as $B = \left[B_1^T, B_2^T\right]^T$, where $B_1 \in \mathbb{R}^{(n-m)\times m}$, $B_2 \in \mathbb{R}^{m\times m}$, and $\det(B_2) \neq 0$.

By the system state transformation [5, 11, 16],

$$Tx = \begin{bmatrix} x_1 \\ x_2 \end{bmatrix}, \quad T = \begin{bmatrix} I_{n-m} & -B_1 B_2^{-1} \\ 0 & B_2^{-1} \end{bmatrix}, \tag{4.38}$$

the system dynamics (4.37) can be rewritten in a regular form,

$$\begin{aligned} \dot{x}_1 &= A_{11}x_1 + A_{12}x_2 + d_u(t, x), \\ \dot{x}_2 &= A_{21}x_1 + A_{22}x_2 + u + d_m(t, x, u), \\ y &= x_1, \end{aligned} \tag{4.39}$$

where $x = \left[x_1^T, x_2^T\right]^T$ denotes the system state ($x_1 \in \mathbb{R}^{n-m}, x_2 \in \mathbb{R}^m$), $y \in \mathbb{R}^{n-m}$ the system output , and $d_u(t, x) \in \mathbb{R}^{n-m}$ and $d_m(t, x, u) \in \mathbb{R}^m$, respectively, the components of the unmatched and matched disturbances. The matched disturbances are usually referred to those lumped into the control channel (see $d_m(t, x, u)$), while the other disturbances are classified as unmatched ones (see $d_u(t, x)$). The transformation (4.38) is of essential importance since it decouples the disturbances into unmatched and matched ones. Otherwise, it is hard to handle the system dynamics (4.37). The system matrices A_{11}, A_{12}, A_{21}, and A_{22} are of proper dimensions.

For the disturbed MIMO system (4.39), we make the following definitions according to the input and output dimensions.

Definition 4.1 (*Input-Dimension-Dominant MIMO Systems*) The system (4.37) is an input-dimension-dominant MIMO system if the dimension of the control input u is equal or larger than the dimension of the system output y, i.e., $m \geq \frac{n}{2}$.

Definition 4.2 (*State-Dimension-Dominant MIMO Systems*) The system (4.37) is an state-dimension-dominant MIMO system if the dimension of the control input u is less than the dimension of the system output y, i.e., $m < \frac{n}{2}$.

In the following subsections, we discuss the time-synchronized control for input-dimension-dominant MIMO systems and state-dimension-dominant MIMO systems, respectively.

4.2.1 Time-Synchronized Control for Input-Dimension-Dominant MIMO Systems

The system matrix A_{21} is assumed to hold the following property.

Assumption 4.4 The row echelon form of the matrix $A_{12} \in \mathbb{R}^{(n-m)\times m}$ $(m \geq \frac{n}{2})$ in the system dynamics (4.39) has pivots in each row.

In Assumption 4.1, the inequality $m \geq \frac{n}{2}$ is necessary due to the following reason.

Lemma 4.5 *For a matrix* $\in \mathbb{R}^{(n-m)\times m}$ $(m \geq \frac{n}{2})$, *the necessary and sufficient condition to have a right inverse is that its row echelon form has pivots in each row (having the maximum number of pivots).*

Proof *(Necessity:)* Given $m < \frac{n}{2}$, define matrices $A \in \mathbb{R}^{(n-m)\times m}$ and $B \in \mathbb{R}^{m\times(n-m)}$. Assume that $AB = I_{n-m}$. It implies that the columns of A span $\mathbb{R}^{n-m}$ while the columns of B are linearly independent. Show that the row echelon form of A has pivots in every row and the row echelon form of B has pivots in every column.

(Sufficiency:) Assume that the row echelon form of A has pivots in every row. The columns of A thus span $\mathbb{R}^{n-m}$, i.e., $Ax = e_i$ has a solution for $\forall i \in \{1, 2, ..., n - m\}$. Denote $b_1, b_2, ..., b_{n-m}$ be the corresponding solutions ($Ab_i = e_i$). Define $B = [b_1, b_2, ..., b_{n-m}]$. Therefore, $AB = I_{n-m}$. □

Lemma 4.6 *There is no right-invertible matrix* $\in \mathbb{R}^{(n-m)\times m}$ *when* $m < \frac{n}{2}$ *(having more rows than columns).*

Proof Given $m < \frac{n}{2}$, define matrices $A \in \mathbb{R}^{(n-m)\times m}$ and $B \in \mathbb{R}^{m\times(n-m)}$. Assume that $AB = I_{n-m}$. This infers that the linear mapping $B : \mathbb{R}^{n-m} \to \mathbb{R}^m$ is injective. Here, a contradiction is derived since $Bx = 0$ has a nontrivial solution. To be more specific, the row echelon form of B must include at least one non-pivot column, contributing to free variables. This contradiction completes the proof. □

Remark 4.8 Similar to Lemma 4.5, one can prove that for a matrix $\in \mathbb{R}^{(n-m)\times m}$ $(m \leq \frac{n}{2})$, the necessary and sufficient condition to have a left inverse is that its row echelon form has pivots in each column (having the maximum number of pivots). The necessity can be addressed following that in the proof of Lemma 4.5, while the sufficiency can be addressed as follows. Assume that the row echelon form of a matrix $A \in \mathbb{R}^{(n-m)\times m}$ $(m \leq \frac{n}{2})$ has pivots in each column. It implies that the columns of A are linearly independent. The rows of A^T are thus likewise linearly independent. The row echelon form of A^T has pivots in each row. Therefore, there exists a matrix $B \in \mathbb{R}^{n-m\times(m)}$ satisfying $A^T B = I_m$. Further, it reads $B^T A = (A^T B)^T = I_m$.

By Lemmas 4.5 and 4.6, the necessity of Assumption 4.1 is explained. The key idea behinds this lies in that we want A_{12} to be right-invertible to ease the control design. However, the case of a state-dimension-dominant MIMO system $(m < \frac{n}{2})$ will be discussed in the next subsection, where the right inverse of A_{12} no longer exists, and a hierarchical method is proposed to stabilize the system dynamics (4.39).

In the following, we first consider the case where there exists no unmatched disturbance $(d_u(t,x) = 0)$. Assume that the matched disturbance satisfies the following conditions.

Assumption 4.5 The matched disturbance is bounded as

$$\|d_m(t,x,u)\| \leq \eta_m \|u\| + h_m(t,x), \tag{4.40}$$

where $\eta_m \geq 0$ is a known positive constant and $h_m(t,x)$ a known positive function.

We construct the following time-synchronized sliding mode

$$s = x_2 + A_{12}^{\dagger} A_{11} x_1 + c A_{12}^{\dagger} \mathrm{sig}_{\mathrm{n}}^{\alpha}(x_1), \tag{4.41}$$

where $c \in \mathbb{R}$ is a positive constant, α a positive constant satisfying $0 < \alpha = \alpha_1/\alpha_2 > 1$ with positive odd integers α_1 and α_2, and $A_{12}^{\dagger}$ the Moore-Penrose pseudo-inverse of the matrix A_{12}.

Remark 4.9 From Assumption 4.4, A_{12} has linearly independent rows (which means the matrix $A_{12}A_{12}^T$ is invertible), its right inverse $A_{12}^{\dagger}$ can be directly calculated by $A_{12}^{\dagger} = A^T(A_{12}A_{12}^T)^{-1}$. In this case, the Moore-Penrose pseudo-inverse of the matrix A_{12} is exactly the right inverse.

For the time-synchronized sliding mode (4.41), we have the following lemma.

Lemma 4.7 *On the manifold $s = 0$ of the sliding mode* (4.41)*, the output of the system* (4.39) *is ratio persistent and converges to the equilibrium time-synchronously.*

Proof On the manifold $s = 0$, Eq. (4.41) becomes

$$x_2 = -A_{12}^{\dagger} A_{11} x_1 - c A_{12}^{\dagger} \mathrm{sig}_{\mathrm{n}}^{\alpha}(x_1). \tag{4.42}$$

Taking (4.42) into the system dynamics (4.39),

$$\begin{aligned} \dot{x}_1 &= A_{11} x_1 - A_{12}\left(A_{12}^{\dagger} A_{11} x_1 + c A_{12}^{\dagger} \mathrm{sig}_{\mathrm{n}}^{\alpha}(x_1)\right) \\ &= -c\,\mathrm{sig}_{\mathrm{n}}^{\alpha}(x_1). \end{aligned} \tag{4.43}$$

From Lemma 3.2, it is concluded that the output y ($y = x_1$) of the system (4.39) is ratio persistent and achieves time-synchronized convergence. □

We aim to build a controller that forces the system state to arrive at time-synchronized sliding manifold (4.41) ($s = 0$). The relevant control design is proposed as follows,

$$u = u^* + \tau^*, \tag{4.44}$$

$$u^* = -\Theta(t, x)\,\mathrm{sign}_{\mathrm{n}}(s), \tag{4.45}$$

$$\Theta(t, x) = \frac{1}{1 - \eta_m}\left(\eta_m \left\|\tau^*\right\| + h_m(t, x) + \varsigma\right), \tag{4.46}$$

$$\begin{aligned} \tau^* = &-A_{21} x_1 - A_{22} x_2 - A_{12}^{\dagger} A_{11} \dot{x}_1 \\ &- c A_{12}^{\dagger}\left((\alpha - 1)\|x_1\|^{\alpha-3} x_1 x_1^T \dot{x}_1 + \|x_1\|^{\alpha-1} \dot{x}_1\right), \end{aligned} \tag{4.47}$$

where ς is a positive control gain, and η_m and $h_m(t, x)$ are a positive constant and a positive function introduced in Assumption 4.4.

Remark 4.10 The controller (4.44) is composite of two components, i.e, u^* and τ^*. The first term u^* directly contributes to the inequality in Lemma 2.5 to achieve finite-time stability. The second term τ^* is to cancel the coupling term in the Lyapunov analysis induced by the time derivative of the sliding mode vector s. Please note that the norm-normalized sign function is utilized in (4.45), which is discontinuous when $s = 0$. As discussed in Chap. 3, Sect. 3.1, it should be understood in the Filippov sense.

The following theorem holds for the proposed time-synchronized controller (4.44).

Theorem 4.3 *Considering the disturbed MIMO system* (4.39) *under Assumptions 4.4 and 4.5, and the time-synchronized control law* (4.44)*, the system state reaches the time-synchronized sliding manifold $s = 0$ in finite time, while the closed-loop* *system output y is ratio persistent on the sliding surface $s = 0$ and achieves time-synchronized convergence.*

Proof Consider a Lyapunov function $V = \frac{1}{2}s^T s$. Differentiating V with respect to time, yields

$$\dot{V} = s^T \left(\dot{x}_2 + A_{12}^{\dagger}A_{11}\dot{x}_1 + cA_{12}^{\dagger}\left((\alpha - 1)\|x_1\|^{\alpha-3}x_1x_1^T\dot{x}_1 + \|x_1\|^{\alpha-1}\dot{x}_1\right)\right). \quad (4.48)$$

Substituting the control law (4.44) into (4.48),

$$\begin{aligned}
\dot{V} &= s^T \Big((A_{21}x_1 + A_{22}x_2 + u + d_m(t, x, u)) + A_{12}^{\dagger}A_{11}\dot{x}_1 \\
&\quad + cA_{12}^{\dagger}\left((\alpha - 1)\|x_1\|^{\alpha-3}x_1x_1^T\dot{x}_1 + \|x_1\|^{\alpha-1}\dot{x}_1\right)\Big) \\
&\le (-\Theta(t, x) + \eta_m \|u\| + h_m(t, x)) \|s\| \\
&\le \left(-\Theta(t, x) + \eta_m \left\|\tau^*\right\| + \eta_m\Theta(t, x) + h_m(t, x)\right) \|s\| \\
&= \left(-(1 - \eta_m)\Theta(t, x) + \eta_m \left\|\tau^*\right\| + h_m(t, x)\right) \|s\| \\
&= \left(-\left(\eta_m \left\|\tau^*\right\| + h_m(t, x) + \varsigma\right) + \eta_m \left\|\tau^*\right\| + h_m(t, x)\right) \|s\| \\
&\le x - \sqrt{2}\varsigma V^{\frac{1}{2}}.
\end{aligned} \quad (4.49)$$

From Lemma 2.5, the system state reaches the sliding manifold $s = 0$ in finite time bounded as

$$T_1 \le \frac{\sqrt{2}}{\varsigma} V^{\frac{1}{2}}(x_0), \quad (4.50)$$

where x_0 denotes the system initial state. While sliding along the time-synchronized sliding mode $s = 0$, from Lemma 4.7, the system output y ($y = x_1$) achieves time-synchronized convergence, with synchronized settling time

$$T_2 = \frac{V_s^{\frac{1-\alpha}{2}}(x_0)}{k(1 - \alpha)}, \quad (4.51)$$

where $V_s = x^T x$.

Therefore, the total synchronized settling time is estimated by $T \le T_1 + T_2$. □

The time-synchronized controller (4.44) includes the term $\alpha \|x_1\|^{\alpha-3} x_1x_1^T\dot{x}_1 + \|x_1\|^{\alpha-1}\dot{x}_1$. A possible singularity exists when $s_1 \ne 0$ and $x_1 \to 0$. (Note that in the case of $s_1 = 0$ and $x_1 \to 0$, $\alpha \|x_1\|^{\alpha-3} x_1x_1^T\dot{x}_1 + \|x_1\|^{\alpha-1}\dot{x}_1$ contains no singularity. This could be verified by taking (4.43) into the controller (4.44).)

To deal with the singularity problem in the case of $s_1 \ne 0$ and $x_1 \to 0$, a switching-based time-synchronized sliding mode as in (3.54) and (4.25) can be employed.

We rewrite the time-synchronized sliding mode in (4.41) as

$$s_1 = x_2 + A_{12}^{\dagger}A_{11}x_1 + cA_{12}^{\dagger}s_s. \quad (4.52)$$

The switching law s_s has the following form,

$$s_s = \begin{cases} \mathrm{sig}_{\mathrm{n}}^{\alpha}(x_1), & \text{if} s^* = 0 \text{ or } s^* \neq 0,\ \|x_1\| > \varepsilon, \\ l_1 x + l_2 \mathrm{sig}_{\mathrm{n}}^{4}(x_1), & \text{if} s^* \neq 0,\ \|x_1\| \leq \varepsilon, \end{cases} \tag{4.53}$$

where ε is a small positive constant, the triggering condition s^* is defined as

$$s^* = x_2 + A_{12}^{\dagger} A_{11} x_1 + c A_{12}^{\dagger} \mathrm{sig}_{\mathrm{n}}^{\alpha}(x_1), \tag{4.54}$$

and l_1 and l_2 are two constant parameters defined as

$$l_1 = \left(\frac{4}{3} - \frac{\alpha}{3}\right) \|\varepsilon\|^{\alpha-1}, \tag{4.55}$$

$$l_2 = \left(\frac{\alpha}{3} - \frac{1}{3}\right) \|\varepsilon\|^{\alpha-4}. \tag{4.56}$$

In addition, the time derivative of s_s has the form

$$\dot{s}_s = \begin{cases} (\alpha - 1)\|x_1\|^{\alpha-3} x_1 x_1^T \dot{x}_1 + \|x_1\|^{\alpha-1} \dot{x}_1, & \text{if} s^* = 0 \text{ or } s^* \neq 0,\ \|x_1\| > \varepsilon, \\ l_1 \dot{x} + 3 l_2 \|x_1\| x_1 x_1^T \dot{x}_1 + l_2 \|x_1\|^3 \dot{x}_1, & \text{if} s^* \neq 0,\ \|x_1\| \leq \varepsilon. \end{cases} \tag{4.57}$$

It is verified that both s_s and $\dot{s}_s$ are continuous.

Accordingly, the time-synchronized controller (4.44) is modified as

$$u = u^* + \tau^*, \tag{4.58}$$

$$u^* = -\Theta(t, x)\, \mathrm{sign}_{\mathrm{n}}(s), \tag{4.59}$$

$$\Theta(t, x) = \frac{1}{1 - \eta_m} \left(\eta_m \left\|\tau^*\right\| + h_m(t, x) + \varsigma\right), \tag{4.60}$$

$$\tau^* = -A_{21} x_1 - A_{22} x_2 - A_{12}^{\dagger} A_{11} \dot{x}_1 - c A_{12}^{\dagger} \dot{s}_s. \tag{4.61}$$

In terms of the non-singular time-synchronized controller (4.45), the following result holds.

Lemma 4.8 *Considering the disturbed MIMO system* (4.39) *under Assumptions 4.4 and 4.5, and the time-synchronized control law* (4.58)*, the system state reaches the time-synchronized sliding manifold $s = 0$ in finite time, while the closed-loop system output y is ratio persistent on the sliding surface $s = 0$ and achieves time-synchronized convergence. No singularity occurs before or after the system reaches the sliding manifold.*

Proof This proof could be suitably addressed following the proof of Theorem 4.3. □

Remark 4.11 For the switching law s_s, the general sliding manifold $s_s = l_1 x_1 + l_2 \mathrm{sig}_{\mathrm{n}}^{4}(x_1)$ only takes place when $s^* \neq 0,\ \|x_1\| \leq \varepsilon$. By selecting a sufficiently small

ε, such that the trajectory of x_1 is governed by $s_s = \mathrm{sig}_{\mathrm{n}}^{\alpha}(x_1)$ rather than $s_s = l_1 x_1 + l_2 \mathrm{sig}_{\mathrm{n}}^{4}(x_1)$.

Despite the fact that the general one does not arise when ε is small enough, the closed-loop convergence property in this case ($s^* \neq 0,\ \|x_1\| \leq \varepsilon$) can still be explored.

Lemma 4.9 *In the case of $s^* \neq 0,\ \|x_1\| \leq \varepsilon$, under the control law* (4.45)*, the closed-loop system* (4.39)*, viewed as a system with the state x_1 and the input Ω, is input-to-state stable, where Ω is defined as $\Omega = d_u(t, x) + A_{12}s_1$.*

Proof In the case of $s^* \neq 0,\ \|q\| \leq \varepsilon$, according to the switching law (4.52), we have

$$
\begin{aligned}
\dot{x}_1 =& A_{11}x_1 + A_{12}\left(s_1 - A_{12}^{\dagger}A_{11}x_1 - cA_{12}^{\dagger}\left(l_1 x + l_2 \mathrm{sig}_{\mathrm{n}}^{4}(x_1)\right)\right) + d_u(t, x) \\
=& -c\left(l_1 x + l_2 \mathrm{sig}_{\mathrm{n}}^{4}(x_1)\right) + d_u(t, x) + A_{12}s_1.
\end{aligned}
\tag{4.62}
$$

Consider a Lyapunov function

$$
V_{ISS} = \frac{1}{2}x_1^T x_1. \tag{4.63}
$$

Its time derivative takes the form

$$
\begin{aligned}
\dot{V}_{ISS} &= -cx_1^T\left(l_1 x_1 + l_2 x_1 \|x_1\|^3\right) + x_1^T\left(d_u(t, x) + A_{12}s_1\right) \\
&= -c\left(l_1 + l_2\|x_1\|^3\right)x_1^T x_1 + x_1^T\left(d_u(t, x) + A_{12}s_1\right).
\end{aligned}
\tag{4.64}
$$

Denoting $L = c\left(l_1 + l_2\|x_1\|^3\right)$, we have

$$
L = c\left(\frac{4}{3} - \frac{\alpha}{3}\right)\|\varepsilon\|^{\alpha-1} + c\left(\frac{\alpha}{3} - \frac{1}{3}\right)\|\varepsilon\|^{\alpha-4}\|x_1\|^3. \tag{4.65}
$$

Noticing $\alpha < 1$ and $\|x_1\| \leq \varepsilon$, we have

$$
L \geq c\left(\frac{4}{3} - \frac{\alpha}{3}\right)\|\varepsilon\|^{\alpha-1} + c\left(\frac{\alpha}{3} - \frac{1}{3}\right)\|\varepsilon\|^{\alpha-1}. \tag{4.66}
$$

Define a positive constant $\lambda = c\|\varepsilon\|^{\alpha-1}$ and a auxiliary vector $\Omega = d_u(t, x) + A_{12}s_1$.

Equation (5.130) becomes

$$
\begin{aligned}
\dot{V}_{ISS} \leq& -\lambda x_1^T x_1 + x_1^T\Omega \\
=& -\lambda(1 - \nu)x_1^T x_1 - \left(\lambda\nu x_1^T x_1 - x_1^T\Omega\right),
\end{aligned}
\tag{4.67}
$$

where $\nu \in (0, 1)$.

Then, for $\lambda \nu x_1^T x_1 - x_1^T \Omega > 0$, we have

$$\dot{V}_{ISS} \leq -\lambda (1-\nu) x_1^T x_1. \tag{4.68}$$

By taking

$$\kappa_1(\delta) = \lambda (1-\nu) \|\delta\|^2, \tag{4.69}$$

$$\kappa_2(\delta) = \frac{\|\delta\|}{\lambda \nu}, \tag{4.70}$$

where the variable $\delta \in \mathbb{R}^n$, we obtain that

$$\dot{V}_{ISS} \leq -\kappa_1 (x_1), \tag{4.71}$$

if $\|x_1\| \geq \kappa_2(\Omega)$. This concludes the proof by showing that, according to Definition 2.5 and Lemma 2.4, the closed-loop system is ISS. The corresponding ISS-Lyapunov function is $V_{ISS} = \frac{1}{2} x_1^T x_1$, with the Lyapunov gain $\kappa_2(\Omega)$. □

Remark 4.12 The time-synchronized controllers (4.44) and (4.45) contain the term $u^* = -\Theta(t, x)\,\text{sign}_\text{n}(s)$ that may hinder real-world applications due to the possible induced frequent chattering. One common solution to alleviate the switching is to replace u^* by its saturated form

$$u^* = \begin{cases} -\Theta(t,x)\text{sign}_\text{n}(s), & \text{if } \|s\| \geq \varepsilon, \\ -\Theta(t,x)\frac{s}{\varepsilon}, & \text{if } \|s\| < \varepsilon, \end{cases} \tag{4.72}$$

where ε is a small constant. This modification is also called as the *boundary layer method*. Under this modified u^*, the original (4.43) becomes

$$\begin{aligned} \dot{x}_1 &= A_{11}x_1 - A_{12}\left(A_{12}^{\dagger}A_{11}x_1 + cA_{12}^{\dagger}\text{sig}_\text{n}^{\alpha}(x_1) - s\right) \\ &= -c\text{sig}_\text{n}^{\alpha}(x_1) + A_{12}s. \end{aligned} \tag{4.73}$$

Considering a Lyapunov candidate $V = \frac{1}{2}x_1^T x_1$, its time derivative is given by,

$$\begin{aligned} \dot{V} &= x_1^T\left(-c\text{sig}_\text{n}^{\alpha}(x_1) + A_{12}s\right) \\ &\leq -cx_1^T\text{sig}_\text{n}^{\alpha}(x_1) + x_1^T A_{12}s \\ &\leq -c\left(\|x_1\|^{\alpha} - \|A_{12}\varepsilon\|\right)\|x_1\|. \end{aligned} \tag{4.74}$$

(Please note that here, we can trivially get $s \leq \varepsilon$ following the proof of Theorem 4.3.) It is observed from (4.74) that $\dot{V} < 0$ when $\|x_1\| > \|A_{12}\varepsilon\|^{\frac{1}{\alpha}}$. Therefore, the system output is bounded by $\|x_1\| > \|A_{12}\varepsilon\|^{\frac{1}{\alpha}}$. That is how this method got its name.

In Theorem 4.3 and Lemma 4.8, the time-synchronized control for the disturbed MIMO system (4.39) is studied (with or without the consideration of singularity avoidance), assuming that no unmatched disturbance exists. When the unmatched disturbance is taken into consideration, the time-synchronized control design in Theorem 4.3 becomes invalid as the Eq. (4.42) contains an additional disturbance term (although under proper assumptions, the system output y can still converge to a bounded region around the origin).

In the sequel, we consider the case where the unmatched disturbance exists, i.e., $d_u(t, x) \neq 0$. The idea is to design a full-order sliding mode, in which x_2 plays as a virtual controller driving x_1 to the origin time-synchronously. Specifically, we propose a desired virtual control input ϕ_2 to substitute x_2 (to further drive x_1 to the origin time-synchronously) and then a controller to make x_2 track the virtual control ϕ_2.

The following necessary assumptions are provided.

Assumption 4.6 The disturbances are bounded as

$$\|d_u(t, x)\| \leq \eta_u \|x\| + h_u, \tag{4.75}$$
$$\|d_m(t, x, u)\| \leq \eta_m \|u\| + h_m(t, x), \tag{4.76}$$

where $\eta_u \geq 0$, $\eta_m \geq 0$, and $h_u \in [0, 1)$ are known positive constants, and $h_m(t, x)$ is a known positive function.

Assumption 4.7 The disturbances are bounded as

$$\left\|\dot{d}_u(t, x)\right\| \leq \kappa_u \|x\| + \vartheta_u, \tag{4.77}$$
$$\left\|\dot{d}_m(t, x, u)\right\| \leq \kappa_m \|u\| + \vartheta_m(t, x), \tag{4.78}$$

where κ_u, κ_m, and ϑ_u are known positive constants, and $\vartheta_m(t, x)$ is a known positive function.

Remark 4.13 Assumptions 5.3 and 5.4 are common in the literature, e.g., similar assumptions are likewise made in [2, 3, 12, 19] to ease the analysis of the control design of uncertain/disturbed systems.

We first define a time-synchronized sliding mode

$$s_1 = \dot{x}_1 + c\,\mathrm{sig}_{\mathrm{n}}^{\alpha}(x_1), \tag{4.79}$$

where c is a positive constant.

Recall Lemma 3.2, on the time-synchronized sliding surface (4.79) ($s_1 = 0$), x_1 achieves time-synchronized convergence with synchronized settling time,

$$T_{m1} = \frac{\|x_1(0)\|^{1-\alpha}}{c(1-\alpha)}. \tag{4.80}$$

Aiming at driving $s_1 \to 0$, a virtual velocity tracking error is defined as

$$z = x_2 - \phi_2, \tag{4.81}$$

where

$$\phi_2 = \phi_{2eq} + \phi_{2n}, \tag{4.82}$$
$$\phi_{2eq} = -A_{12}^{\dagger}\left(A_{11}x_1 + c\mathrm{sig}_{\mathrm{n}}^{\alpha}(x_1)\right), \tag{4.83}$$
$$\dot{\phi}_{2n} = -A_{12}^{\dagger}\Theta_1(x)\mathrm{sign}_{\mathrm{n}}(s_1), \tag{4.84}$$
$$\Theta_1(x) = \kappa_u\|x\| + \vartheta_u + \varsigma_1, \tag{4.85}$$

where ς_1 is a positive constant.

We will prove that s_1 converges to the origin in finite time when $z = \dot{z} = 0$ (the desired virtual velocity ϕ_2 can be used to replace x_2).

Lemma 4.10 *In the case of $z = \dot{z} = 0$, the system dynamics* (4.39) *arrives at the sliding manifold $s_1 = 0$ in finite time.*

Proof Since $x_2 = z + \phi_2$, we have

$$\dot{x}_1 = A_{11}x_1 + A_{12}(\phi_2 + z) + d_u(t, x). \tag{4.86}$$

Then, s_1 becomes

$$\begin{aligned} s_1 &= A_{11}x_1 + A_{12}(\phi_2 + z) + d_u(t, x) + c\mathrm{sig}_{\mathrm{n}}^{\alpha}(x_1) \\ &= A_{11}x_1 + A_{12}\phi_{2eq} + A_{12}\phi_{2n} + A_{12}z + d_u(t, x) + c\mathrm{sig}_{\mathrm{n}}^{\alpha}(x_1) \\ &= A_{12}\phi_{2n} + A_{12}z + d_u(t, x). \end{aligned} \tag{4.87}$$

Construct a Lyapunov function $V_1 = \frac{1}{2}s_1^T s_1$. Its time derivative is

$$\begin{aligned} \dot{V}_1 &= s_1^T\left(A_{12}\dot{\phi}_{2n} + A_{12}\dot{z} + \dot{d}_u\right) \\ &= -\Theta_1 s_1^T A_{12}A_{12}^{\dagger}\mathrm{sign}_{\mathrm{n}}(s_1) + s_1^T\dot{d}_u + s_1^T A_{12}\dot{z} \\ &\le -\Theta_1\|s_1\| + \|s_1\|\left\|\dot{d}_u\right\| + s_1^T A_{12}\dot{z} \\ &\le -\Theta_1\|s_1\| + \|s_1\|(\kappa_u\|x\| + \vartheta_u) + s_1^T A_{12}\dot{z} \\ &\le -\varsigma_1\|s_1\| + s_{21}^T A_{12}\dot{z}. \end{aligned} \tag{4.88}$$

When $z = \dot{z} = 0$, the inequality (4.88) becomes

$$\dot{V}_1 \le -\varsigma_1\|s_1\| \le -\sqrt{2}\varsigma_1 V_1^{\frac{1}{2}}. \tag{4.89}$$

From Lemma 2.5, the system state reaches the sliding manifold $s_1 = 0$ in finite time bounded by

$$T_{m2} \leq \frac{\sqrt{2}}{\varsigma} V_1^{\frac{1}{2}}(x_0). \tag{4.90}$$

The proof ends here. □

In the follow design, we try to drive z to the origin in synchronized settling time, by defining a second time-synchronized sliding mode,

$$s_2 = \dot{z} + c\mathrm{sig}_{\mathrm{n}}^{\alpha}(z). \tag{4.91}$$

Based on Lemma 3.2, on the surface (4.91) ($s_2 = 0$), z achieves time-synchronized convergence with synchronized settling time,

$$T_{m3} = \frac{\|z(0)\|^{1-\alpha}}{c(1-\alpha)}. \tag{4.92}$$

Based on the sliding mode (4.91), a full-order sliding mode controller is proposed,

$$u = u^* + \tau^*, \tag{4.93}$$
$$\dot{u}^* = -\Theta_2 \mathrm{sign}_{\mathrm{n}}(s_2), \tag{4.94}$$
$$\begin{aligned}\tau^* = &-\Big(A_{21}x_1 + A_{22}x_2 + A_{12}^{\dagger}(A_{11} + \theta(x_1))(A_{11}x_1 + A_{12}x_2) \\ &+ A_{12}^{\dagger}\Theta_1(x)\mathrm{sign}_{\mathrm{n}}(s_1) + c\mathrm{sig}_{\mathrm{n}}^{\alpha}(z)\Big),\end{aligned} \tag{4.95}$$
$$\begin{aligned}\Theta_2 = &\,\kappa_m \left\|\tau^*\right\| + \vartheta_m(t,x) + \left\|A_{12}^{\dagger}(A_{11} + \theta(x_1))\right\|(\kappa_u\|x\| + \vartheta_u) \\ &+ \left\|A_{12}^{\dagger}\dot{\theta}(x_1)\right\| \|\eta_u\|x\| + h_u\| + \kappa_m \left\|\int_0^t \Theta_2 \mathrm{sign}_{\mathrm{n}}(s_2)\,dt\right\| + \varsigma_2,\end{aligned} \tag{4.96}$$

where ς_2 is a positive constant, and a function and a function $\theta(x_1) : \mathbb{R}^{n-m} \to \mathbb{R}^{n-m} \times \mathbb{R}^{n-m}$ is defined as

$$\theta(x_1) = c(\alpha - 1)\|x_1\|^{\alpha-3} x_1 x_1^T + c\|x_1\|^{\alpha-1} I_{n-m}. \tag{4.97}$$

The controller (4.93) is quasi-continuous due to the discontinuity at $s_1 = 0$ induced by τ^*. It should be thus understood in the Filippov sense [7].

Theorem 4.4 *Considering the disturbed input-dimension-dominant MIMO system* (4.39) *under Assumptions 4.3, 4.4, 4.6, and 4.7, and the full-order time-synchronized control law* (4.93), *the system state reaches the time-synchronized sliding manifolds* $s_1 = 0$ *and* $s_2 = 0$ *in finite time, while the closed-loop system output* y *is ratio persistent on the sliding surface* $s_1 = 0$ *and achieves time-synchronized convergence.*

Proof The sliding mode s_2 can be expended as,

$$\begin{aligned} s_2 &= \dot{x}_2 - \dot{\phi}_2 + c\mathrm{sig}_{\mathrm{n}}^{\alpha}(z) \\ &= A_{21}x_1 + A_{22}x_2 + u + d_m \\ &\quad + A_{12}^{\dagger}\left(A_{11} + \theta\left(x_1\right)\right)\left(A_{11}x_1 + A_{12}x_2 + d_u(t,x)\right)) \\ &\quad + A_{12}^{\dagger}\Theta_1 \mathrm{sign}_{\mathrm{n}}\left(s_1\right) + c\mathrm{sig}_{\mathrm{n}}^{\alpha}(z). \end{aligned} \tag{4.98}$$

Taking the full-order controller (4.93) into (4.98), the sliding mode vector and its derivative become,

$$s_2 = u^* + A_{12}^{\dagger}\left(A_{11} + \theta\left(x_1\right)\right) d_u(t,x) + d_m, \tag{4.99}$$

$$\dot{s}_2 = \dot{u}^* + A_{12}^{\dagger}\left(A_{11} + \theta\left(x_1\right)\right) \dot{d}_u(t,x) + A_{12}^{\dagger}\dot{\theta}\left(x_1\right) d_u(t,x) + \dot{d}_m. \tag{4.100}$$

Remark 4.14 The derivative of s_2 in (4.100) can be acquired by the following treatment,

$$\begin{aligned} \dot{s}_2 &= \dot{u}^* + A_{12}^{\dagger}A_{11}\dot{d}_u(t,x) + \frac{\mathrm{d}}{\mathrm{d}t}A_{12}^{\dagger}\theta\left(x_1\right) d_u(t,x) + \dot{d}_m \\ &= A_{12}^{\dagger}\frac{\left(\theta\left(x_1\right) + \Delta\theta\left(x_1\right)\right)\left(d_u(t,x) + \Delta d_u(t,x)\right) - \theta\left(x_1\right) d_u(t,x)}{\Delta t} \\ &\quad + \dot{u}^* + A_{12}^{\dagger}A_{11}\dot{d}_u(t,x) + \dot{d}_m \\ &= A_{12}^{\dagger}\frac{\theta\left(x_1\right)\Delta d_u(t,x) + \Delta\theta\left(x_1\right) d_u(t,x)}{\Delta t} + \dot{u}^* + A_{12}^{\dagger}A_{11}\dot{d}_u(t,x) + \dot{d}_m \\ &= A_{12}^{\dagger}\theta\left(x_1\right)\dot{d}_u(t,x) + \dot{\theta}\left(x_1\right) d_u(t,x) + \dot{u}^* + A_{12}^{\dagger}A_{11}\dot{d}_u(t,x) + \dot{d}_m \\ &= \dot{u}^* + A_{12}^{\dagger}\left(A_{11} + \theta\left(x_1\right)\right)\dot{d}_u(t,x) + A_{12}^{\dagger}\dot{\theta}\left(x_1\right) d_u(t,x) + \dot{d}_m. \end{aligned} \tag{4.101}$$

On the other hand, define $f = \theta\left(x_1\right) d_u(t,x)$, and the manipulation in (4.100) can also be interpreted as follows,

$$\begin{aligned} \frac{\mathrm{d}}{\mathrm{d}t}A_{12}^{\dagger}f &= A_{12}^{\dagger}\left(\frac{\partial f}{\partial\theta\left(x_1\right)}\frac{\partial\theta\left(x_1\right)}{\partial t} + \frac{\partial f}{\partial d_u(t,x)}\frac{\partial d_u(t,x)}{\partial t}\right) \\ &= A_{12}^{\dagger}\left(\frac{\partial f}{\partial\,\mathrm{vec}\left(\theta\left(x_1\right)\right)}\frac{\partial\,\mathrm{vec}\left(\theta\left(x_1\right)\right)}{\partial t} + \frac{\partial f}{\partial d_u(t,x)}\frac{\partial d_u(t,x)}{\partial t}\right) \\ &= A_{12}^{\dagger}\left(\left(\left(d_u(t,x)\right)^T \otimes I_{n-m}\right)\mathrm{vec}\left(\dot{\theta}\left(x_1\right)\right) + \theta\left(x_1\right)\dot{d}_u(t,x)\right) \\ &= A_{12}^{\dagger}\left(\dot{\theta}\left(x_1\right) d_u(t,x) + \theta\left(x_1\right)\dot{d}_u(t,x)\right), \end{aligned} \tag{4.102}$$

where vec (•) denotes the vectorization form of a matrix and ⊗ the kronecker product. Clearly, the second method less straightforward but yields the same result.

Consider a Lyapunov function $V_2 = \frac{1}{2}s_2^T s_2$. From (4.93) and (4.100), its time derivative takes the form,

$$
\begin{aligned}
\dot{V}_2 &= s_2^T \left(\dot{u}^* + A_{12}^{\dagger} \left(A_{11} + \theta \left(x_1 \right) \right) \dot{d}_u(t, x) + A_{12}^{\dagger} \dot{\theta} \left(x_1 \right) d_u(t, x) + \dot{d}_m \right) \\
&\le -\Theta_2 \left\| s_2 \right\| + \left\| s_2 \right\| \left\| \dot{d}_m \right\| + \left\| s_2 \right\| \left\| A_{12}^{\dagger} \left(A_{11} + \theta \left(x_1 \right) \right) \right\| \left\| \dot{d}_u \right\| \\
&\quad + \left\| s_2 \right\| \left\| A_{12}^{\dagger} \dot{\theta} \left(x_1 \right) \right\| \left\| \eta_u \left\| x \right\| + h_u \right\| \\
&\le -\left(\Theta_2 - \kappa_m \left\| u \right\| - \vartheta_m(t, x) - \left\| A_{12}^{\dagger} \left(A_{11} + \theta \left(x_1 \right) \right) \right\| \left(\kappa_u \left\| x \right\| + \vartheta_u \right) \right. \\
&\quad \left. - \left\| A_{12}^{\dagger} \dot{\theta} \left(x_1 \right) \right\| \left\| \eta_u \left\| x \right\| + h_u \right\| \right) \left\| s_2 \right\| \\
&\le -\left\| s_2 \right\| \left(\Theta_2 - \kappa_m \left\| \tau^* \right\| - \kappa_m \left\| \int_0^t \Theta_2 \mathrm{sign_n} \left(s_2 \right) dt \right\| - \vartheta_m(t, x) \right. \\
&\quad \left. - \left\| A_{12}^{\dagger} \left(A_{11} + \theta \left(x_1 \right) \right) \right\| \left(\kappa_u \left\| x \right\| + \vartheta_u \right) - \left\| A_{12}^{\dagger} \dot{\theta} \left(x_1 \right) \right\| \left\| \eta_u \left\| x \right\| + h_u \right\| \right) \\
&\le -\sqrt{2} \varsigma_2 V_2^{\frac{1}{2}}.
\end{aligned} \tag{4.103}
$$

According to Lemma 2.5, the system state reaches the sliding manifold $s_2 = 0$ in finite time bounded by

$$
T_{m4} \le \frac{\sqrt{2}}{\varsigma} V_2^{\frac{1}{2}} \left(x_0 \right). \tag{4.104}
$$

We can now derive the total synchronized settling time to achieve time-synchronized convergence,

$$
T_m \le T_{m1} + T_{m2} + T_{m3} + T_{m4}, \tag{4.105}
$$

where T_{m1} and T_{m3} are the synchronized settling time instants defined in (4.80) and (4.92). □

Remark 4.15 The full-order controller (4.93) involves the term $\theta\left(x_1\right)$. A possible singularity exists when $s_1 \neq 0$ and $x_1 \to 0$. To eliminate this potential threat, a switching-based time-synchronized sliding mode as in (3.54) and (4.25) can be applied to replace (4.79), and the time-synchronized convergence result still holds without singularities. Detailed analysis is omitted here, which is similar to the analysis in Lemma 4.8.

Remark 4.16 The total synchronized settling time estimation in (4.105) is less conservative, in view of the fact the T_{m1} and T_{m3} (defined in (4.80) and (4.92), respectively), are not estimations but the exact value of the synchronized settling time of the time-synchronized sliding modes, which are in line with Theorem 2.2 and Lemma 3.2.

4.2.2 Semi-time-Synchronized Control for State-Dimension-Dominant MIMO Systems

We have proposed several time-synchronized controllers for the state-dimension-dominant MIMO systems (4.39) under Assumption 4.4 ($m \geq \frac{n}{2}$). In what follows, we explore time-synchronized control for the state-dimension-dominant MIMO system (the case with $m < \frac{n}{2}$). It means that the right inverse of A_{12} does not exist according to Lemma 4.6, bringing substantial challenges to the time-synchronized control design.

Not all the state-dimension-dominant MIMO systems, actually, can be driven to the equilibrium time-synchronously. Even for those can "achieve time-synchronized convergence", only a subset of the elements of the system output y are able to reach the origin at the same time. Although only semi-time-synchronized convergence is guaranteed in this subsection, there are still d_1 ($d_1 \leq m$) elements of the system output able to achieve time-synchronized convergence, where d_1 is the largest number in the controllability indices (named the controllability index) of the system $\dot{x}_1 = A_{11}x_1 + A_{12}x_2 + d_u(t, x)$, by proper state transformations treating x_2 as the control input.

In this subsection, we call this kind of convergence property "*Semi-Time-Synchronized Convergence*", where a subset of the elements of the system state arrive at the origin time-synchronously.

The following assumption on the unmatched disturbance makes semi-time-synchronized control for state-dimension-dominant MIMO systems possible.

Assumption 4.8 The unmatched disturbance satisfies $d_u(t, x) = A_{12}\omega(t, x)$, where $\omega(t, x)$ is an unknown function characterizing the disturbances. In addition, $\omega(t, x)$ meets the following constraints,

$$\|\omega(t, x)\| \leq \overline{\eta}_u \|x\| + \overline{h}_u, \tag{4.106}$$

$$\|\dot{\omega}(t, x)\| \leq \bar{\kappa}_u \|x\| + \bar{\vartheta}_u, \tag{4.107}$$

where $\bar{\kappa}_u$ and $\bar{\vartheta}_u$ are two known positive constants.

Inspired by the proposal in [6], state transformations are taken to process the system (4.39). In summary, the state transformations are given by

$$x_1 = \Gamma_1^{-1}\Gamma_2 z, \tag{4.108}$$

$$z = \Gamma_2^{-1}\Gamma_1 x_1, \tag{4.109}$$

where z is the transformed state vector, and Γ_1 and Γ_2 are transformation matrices that will be introduced later. By this state transformation process, we extract an "invertable part" from A_{12}.

The following block controllable form is derived by the state transformation $x' = \Gamma_1 x_1$, where Γ_1 is a constant matrix (for details, refer to [15, 17, 18]),

$$\begin{bmatrix} \dot{x}'_r \\ \dot{x}'_{r-1} \\ \vdots \\ \dot{x}'_3 \\ \dot{x}'_2 \\ \dot{x}'_1 \end{bmatrix} = \begin{bmatrix} A'_{r,r} & B'_{r,r-1} & \cdots & 0 & 0 & 0 \\ A'_{r-1,r} & A'_{r-1,r-1} & \cdots & 0 & 0 & 0 \\ \vdots & \vdots & \ddots & \vdots & \vdots & \vdots \\ A'_{3,r} & A'_{3,r-1} & \cdots & A'_{3,3} & B'_{3,2} & 0 \\ A'_{2,r} & A'_{2,r-1} & \cdots & A'_{2,3} & A'_{2,2} & B'_{2,1} \\ A'_{1,r} & A'_{1,r-1} & \cdots & A'_{1,3} & A'_{1,2} & A'_{1,1} \end{bmatrix} \begin{bmatrix} x'_r \\ x'_{r-1} \\ \vdots \\ x'_3 \\ x'_2 \\ x'_1 \end{bmatrix} + \begin{bmatrix} 0 \\ 0 \\ \vdots \\ 0 \\ 0 \\ B_{1,0}(x_2+\omega) \end{bmatrix}. \tag{4.110}$$

The following dimensions can be defined,

$$d_1 = \text{rank}\left(B'_{1,0}\right) = \dim\left(x'_1\right), \tag{4.111}$$

$$d_i = \text{rank}\left(B'_{i,i-1}\right) = \dim\left(x'_i\right),\ i = 2, \cdots, r, \tag{4.112}$$

where clearly, $\sum_{i=1}^{r} d_i = n - m$. Equation (4.111) implies that it is possible to utilize the right inverse of $B'_{1,0}$. And also, from rank $\left(B'_{i,i-1}\right) = \dim\left(x'_i\right),\ i = 2, \cdots, r$, in (4.112) (i.e., the dimension of the subsystem state in each block is the same as the dimension of the corresponding fictitious input in this block), the right inverse of $B'_{i,i-1}$ exists, although its inverse will not be used in the subsequent proposal.

Remark 4.17 The state transformation $x' = \Gamma_1 x_1$ involves $r-1$ steps [11]. The block controllable form has r blocks, where the dimensions of the control vector and the subsystem state coincide in each block [15, 17, 18]. The matrices $B'_{i,i-1}$ are of full rank as shown in (4.111) and (4.112), and $(d_1, ..., d_r)$ is the controllability indices (also called the controllability index vector). Moreover, d_1 is the controllability index, since $d_1 \geq d_2 \geq ... \geq d_r$.

From the block controllable form in (4.110), we get the decoupled block controllable form by another state transformation $x' = \Gamma_2 z$ (for details, refer to [11]),

$$\begin{bmatrix} \dot{z}_r \\ \dot{z}_{r-1} \\ \vdots \\ \dot{z}_3 \\ \dot{z}_2 \\ \dot{z}_1 \end{bmatrix} = \begin{bmatrix} B_r & B'_{r,r-1} & \cdots & 0 & 0 & 0 \\ 0 & N_{r-1} & \cdots & 0 & 0 & 0 \\ \vdots & \vdots & \ddots & \vdots & \vdots & \vdots \\ 0 & 0 & \cdots & N_3 & B'_{2,2} & 0 \\ 0 & 0 & \cdots & 0 & N_2 & B'_{2,1} \\ A''_{1,r} & A''_{1,r-1} & \cdots & A''_{1,3} & A''_{1,2} & A''_{1,1} \end{bmatrix} \begin{bmatrix} z_r \\ z_{r-1} \\ \vdots \\ z_3 \\ z_2 \\ z_1 \end{bmatrix} + \begin{bmatrix} 0 \\ 0 \\ \vdots \\ 0 \\ B_{1,0}(x_2+\omega) \end{bmatrix}, \tag{4.113}$$

where the transformation matrix Γ_2 takes the form,

$$\Gamma_2 = \begin{bmatrix} I & 0 & 0 & \cdots & 0 & 0 \\ -K_{r-1,r} & I & 0 & \cdots & 0 & 0 \\ -K_{r-2,r} & -K_{r-2,r-1} & I & \cdots & 0 & 0 \\ \vdots & \vdots & \vdots & \ddots & \vdots & \vdots \\ -K_{2,r} & -K_{2,r-1} & -K_{2,r-2} & \cdots & I & 0 \\ -K_{1,r} & -K_{1,r-1} & -K_{1,r-2} & \cdots & -K_{1,2} & I \end{bmatrix}, \tag{4.114}$$

where in addition, for $i = 1, \cdots, r-1$,

$$K_{i,i+1} = B_{i+1,i}^{\prime\dagger}\left(K_{i+1,i+2}B_{i+2,i+1}^{\prime} + A_{i+1,i+1}^{\prime} - N_{i+1}\right), \tag{4.115}$$

for $j = i+2, \cdots, r,\ i = 1, \cdots, r-2$,

$$K_{i,j} = B_{i+1,i}^{\prime\dagger}\left(K_{i+1,j}N_j + A_{i+1,j}^{\prime} + K_{i+1,j+1}B_{j+1,j}^{\prime} - \sum_{k=j-1}^{i+1} A_{i+1,k}^{\prime}K_{k,j}\right), \tag{4.116}$$

and

$$K_{i,j} = 0,\ \forall i \notin \{1, \cdots, r-1\} \text{ and } j \notin \{2, \cdots, r\}, \tag{4.117}$$

$$B_{i,j}^{\prime} = 0,\ \forall i \notin \{1, \cdots, r\} \text{ and } j \notin \{0, \cdots, r-1\}. \tag{4.118}$$

Based on the above, we rewrite the decoupled block control form as

$$\dot{z}_1 = \sum_{i=1}^{r} A_{1,i}^{\prime\prime} z_i + B_{1,0}\left(x_2 + \omega\right), \tag{4.119}$$

$$\dot{z}_i = N_i z_i + B_{i,i-1}^{\prime} z_{i-1},\quad i = 2, \ldots, r, \tag{4.120}$$

where N_i is a matrix defined as $N_i = -\ell_i I_{\pi_i}$, ℓ_i is a positive constant satisfying $0 < \ell_r < \ldots < \ell_2$, and π_i denotes the dimension of z_i.

Remark 4.18 The recursive structure in (4.119) and (4.120) explains why only semi-time-synchronized convergence rather than time-synchronized one can be achieved. When a time-synchronized controller is suitably proposed to force z_1 to the origin time-synchronously, $z_2, \ldots, z_r$ will arrive at the origin one after another asymptotically. Therefore, time-synchronized convergence is concluded only in d_1 dimensions.

A time-synchronized sliding mode is constructed to force $z_1 \to 0$,

$$s_1 = \dot{z}_1 + c\,\mathrm{sig}_{\mathrm{n}}^{\alpha}\left(z_1\right), \tag{4.121}$$

where c is a positive constant. Recall Lemma 3.2, on the surface (4.121) ($s_1 = 0$), z_1 achieves time-synchronized convergence with synchronized settling time,

$$T_{m1} = \frac{\|z_1(0)\|^{1-\alpha}}{c(1-\alpha)}. \tag{4.122}$$

A virtual velocity tracking error is then defined as

$$\mu = x_2 - \phi_2, \tag{4.123}$$

where ϕ_2 is introduced as follows to drive $s_1 \to 0$,

$$\phi_2 = \phi_{2eq} + \phi_{2n} \tag{4.124}$$

$$\phi_{2eq} = -B_{1,0}^{\dagger}\left(\sum_{i=1}^{r} A_{1,i}'' z_i + c\mathrm{sig}_{\mathrm{n}}^{\alpha}(z_1)\right) \tag{4.125}$$

$$\dot{\phi}_{2n} = -B_{1,0}^{\dagger}\Theta_1(x)\mathrm{sign}_{\mathrm{n}}(s_1) \tag{4.126}$$

$$\Theta_1(x) = \left(\bar{\kappa}_u \|x\| + \bar{\vartheta}_u\right)\left\|B_{1,0}\right\| + \varsigma_1 \tag{4.127}$$

where ς_1 is a predefined positive constant, and κ_u and ϑ_u are introduced in Assumption 4.7.

Lemma 4.11 *In the case of $\mu = \dot{\mu} = 0$, the system dynamics* (4.119) *arrives at the sliding manifold $s_1 = 0$ in finite time.*

Proof Since $x_2 = \mu + \phi_2$, based on (4.124), we have

$$\begin{aligned}\dot{z}_1 &= \sum_{i=1}^{r} A_{1,i}'' z_i + B_{1,0}(\mu + \phi_2 + \omega) \\ &= -c\mathrm{sig}_{\mathrm{n}}^{\alpha}(z_1) + B_{1,0}\phi_{2n} + B_{1,0}\omega + B_{1,0}\mu.\end{aligned} \tag{4.128}$$

Taking (4.128) into (4.121), s_1 and its derivative become

$$s_1 = B_{1,0}\phi_{2n} + B_{1,0}\omega + B_{1,0}\mu \tag{4.129}$$

$$\dot{s}_1 = -\Theta_1(x)\mathrm{sign}_{\mathrm{n}}(s_1) + B_{1,0}\dot{\omega} + B_{1,0}\dot{\mu}. \tag{4.130}$$

Consider a Lyapunov function $V_1 = \frac{1}{2}s_1^T s_1$. Applying Assumption 4.8, its time derivative is

$$\begin{aligned}\dot{V}_1 &= s_1^T\left(-\Theta_1(x)\mathrm{sign}_{\mathrm{n}}(s_1) + B_{1,0}\dot{\omega} + B_{1,0}\dot{\mu}\right) \\ &\le -\Theta_1(x)\|s_1\| + \left\|B_{1,0}\right\|\left(\bar{\kappa}_u\|x\| + \bar{\vartheta}_u\right)\|s_1\| + s_1^T B_{1,0}\dot{\mu} \\ &\le -\zeta_1\|s_1\| + s_1^T B_{1,0}\dot{\mu}.\end{aligned} \tag{4.131}$$

When $\mu = \dot{\mu} = 0$, the inequality (4.131) becomes

$$\dot{V}_1 \le -\varsigma_1\|s_1\| \le -\sqrt{2}\varsigma_1 V_1^{\frac{1}{2}}. \tag{4.132}$$

From Lemma 2.5, the system state reaches the sliding manifold $s_1 = 0$ in finite time bounded by

$$T_{m2} \le \frac{\sqrt{2}}{\varsigma} V_1^{\frac{1}{2}}(x_0). \tag{4.133}$$

□

Next, we aim to drive $\mu \to 0$ in synchronized settling time, using another time-synchronized sliding mode,

$$s_2 = \dot{\mu} + c\mathrm{sig}_{\mathrm{n}}^{\alpha}(\mu). \tag{4.134}$$

From Lemma 3.2, on the surface 4.91 ($s_2 = 0$), μ achieves time-synchronized convergence with synchronized settling time,

$$T_{m3} = \frac{\|\mu(0)\|^{1-\alpha}}{c(1-\alpha)}. \tag{4.135}$$

Based on the sliding mode (4.134), a full-order sliding mode controller is proposed,

$$u = u^* + \tau^*, \tag{4.136}$$

$$\dot{u}^* = -\Theta_2 \mathrm{sign}_{\mathrm{n}}(s_2), \tag{4.137}$$

$$\begin{aligned}\tau^* = -\Bigg(& A_{21}x_1 + A_{22}x_2 + B_{1,0}^{\dagger}\sum_{i=2}^{r} A_{1,i}''\left(N_i z_i + B_{i,i-1}' z_{i-1}\right) \\ & + B_{1,0}^{\dagger}\left(A_{1,1}'' + \theta(z_1)\right)\left(\sum_{i=2}^{r} A_{1,i}'' + B_{1,0}x_2\right) + B_{1,0}^{\dagger}\Theta_1(x)\mathrm{sign}_{\mathrm{n}}(s_1) + c\mathrm{sig}_{\mathrm{n}}^{\alpha}(\mu)\Bigg),\end{aligned} \tag{4.138}$$

$$\begin{aligned}\Theta_2 = {} & \kappa_m \left\|\tau^*\right\| + \vartheta_m(t,x) + \left\|B_{1,0}^{\dagger}\left(A_{1,1}'' + \theta(z_1)\right)B_{1,0}\right\|\left(\bar{\kappa}_u \|x\| + \bar{\vartheta}_u\right) \\ & + \left\|B_{1,0}^{\dagger}\dot{\theta}(z_1)\right\|\left\|\bar{\eta}_u \|x\| + \bar{h}_u\right\| + \kappa_m \left\|\int_0^t \Theta_2 \mathrm{sign}_{\mathrm{n}}(s_2)\, dt\right\| + \varsigma_2,\end{aligned} \tag{4.139}$$

where ς_2 is a positive constant, and a function and a function $\theta(z_1) : \mathbb{R}^{d_1} \to \mathbb{R}^{d_1} \times \mathbb{R}^{d_1}$ is defined as

$$\theta(z_1) = c(\alpha - 1)\|z_1\|^{\alpha-3} z_1 z_1^T + c\|z_1\|^{\alpha-1} I_{d_1}. \tag{4.140}$$

The controller (4.136) is quasi-continuous due to the discontinuity at $s_1 = 0$ induced by τ^*. It should be thus understood in the Filippov sense [7].

Theorem 4.5 *Considering the disturbed state-dimension-dominant MIMO system* (4.39) *under Assumptions 4.3, 4.6 and 4.8, and the full-order sliding-mode control law* (4.136), *the system state reaches the semi-time-synchronized sliding manifold $s_2 = 0$ and $s_1 = 0$ successively in finite time. The closed-loop system output z achieves semi-time-synchronized convergence. Specifically, z_1 is ratio persistent on the sliding surface $s_1 = 0$ and achieves time-synchronized convergence, while z_i, $i \in \{2, ..., r\}$ achieves asymptotic convergence.*

Proof The sliding mode s_2 can be expended as,

$$\begin{aligned}
s_2 &= \dot{x}_2 - \dot{\phi}_2 + c\mathrm{sig}_{\mathrm{n}}^{\alpha}(\mu) \\
&= A_{21}x_1 + A_{22}x_2 + u + d_m \\
&\quad + B_{1,0}^{\dagger}\left(\sum_{i=2}^{r} A_{1,i}''\left(N_i z_i + B_{i,i-1}' z_{i-1}\right) + \left(A_{1,1}'' + \theta(z_1)\right)\left(\sum_{i=1}^{r} A_{1,i}'' z_i + B_{1,0}(x_2 + \omega)\right)\right) \\
&\quad + B_{1,0}^{\dagger}\Theta_1(x)\mathrm{sign}_{\mathrm{n}}(s_1) + c\mathrm{sig}_{\mathrm{n}}^{\alpha}(\mu).
\end{aligned} \tag{4.141}$$

Taking the full-order controller (4.136) into (4.141), the sliding mode vector and its derivative become,

$$s_2 = u^* + B_{1,0}^{\dagger}\left(A_{1,1}'' + \theta(z_1)\right) B_{1,0}\omega + d_m, \tag{4.142}$$

$$\dot{s}_2 = \dot{u}^* + B_{1,0}^{\dagger}\left(A_{1,1}'' + \theta(z_1)\right) B_{1,0}\dot{\omega} + B_{1,0}^{\dagger}\dot{\theta}(z_1) B_{1,0}\omega + \dot{d}_m. \tag{4.143}$$

(The derivation in (4.143) can be obtained following the discussion in Remark 4.14.)

Consider a Lyapunov function $V_2 = \frac{1}{2}s_2^T s_2$. Based on (4.136) and (4.143), the time derivative of V_2 becomes,

$$\begin{aligned}
\dot{V}_2 &= s_2^T\left(\dot{u}^* + B_{1,0}^{\dagger}\left(A_{1,1}'' + \theta(z_1)\right) B_{1,0}\dot{\omega} + B_{1,0}^{\dagger}\dot{\theta}(z_1) B_{1,0}\omega + \dot{d}_m\right) \\
&\le -\Theta_2\|s_2\| + \|s_2\|\left\|\dot{d}_m\right\| + \|s_2\|\left\|B_{1,0}^{\dagger}(A_{11} + \theta(z_1)) B_{1,0}\right\|\left(\bar{\kappa}_u\|x\| + \bar{\vartheta}_u\right) \\
&\quad + \left\|B_{1,0}^{\dagger}\dot{\theta}(z_1)\right\|\left\|\bar{\eta}_u\|x\| + \bar{h}_u\right\|\|s_2\| \\
&\le -\left(\Theta_2 - \kappa_m\|u\| - \vartheta_m(t,x) - \left\|B_{1,0}^{\dagger}(A_{11} + \theta(z_1)) B_{1,0}\right\|\left(\bar{\kappa}_u\|x\| + \bar{\vartheta}_u\right)\right. \\
&\quad \left. - \left\|B_{1,0}^{\dagger}\dot{\theta}(z_1)\right\|\left\|\bar{\eta}_u\|x\| + \bar{h}_u\right\|\right)\|s_2\| \\
&\le -\left(\Theta_2 - \kappa_m\left\|\tau^*\right\| - \kappa_m\left\|\int_0^t \Theta_2\mathrm{sign}_{\mathrm{n}}(s_2)\,dt\right\| - \vartheta_m(t,x)\right. \\
&\quad \left. - \left\|B_{1,0}^{\dagger}(A_{11} + \theta(z_1)) B_{1,0}\right\|\left(\bar{\kappa}_u\|x\| + \bar{\vartheta}_u\right) - \left\|B_{1,0}^{\dagger}\dot{\theta}(z_1)\right\|\left\|\bar{\eta}_u\|x\| + \bar{h}_u\right\|\right)\|s_2\| \\
&\le -\sqrt{2}\varsigma_2 V_2^{\frac{1}{2}}.
\end{aligned} \tag{4.144}$$

According to Lemma 2.5, the system state reaches the sliding manifold $s_2 = 0$ in finite time bounded by

$$T_{m4} \le \frac{\sqrt{2}}{\varsigma} V_2^{\frac{1}{2}}(x_0). \tag{4.145}$$

We can now derive the total synchronized settling time for z_1 to achieve semi-time-synchronized convergence,

$$T_m \le T_{m1} + T_{m2} + T_{m3} + T_{m4}, \tag{4.146}$$

where T_{m1} and T_{m3} are the synchronized settling time instants introduced in (4.122) and (4.135). The subsequent state z_i, $i \in \{2, ..., r\}$ achieves asymptotic convergence successively as discussed in Remark 4.18. □

Remark 4.19 Similar to the discussion in Remark 4.16, the total synchronized settling time estimation in (4.146) is less conservative due to the fact the T_{m1} and T_{m3} are the exact value of the synchronized settling time in accordance with Theorem 2.2 and Lemma 3.2.

Remark 4.20 There are many graceful results in the literature focus on finite-time control of MIMO systems (under different assumptions, e.g., with or without matched/unmatched disturbances and uncertainties), although time-synchronized convergence is not investigated therein. One of the widely-applied techniques is to construct the sliding mode s with the function $\mathrm{sign_c}$, e.g.,

$$s = c_1 x_1 + c_2 x_2 + c_3 \mathrm{sign_c}(x_1), \tag{4.147}$$

as in [8], Theorem 1. The sliding model (4.147) can be rewritten in the form of differential inclusions and understood in the Filippov sense [7]. However, in this case, we should be very careful when differentiating a Lyapunov function candidate. For example, in view of a Lyapunov candidate $V = \frac{1}{2}s^T s$, its time derivative is given by

$$\dot{V} = s^T \frac{\mathrm{d}}{\mathrm{d}t}\left(c_1 x_1 + c_2 x_2 + c_3 \mathrm{sign_c}(x_1)\right). \tag{4.148}$$

In general

$$\lim_{x_1 \to 0} c_3 \frac{\mathrm{d}}{\mathrm{d}t} \mathrm{sign_c}(x_1) \neq 0, \tag{4.149}$$

which certainly complicates the stability analysis since differentiating $\mathrm{sign_c}(x_1)$ with respect to x_1 is two times the Dirac delta function $2\delta(x_1)$, i.e.,

$$\frac{\partial}{\partial x_1} \mathrm{sign_c}(x_1) = 2\delta(x_1), \tag{4.150}$$

where

$$\delta(x_1) = \begin{cases} +\infty, & x_1 = 0 \\ 0, & x_1 \neq 0 \end{cases}, \quad \int_{-\infty}^{\infty} \delta(x_1)\mathrm{d}x_1 = 1. \tag{4.151}$$

This problem could be avoided by reconstructing the sliding mode s by the continuous functions $\mathrm{sig_c^\alpha}$ (to achieve finite-time convergence) and $\mathrm{sig_n^\alpha}$ (to achieve time-synchronized convergence), and by using the (full-order) sliding-mode control techniques introduced in this chapter.

4.3 Summary

This chapter has introduced various robust time-synchronized controllers for the affine system, the Euler-Lagrange system, and the general MIMO system. A multivariable super-twisting observer has been integrated into the time-synchronized control scheme to accurately compensate the external matched disturbance or the model uncertainty (that can be lumped into disturbances). To realize time-synchronized convergence for a class of general MIMO systems under the unmatched disturbance is non-trivial. It has been shown that only semi-time-synchronized convergence can be achieved if the target MIMO system is classified as the state-dimension-dominant MIMO system. Discussions on the recent predefined-time control have also been addressed, where several comparative simulations have been conducted to showcase the merits attained by the proposed time-synchronized controllers.

Acknowledgements The authors would like to extend their gratitude towards the support and fruitful discussions of Dr. J. Qian from Stony Brook University, New York, Prof. L. Xie and Dr. K. Cao from Nanyang Technological University, and Dr. Z. Hou from National University of Singapore.

References

1. Ahmed, J., Coppola, V.T., Bernstein, D.S.: Adaptive asymptotic tracking of spacecraft attitude motion with inertia matrix identification. Journal of Guidance, Control, and Dynamics **21**(5), 684–691 (1998)
2. Basin, M.: Finite- and fixed-time convergent algorithms: Design and convergence time estimation. Annual Reviews in Control **48**, 209–221 (2019). https://doi.org/10.1016/j.arcontrol.2019.05.007
3. Basin, M., Panathula, C.B., Shtessel, Y.: Multivariable continuous fixed-time second-order sliding mode control: Design and convergence time estimation. IET Control Theory and Applications **11**(8), 1104–1111 (2017)
4. Basin, M., Rodriguez-Ramirez, P., Garza-Alonso, A.: Continuous Fixed-Time Convergent Super-Twisting Algorithm in Case of Unknown State and Disturbance Initial Conditions. Asian Journal of Control **21**(1), 323–338 (2019)
5. Decarlo, R.A., Zak, S.H., Matthews, G.P.: Variable Structure Control of Nonlinear Multivariate Systems: A Tutorial. Proceedings of the IEEE **76**(3), 212–232 (1988). https://doi.org/10.1109/5.4400
6. Feng, Y., Zhou, M., Zheng, X., Han, F., Yu, X.: Full-order terminal sliding-mode control of MIMO systems with unmatched uncertainties. Journal of the Franklin Institute **355**(2), 653–674 (2018)
7. Filippov, A.F.: Differential Equations with Discontinuous Right-Hand Side. Dordrecht, The Netherlands: Kluwer (1988)
8. Huang, J., Sun, L., Han, Z., Liu, L.: Adaptive terminal sliding mode control for nonlinear differential inclusion systems with disturbance. Nonlinear Dynamics **72**(1–2), 221–228 (2013)
9. Li, B., Hu, Q., Yang, Y., Postolache, O.A.: Finite-time disturbance observer based integral sliding mode control for attitude stabilisation under actuator failure. IET Control Theory and Applications **13**(1), 50–58 (2019). https://doi.org/10.1049/iet-cta.2018.5477
10. Li, D., Ma, G., Li, C., He, W., Mei, J., Ge, S.S.: Distributed Attitude Coordinated Control of Multiple Spacecraft with Attitude Constraints. IEEE Transactions on Aerospace and Electronic Systems **54**(5), 2233–2245 (2018)

11. Loukianov, A.G., Utkin, V.I.: Time-varying linear system decomposed control. In: Proceedings of the 1998 American Control Conference. ACC (IEEE Cat. No. 98CH36207), vol. 5, pp. 2884–2888. IEEE (1998)
12. Nagesh, I., Edwards, C.: A multivariable super-twisting sliding mode approach. Automatica **50**(3), 984–988 (2014). https://doi.org/10.1016/j.automatica.2013.12.032
13. Pal, A.K., Kamal, S., Nagar, S.K., Bandyopadhyay, B., Fridman, L.: Design of controllers with arbitrary convergence time. Automatica **112**, 108710 (2020)
14. Precup, R.E., Tomescu, M.L.: Stable fuzzy logic control of a general class of chaotic systems. Neural Computing and Applications **26**(3), 541–550 (2015)
15. Rao, S., Brandtstädter, H., Utkin, V., Buss, M.: Generalized Block Control Principle. IFAC Proceedings Volumes **38**(1), 229–234 (2005)
16. Utkin, V., Guldner, J., Shijun, M.: Sliding mode control in electro-mechanical systems. CRC press (1999)
17. Utkin, V.I., Chen, D.S., Chang, H.C.: Block control principle for mechanical systems. J. Dyn. Sys., Meas., Control **122**(1), 1–10 (2000)
18. Utkin, V.I., Drakunov, S.V., Izosimov, D.E., Lukyanov, A.G., Utkin, V.A.: A hierarchical principle of the control system decomposition based on motion separation. IFAC Proceedings Volumes **17**(2), 1553–1558 (1984)
19. Yin, Z., Suleman, A., Luo, J., Wei, C.: Appointed-time prescribed performance attitude tracking control via double performance functions. Aerospace Science and Technology **93**, 105337 (2019). https://doi.org/10.1016/j.ast.2019.105337
20. Zhou, B.: Finite-time stability analysis and stabilization by bounded linear time-varying feedback. Automatica **121**, 109191 (2020)

Chapter 5
Fixed-Time-Synchronized Control with Singularity Avoidance

Abstract In this chapter, we move forward to fixed-time-synchronized control for simple affine systems, where global time-synchronized stability is guaranteed and the upper bound of the synchronized settling time is independent of the initial system state. Extensions to a class of more general MIMO systems are provided under the consideration of matched and unmatched disturbances. The general MIMO system itself is classified into two specific types according to its state and input dimensions, i.e., the input-dimension-dominant MIMO system and the state-dimension-dominant MIMO system. It is noteworthy that the fixed-time-synchronized control for the state-dimension-dominant MIMO system is in general more complicated due to the vanishing of the right inverse of a crucial system matrix. In addition, in this case, only semi-fixed-time-synchronized convergence is achievable, inferring only a subset of the state elements, rather than all of them, converge simultaneously. The important related matters of singularity avoidance are likewise considered in the control design in this chapter.

5.1 Fixed-Time-Synchronized Control for Affine Systems with Singularity Avoidance

In this subsection, we investigate singularity-free fixed-time-synchronized control problems for simple affine systems, specifically, first-order and second-order affine systems. In addition, the explicit fixed-time-synchronized analytical solution is drawn for the first-order affine system, and a switching fixed-time-synchronized sliding mode is proposed for the second-order affine system.

D. Li et al., *Time-Synchronized Control: Analysis and Design*,
https://doi.org/10.1007/978-981-16-3089-7_5

5.1.1 Control Design for First-Order Systems

We first consider the following extremely simple first-order system in $\mathbb{R}^n$,

$$\dot{x} = u, \tag{5.1}$$

where $x = [x_1, x_2, \dots, x_n]^T \in \mathbb{R}^n$ is the state vector,

Theorem 5.1 *Considering the system (5.1), under the fixed-time-synchronized control law*

$$u = -\alpha \mathrm{sig}_{\mathrm{n}}^{p}(x) - \beta \mathrm{sig}_{\mathrm{n}}^{g}(x), \tag{5.2}$$

where $k_1,\ k_2 > 0,\ g > 1$, and $p \in (0, 1)$, the closed-loop system is fixed-time-synchronized stable. The system state x is ratio persistent. The synchronized settling time is bounded by

$$T(x_0) \le \frac{1}{k_1(1-p)} + \frac{1}{k_1(g-1)}. \tag{5.3}$$

Proof Consider a Lyapunov function candidate $V = x^T x = \|x\|^2$, and differentiate V with respect to time. It yields

$$\begin{aligned}\dot{V} &= -2x^T \left(k_1 \mathrm{sig}_{\mathrm{n}}^{p}(x) + k_2 \mathrm{sig}_{\mathrm{n}}^{g}(x)\right) \\ &= -2k_1 \|x\|^{p+1} - 2k_2 \|x\|^{g+1} \\ &= -2k_1 V^{\frac{p+1}{2}} - 2k_2 V^{\frac{g+1}{2}}.\end{aligned} \tag{5.4}$$

According to Theorem 2.3, the system state x achieves fixed-time-synchronized stabilization, with synchronized settling time

$$T(x_0) \le \frac{1}{k_1(1-p)} + \frac{1}{k_2(g-1)}. \tag{5.5}$$

□

Next, by modifying the controller (5.2), the analytical solution of the fixed-time-synchronized stable system (5.1) is obtained.

Theorem 5.2 *Considering the system (5.1) under the fixed-time-synchronized control law*

$$u = -k_1 \mathrm{sig}_{\mathrm{n}}^{\eta}(x), \tag{5.6}$$

where $k_1 > 0,\ g > 1,\ p \in (0, 1)$, and

$$\eta = \begin{cases} g, \text{ if } \|x\| > 1, \\ p, \text{ if } \|x\| \leq 1, \end{cases} \tag{5.7}$$

the closed-loop system is fixed-time-synchronized stable. The system state x is ratio persistent. The synchronized settling time has an explicit form

$$T(x_0) = \begin{cases} \frac{1 - V^{\frac{1-g}{2}}(x_0)}{k_1(g-1)} + \frac{1}{k_1(1-p)}, & \text{if } \|x_0\| > 1, \\ \frac{V^{\frac{1-p}{2}}(x_0)}{k_1(1-p)}, & \text{if } \|x_0\| \leq 1, \end{cases} \tag{5.8}$$

where $V = x^T x$. The analytical solution of the closed-loop system (5.1) takes the form:

(1) in the case of $\|x_0\| > 1$,

$$x = \begin{cases} \left(\|x_0\|^{1-g} - \vartheta_g t\right)^{\frac{1}{1-g}} \frac{x_0}{\|x_0\|}, & \text{if } t < T_1, \\ \left(k_1^{\frac{p-1}{2}} - \vartheta_p (t - T_1)\right)^{\frac{1}{1-p}} \frac{x_0}{\|x_0\|}, & \text{if } t \geq T_1; \end{cases} \tag{5.9}$$

and (2) in the case of $\|x_0\| \leq 1$,

$$x = \left(\|x_0\|^{1-p} - \vartheta_p t\right)^{\frac{1}{1-p}} \frac{x_0}{\|x_0\|}, \tag{5.10}$$

where $T_1 = (1 - V^{\frac{1-g}{2}}(x_0)) / (k_1(g-1))$, $\vartheta_p = k_1(1-p)$ and $\vartheta_g = k_1(1-g)$.

Proof Considering a Lyapunov function $V = x^T x$ and analyzing its time derivative, the proof is then readily derived from Theorem 2.4. □

Remark 5.1 Despite the switching power η, it can be observed that the fixed-time synchronized control law (5.6) is continuous by verifying the controller (5.6) at the point $\|x\| = 1$.

5.1.2 Control Design for Second-Order Systems

In this subsection, to demonstrate fixed-time-synchronized stability and to consider a class of more practical systems, a fixed-time-synchronized control law is designed for second-order systems. Before moving forward, a useful lemma is first proposed.

Lemma 5.1 *Consider the following system dynamics*

$$\dot{x} = -\alpha \text{sig}_n^p(x) - \beta \text{sig}_n^g(x), \tag{5.11}$$

where $x \in \mathbb{R}^n$ is the state vector, α, β, p and g are positive constants, satisfying $0 < p = p_1^*/p_2^* < 1$ and $1 < g = g_1^*/g_2^*$, with positive odd integers p_1^*, p_2^*, g_1^* and

g_2^*. *The closed-loop system is ratio persistent and fixed-time-synchronized stable with the synchronized settling time*

$$T_h \leq \frac{2^{\frac{1-p}{2}}}{\alpha\left(1-p\right)} + \frac{2^{\frac{1-g}{2}}}{\beta\left(g-1\right)}. \tag{5.12}$$

Proof Consider a Lyapunov function $V_h = \frac{1}{2}x^T x$. Taking (5.11) into the derivative of V_h, yields

$$\begin{aligned} V_h &= -\alpha x^T \mathrm{sig}_{\mathrm{n}}^p(x) - \beta q^T \mathrm{sig}_{\mathrm{n}}^g(x) \\ &= -2^{\frac{1+p}{2}} \alpha V_h^{\frac{1+p}{2}} - 2^{\frac{1+g}{2}} \beta V_h^{\frac{1+g}{2}}. \end{aligned} \tag{5.13}$$

It can be verified that x is ratio persistent. According to Theorem 2.3, we know that the system (5.11) is fixed-time-synchronized stable with the synchronized settling time as shown in (5.12). □

As we will present in what follows, the first-order dynamics (5.11) and its corresponding results given in Lemma 5.1 actually serve as a basis of fixed-time-synchronized control for more practical system dynamics.

Consider the following second-order affine system

$$\ddot{q} = f\left(q, \dot{q}\right) + b\left(q, \dot{q}\right) u, \tag{5.14}$$

where $q \in \mathbb{R}^n$ and $\dot{q} \in \mathbb{R}^n$, respectively, denote the general position and velocity vectors, $u \in \mathbb{R}^n$ the control input, and $f\left(q, \dot{q}\right) \in \mathbb{R}^n$ and $b\left(q, \dot{q}\right) \in \mathbb{R}^{n \times n}$ ($\det\left(b\left(q, \dot{q}\right)\right) \neq 0$) are the known parts. The second-order dynamics (5.14) can be rewritten as an Euler-Lagrange system for applications to mechanical systems, where q stands for the vector of the Lagrangian coordinates.

We design a sliding-mode manifold

$$s = \dot{q} + \alpha_1 \mathrm{sig}_{\mathrm{n}}^{p_1}\left(q\right) + \beta_1 \mathrm{sig}_{\mathrm{n}}^{g_1}\left(q\right), \tag{5.15}$$

where α_1 and β_1 are positive constants, and p_1 and g_1 are defined as $0 < p_1 = p_1^*/p_2^* < 1$ and $g_1 = g_1^*/g_2^* > 1$, with positive odd integers p_1^*, p_2^*, g_1^* and g_2^*.

The following fixed-time-synchronized control law is then proposed

$$\begin{aligned} u = &-b^{-1}\left(q, \dot{q}\right)\big(\alpha_2 \mathrm{sig}_{\mathrm{n}}^{p_2}\left(s\right) + \beta_2 \mathrm{sig}_{\mathrm{n}}^{g_2}\left(s\right) \\ &+ \rho_1 q q^T \dot{q} + \rho_2 \dot{q} + f\left(q, \dot{q}\right)\big), \end{aligned} \tag{5.16}$$

where α_2, β_2, p_2 and g_2 are positive constants, satisfying $0 < p_2 = p_3^*/p_4^* < 1$ and $g_2 = g_3^*/g_4^* > 1$, with positive odd integers p_3^*, p_4^*, g_3^* and g_4^*, and variables ρ_1 and ρ_2 take the forms

$$\rho_1 = \alpha_1 (p_1 - 1) \|q\|^{p_1-3} + \beta_1 (g_1 - 1) \|q\|^{g_1-3}, \tag{5.17}$$

$$\rho_2 = \alpha_1 \|q\|^{p_1-1} + \beta_1 \|q\|^{g_1-1}. \tag{5.18}$$

In light of the controller (5.16), we propose the following theorem.

Theorem 5.3 *Considering the second-order affine system governed by (5.14), under the fixed-time-synchronized control law (5.16), the closed-loop system is fixed-time-synchronized stable. The sliding-mode variable s is ratio persistent, while the closed-loop system state q is ratio persistent on the sliding surface $s = 0$.*

Proof Consider a Lyapunov function $V_{h1} = \frac{1}{2}s^T s$. Taking the controller (5.16) into the derivative of V_{h1}, we have

$$\begin{aligned} \dot{V}_{h1} &= s^T \left(f(q, \dot{q}) + b(q, \dot{q}) u + \rho_1 q q^T \dot{q} + \rho_2 \ddot{q}\right) \\ &= -\alpha_2 s^T \mathrm{sig}_{\mathrm{n}}^{p_2}(s) - \beta_2 s^T \mathrm{sig}_{\mathrm{n}}^{g_2}(s) \\ &= -2^{\frac{1+p_2}{2}} \alpha_2 V_{h1}^{\frac{1+p_2}{2}} - 2^{\frac{1+g_2}{2}} \beta_2 V_{h1}^{\frac{1+g_2}{2}}. \end{aligned} \tag{5.19}$$

Next, taking (5.16) into the derivative of (5.15), we can verify that s is ratio persistent. Based on Theorem 2.3, the sliding mode variable s is fixed-time-synchronized stable with the synchronized settling time

$$T_{h1} \leq \frac{2^{\frac{1-p_2}{2}}}{\alpha_2 (1 - p_2)} + \frac{2^{\frac{1-g_2}{2}}}{\beta_2 (g_2 - 1)}. \tag{5.20}$$

Further, on the sliding manifold $s = 0$, we have

$$\dot{q} = -\alpha_1 \mathrm{sig}_{\mathrm{n}}^{p_1}(q) - \beta_1 \mathrm{sig}_{\mathrm{n}}^{g_1}(q). \tag{5.21}$$

Based on Lemma 5.1, after the system state arrives at the sliding surface, the closed-loop state q is ratio persistent and fixed-time-synchronized stable with the synchronized settling time

$$T_{h2} \leq \frac{2^{\frac{1-p_1}{2}}}{\alpha_1 (1 - p_1)} + \frac{2^{\frac{1-g_1}{2}}}{\beta_1 (g_1 - 1)}. \tag{5.22}$$

The total synchronized settling time of the state q takes the form $T_{hm} \leq T_{h1} + T_{h2}$, which concludes the proof by showing that under the controller (5.16), the system (5.14) is fixed-time-synchronized stable. □

The fixed-time-synchronized control law (5.16), however, has negative powers -3 and -1 in ρ_1 and ρ_2 (see (5.17) and (5.18)), causing a possible singularity problem at the point $q = 0$. To avoid the singularity, we propose a switching sliding mode manifold as

$$s = \dot{q} + s_s. \tag{5.23}$$

The switching law s_s and its derivative take the forms,

$$s_s = \begin{cases} \alpha_1 \mathrm{sig}_{\mathrm{n}}^{p_1}(q) + \beta_1 \mathrm{sig}_{\mathrm{n}}^{g_1}(q), & \text{if } s^* = 0 \text{ or } s^* \neq 0,\ \|q\| > \varepsilon, \\ l_1 q + l_2 \mathrm{sig}_{\mathrm{n}}^{4}(q), & \text{if } s^* \neq 0,\ \|q\| \leq \varepsilon, \end{cases} \tag{5.24}$$

$$\dot{s}_s = \begin{cases} \rho_1 q q^T \dot{q} + \rho_2 \dot{q}, & \text{if } s^* = 0 \text{ or } s^* \neq 0,\ \|q\| > \varepsilon, \\ l_1 \dot{q} + 3 l_2 \|q\| q q^T \dot{q} + l_2 \|q\|^3 \dot{q}, & \text{if } s^* \neq 0,\ \|q\| \leq \varepsilon, \end{cases} \tag{5.25}$$

where ε is a small positive constant, the auxiliary variable s^* takes the form

$$s^* = \dot{q} + \alpha_1 \mathrm{sig}_{\mathrm{n}}^{p_1}(q) + \beta_1 \mathrm{sig}_{\mathrm{n}}^{g_1}(q), \tag{5.26}$$

constants l_1 and l_2 take the forms

$$l_1 = \alpha_1 \left(\frac{4}{3} - \frac{p_1}{3} \right) \|\varepsilon\|^{p_1 - 1} + \beta_1 \left(\frac{4}{3} - \frac{g_1}{3} \right) \|\varepsilon\|^{g_1 - 1}, \tag{5.27}$$

$$l_2 = \alpha_1 \left(\frac{p_1}{3} - \frac{1}{3} \right) \|\varepsilon\|^{p_1 - 4} + \beta_1 \left(\frac{g_1}{3} - \frac{1}{3} \right) \|\varepsilon\|^{g_1 - 4}, \tag{5.28}$$

and the corresponding powers are redefined as

$$\frac{1}{2} < p_1 = p_1^* / p_2^* < 1, \tag{5.29}$$

$$1 < g_1 = g_1^* / g_2^* < 4, \tag{5.30}$$

with positive odd integers p_1^*, p_2^*, g_1^* and g_2^* (the rationality of the value ranges of the parameters p_1 and g_1 will be detailed in Appendixes 1 and 2 at the end of this chapter).

It could be verified that the forms of l_1 and l_2 ensure the continuity of the switching algorithm s_s (5.24) and $\dot{s}_s$ (5.25).

Based on the proposed sliding manifold (5.23), we rewrite the control law (5.16) as follows,

$$u = -b^{-1}(q, \dot{q}) \left(\alpha_2 \mathrm{sig}_{\mathrm{n}}^{p_2}(s) + \beta_2 \mathrm{sig}_{\mathrm{n}}^{g_2}(s) + \dot{s}_s + f(q, \dot{q}) \right), \tag{5.31}$$

where the corresponding powers are defined as $0 < p_2 = p_3^* / p_4^* < 1$ and $g_2 = p_3^* / p_4^* > 1$ with positive odd integers p_3^*, p_4^*, g_3^* and g_4^*.

The main result of this subsection is proposed.

Theorem 5.4 *Considering a second-order system governed by (5.14), under the singularity-free fixed-time-synchronized control law (5.31), the closed-loop system is fixed-time-synchronized stable. The sliding-mode variable s is ratio persistent, while the closed-loop system state q is ratio persistent on the sliding surface* $s = 0$.

Proof Consider a Lyapunov function $V_{s1} = \frac{1}{2}s^T s$. Using the control law (5.31), the derivative of V_{s1} can be written as

$$\dot{V}_{s1} = -2^{\frac{1+p_2}{2}}\alpha_2 V_{s1}^{\frac{1+p_2}{2}} - 2^{\frac{1+g_2}{2}}\beta_2 V_{s1}^{\frac{1+g_2}{2}}. \tag{5.32}$$

Again, it can be verified that the variable s is ratio persistent. Based on Theorem 2.3, we know s is fixed-time-synchronized stable. The corresponding synchronized settling time takes the form

$$T_{s1} \leq \frac{2^{\frac{1-p_2}{2}}}{\alpha_2(1-p_2)} + \frac{2^{\frac{1-g_2}{2}}}{\beta_2(g_2-1)}. \tag{5.33}$$

After the system state q reaches the sliding manifold $s = 0$, we have $\dot{q} = -\alpha_1 \mathrm{sig}_\mathrm{n}^{p_1}(q) - \beta_1 \mathrm{sig}_\mathrm{n}^{g_1}(q)$, which leads to the ratio persistence of the state q. According to Lemma 5.1, we know that the closed-loop state is fixed-time-synchronized stable with the synchronized settling time

$$T_{s2} \leq \frac{2^{\frac{1-p_1}{2}}}{\alpha_1(1-p_1)} + \frac{2^{\frac{1-g_1}{2}}}{\beta_1(g_1-1)}, \tag{5.34}$$

which makes the total setting time of the closed-loop system state q bounded by $T_{sm} \leq T_{s1} + T_{s2}$, and this concludes the proof. □

Next, the analysis of the singularity problem is given, followed by some further discussion of the stability analysis of the second-order dynamics (5.14) under the fixed-time-synchronized control law (5.31).

Lemma 5.2 *Using the proposed control law (5.31) with the switching sliding manifold (5.23), the singularity problem can be avoided.*

Proof For details, refer to Appendix 1 at the end of this chapter. □

Actually, in light of the switching law (5.24), $s_s = l_1 q + l_2 \mathrm{sig}_\mathrm{n}^4(q)$ only takes place when $s^* \neq 0$, $\|q\| \leq \varepsilon$. We here further discuss the system convergence property before the sliding manifold s converges to the origin ($0 \leq t < T_{s1}$) while $\|q\| \leq \varepsilon$.

Lemma 5.3 *In the case of $s^* \neq 0$, $\|q\| \leq \varepsilon$, under the control law (5.31) and the switching law (5.24), the closed-loop system (5.14), viewed as a system with the state q and the input s, is input-to-state stable.*

Proof For details, refer to Appendix 2 at the end of this chapter. □

Remark 5.2 By choosing a small enough ε, the state x will converge along the terminal sliding manifold rather than the general sliding manifold to guarantee finite time convergence, i.e., the case of $s^* \neq 0$, $\|x\| \leq \varepsilon$ as in Lemma 5.3 will not occur during actual convergence. The rationale is similar to the discussion in Remark 4 from [9], which is omitted here.

5.1.3 Numerical Results

In this subsection, we take the attitude stabilization of a rigid spacecraft as an example, as some highly critical phases of space missions have very strict requirements for the convergence performance of the spacecraft.

First, the spacecraft attitude dynamics are briefly recalled.

We define the cross-product skew symmetric matrix $z^{\times}$ associated with a vector $z = (z_1, z_2, z_3)^T \in \mathbb{R}^3$ as

$$z^{\times} = \begin{bmatrix} 0 & -z_3 & z_2 \\ z_3 & 0 & -z_1 \\ -z_2 & z_1 & 0 \end{bmatrix}. \tag{5.35}$$

Then, utilizing the modified Rodrigues parameters (MRPs) [1], the attitude dynamics with respect to an inertial frame can be represented as

$$\dot{\sigma} = G(\sigma)\,\omega, \tag{5.36}$$

$$J\dot{\omega} = -\omega^{\times} J\omega + \tau, \tag{5.37}$$

where $\sigma \overset{\text{def}}{=} [\sigma_1, \sigma_2, \sigma_3]^T = \tan(\Phi/4)\, e \in \mathbb{R}^3$ denotes the MRPs, $\Phi \in \mathbb{R}$ denotes Euler angle, $e \in \mathbb{R}^3$ the Euler axis, $\omega \overset{\text{def}}{=} [\omega_1, \omega_2, \omega_3]^T \in \mathbb{R}^3$ the angular velocity, $J \in \mathbb{R}^{3\times 3}$ the inertia matrix, $\tau \in \mathbb{R}^3$ the control torque, and $G(\sigma) \in \mathbb{R}^{3\times 3}$ is given by

$$G(\sigma) = \frac{1}{2}\left(\frac{1 - \sigma^T\sigma}{2} I_3 + \sigma^{\times} + \sigma\sigma^T\right). \tag{5.38}$$

Then, (5.36) and (5.37) can be rewritten as an Euler-Lagrange equation, that is,

$$M(\sigma)\ddot{\sigma} + C(\sigma, \dot{\sigma})\dot{\sigma} = u(t), \tag{5.39}$$

where the relevant parameters are rewritten as

$$M(\sigma) = G^{-T}(\sigma)\, J G^{-1}(\sigma), \tag{5.40}$$

$$C(\sigma, \dot{\sigma}) = -G^{-T} J G^{-1}\dot{G}G^{-1} - G^{-T}(J\omega)^{\times} G^{-1}, \tag{5.41}$$

$$u = G^{-T}(\sigma)\,\tau. \tag{5.42}$$

Therefore, according to the Euler-Lagrange system (5.39), we can rewrite the fixed-time-synchronized control law (5.31) for the spacecraft attitude stabilization as

$$u = -M(\sigma)\left(\alpha_2 \mathrm{sig}_{\mathrm{n}}^{p_2}(s) + \beta_2 \mathrm{sig}_{\mathrm{n}}^{g_2}(s) + \dot{s}_s\right) + C(\sigma, \dot{\sigma})\,\dot{\sigma}. \tag{5.43}$$

In what follows, to illustrate the features of fixed-time-synchronized stability, we apply the fixed-time-synchronized control law (5.43) to the spacecraft attitude stabilization. Moreover, to offer a better comparison, we here introduce a fixed-time control law for the Euler-Lagrange systems.

Lemma 5.4 *Considering the system governed by (5.39), under the following fixed-time control law:*

$$\bar{u} = -M(\sigma)\left(\alpha_2 \text{sig}_{\text{c}}^{p_2}(\bar{s}) + \beta_2 \text{sig}_{\text{c}}^{g_2}(\bar{s}) + \dot{\bar{s}}_s\right) + C(\sigma, \dot{\sigma})\dot{\sigma}. \tag{5.44}$$

The sliding manifold $\bar{s} = \dot{\sigma} + \bar{s}_s$, *the trigger sliding-mode variable* $\bar{s}^* = \dot{\sigma} + \alpha_1 \text{sig}_{\text{c}}^{p_1}(\sigma) + \beta_1 \text{sig}_{\text{c}}^{g_1}(\sigma)$, *and the switching law take the forms*

$$\bar{s}_{s,i} = \begin{cases} \alpha_1 \text{sig}_{\text{c}}^{p_1}(\sigma_i) + \beta_1 \text{sig}_{\text{c}}^{g_1}(\sigma_i), & \text{if } \bar{s}^* = 0, \\ & \text{or } \bar{s}^* \neq 0, |\sigma_i| > \varepsilon, \\ l_3\sigma_i + l_4 \text{sig}_{\text{c}}^2(\sigma_i), & \text{if } \bar{s}^* \neq 0, |\sigma_i| \leq \varepsilon, \end{cases}$$

$$\dot{\bar{s}}_{s,i} = \begin{cases} \alpha_1 p_1 |\sigma_i|^{p_1 - 1}\dot{\sigma}_i + \beta_1 g_1 |\sigma_i|^{g_1 - 1}\dot{\sigma}_i, & \text{if } \bar{s}^* = 0, \\ & \text{or } \bar{s}^* \neq 0, |\sigma_i| > \varepsilon, \\ l_3\dot{\sigma} + l_4 |\sigma_i| \dot{\sigma}_i, & \text{if } \bar{s}^* \neq 0, |\sigma_i| \leq \varepsilon, \end{cases}$$

where σ_i, $\bar{s}_{s,i}$ *and* $\dot{\bar{s}}_{s,i}$ *denote the* i*th element of* σ, $\bar{s}_s$ *and* $\dot{\bar{s}}_s$, $i = 1, 2, ..., n$. l_3 *and* l_4 *take the following forms to guarantee the continuity of* $\bar{s}_s$ *and* $\dot{\bar{s}}_s$:

$$l_3 = (2 - p_1)\alpha_1 \varepsilon^{p_1 - 1} + (2 - g_1)\beta_1 \varepsilon^{g_1 - 1}, \tag{5.45}$$

$$l_4 = (p_1 - 1)\alpha_1 \varepsilon^{p_1 - 2} + (g_1 - 1)\beta_1 \varepsilon^{g_1 - 2}. \tag{5.46}$$

The closed-loop system state converges to the origin in fixed time.

Proof The proof can be given following the procedure in [9]. □

In Table 5.1, the simulation parameters for the fixed-time-synchronized control law (5.43) and the fixed-time control law (5.44) are provided, where the control parameters for both the controllers are identical for fairness of comparison.

Table 5.1 Simulation parameters: case 1

Parameters	Values	Parameters	Values
α_1	0.1	p_1	0.65
α_2	0.05	p_2	0.65
β_1	0.1	g_1	1.1
β_2	0.05	g_2	1.1
σ_0	$[0.06, -0.03, 0.01]^T$	ε	0.0001
J	[12, 0.4, 0.2; 0.4, 10, 0.6; 0.2, 0.6, 11] (kg m^2)		

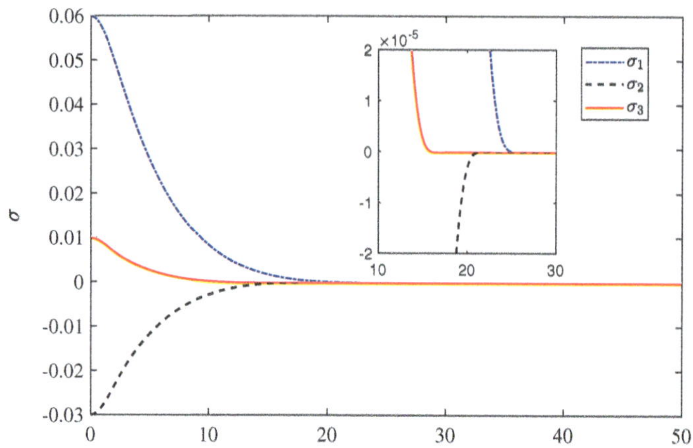

(a)

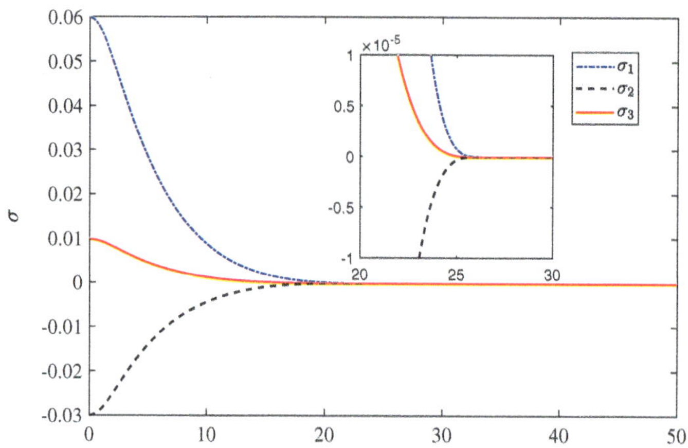

(b)

Fig. 5.1 System state σ: **a** performance of the fixed-time control law (5.44); **b** performance of the fixed-time-synchronized control law (5.43)

Using the fixed-time control law (5.44) and the fixed-time-synchronized control law (5.43), the corresponding closed-loop system states are respectively shown in Fig. 5.1a, b. In Fig. 5.1a, it explicitly indicates that under the fixed-time control law (5.44), three state elements σ_1, σ_2 and σ_3 converge to the origin separately at different time instants, while as shown in Fig. 5.1b, the fixed-time-synchronized control law (5.43) synchronously drives all the state elements to the equilibrium. For greater clarity, the norms of the state elements governed by two control laws are given in Fig. 5.2a, b. In Fig. 5.2a, the state element σ_3 arrive at the origin first at $t = 16$ s, and

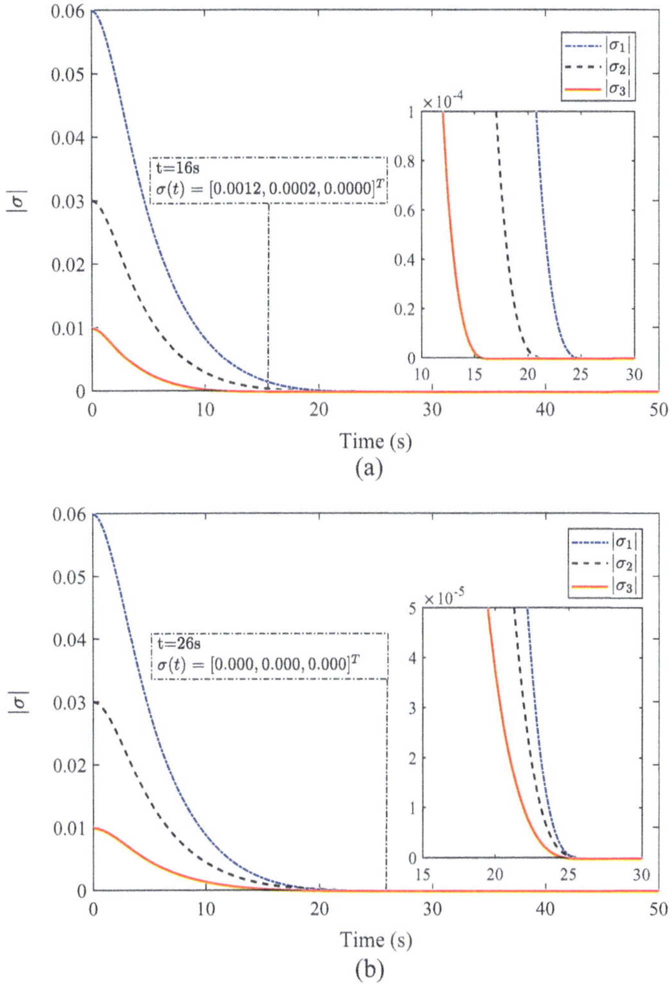

Fig. 5.2 Norm of the system state σ: **a** performance of the fixed-time control law (5.44); **b** performance of the fixed-time-synchronized control law (5.43)

then the other two elements catch up at $t = 22$ s and $t = 26$ s. On the contrary, all the state elements in Fig. 5.2b, reach the origin synchronously at $t = 26$ s.

Moreover, in Fig. 5.3, it demonstrates that the state trajectory generated by the fixed-time-synchronized control law (5.43) is ratio persistent, making the whole trajectory shorter compared with that governed by the fixed-time control law (5.44), as it is a straight line between the start point and the end point.

Specifically, the property of ratio persistence is further shown in Fig. 5.4a, where we have

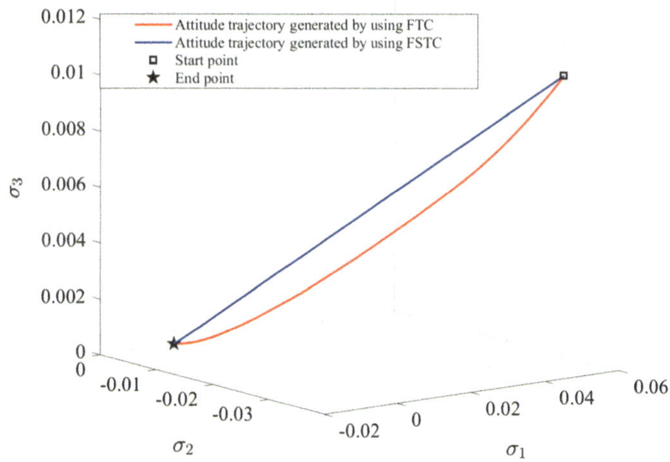

Fig. 5.3 Comparison of the attitude trajectories under the fixed-time control (FTC) law (5.44) and the fixed-time-synchronized control (FTSC) law (5.43) (Please note that these trajectories are not the real movements of the spacecraft attitude, as modified Rodrigues parameters are applied in the system dynamics)

$$\frac{\sigma_1(t)}{\sigma_2(t)} = \frac{\sigma_1(0)}{\sigma_2(0)} = -2, \tag{5.47}$$

$$\frac{\sigma_1(t)}{\sigma_3(t)} = \frac{\sigma_1(0)}{\sigma_3(0)} = 6, \tag{5.48}$$

$$\frac{\sigma_2(t)}{\sigma_3(t)} = \frac{\sigma_2(0)}{\sigma_3(0)} = -3. \tag{5.49}$$

It also evidently approves declared fixed-time-synchronized stability under the fixed-time-synchronized control law (5.43). Finally, the relevant control inputs of the fixed-time-synchronized control law (5.43) are given in Fig. 5.4b. One may wonders why the ratios in Fig. 5.4a do not run into a singularity (0 divided by 0) after the state elements converge to the origin. It is because the simulation software has a limit on the numerical resolution (e.g., by default, MATLAB®uses 16 digits of precision). Thus, although σ_i and σ_j can reach the origin theoretically, they might be respectively represented as $\sigma_i = 6 * 10^{-15}$ and $\sigma_j = 1 * 10^{-15}$ in the simulation software, making their ratio remains to be a constant with the value of 6.

Next, for a more comprehensive comparison, we choose another set of parameters for simulation as shown in Table 5.2. Basically, the power parameters p_1, p_2, g_1 and g_2 are re-selected (as they are the crucial ones to affect the performance), while the other parameters remain the same.

Here, we directly plot the norms of the state elements under two controllers in Fig. 5.5a, b. Again, Fig. 5.5a shows that the time-synchronized convergence property cannot be expected from the fixed-time control law (5.44). The main reason is that

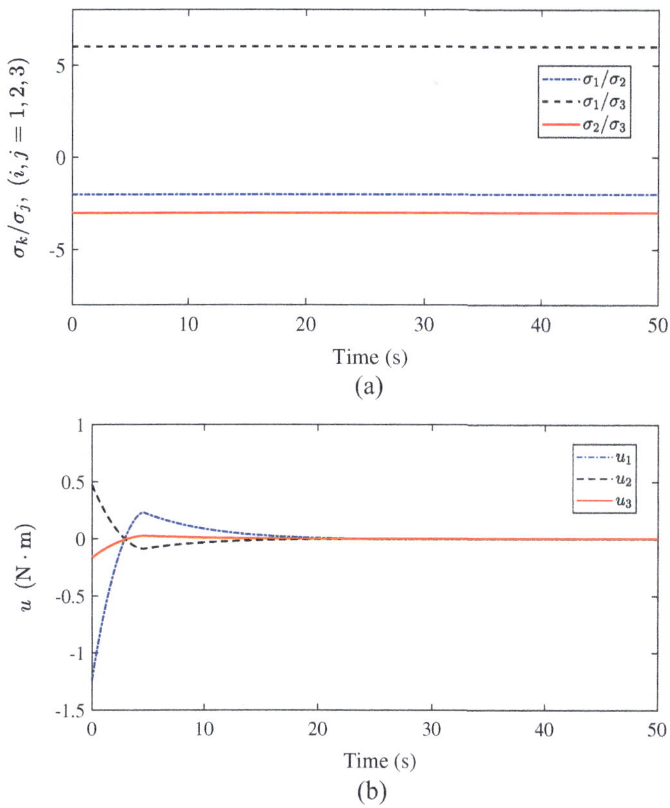

Fig. 5.4 Performance of the fixed-time-synchronized control law (5.43): **a** ratio of the elements of the system state σ; **b** control inputs

Table 5.2 Simulation parameters: case 2

Parameters	Values	Parameters	Values
α_1	0.1	p_1	0.7
α_2	0.05	p_2	0.7
β_1	0.1	g_1	1.2
β_2	0.05	g_2	1.2
σ_0	$[0.06, -0.03, 0.01]^T$	ε	0.0001
J	[12, 0.4, 0.2; 0.4, 10, 0.6; 0.2, 0.6, 11]		

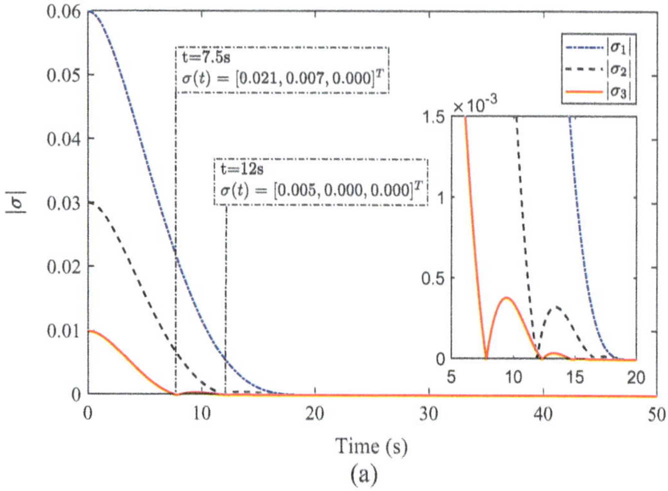

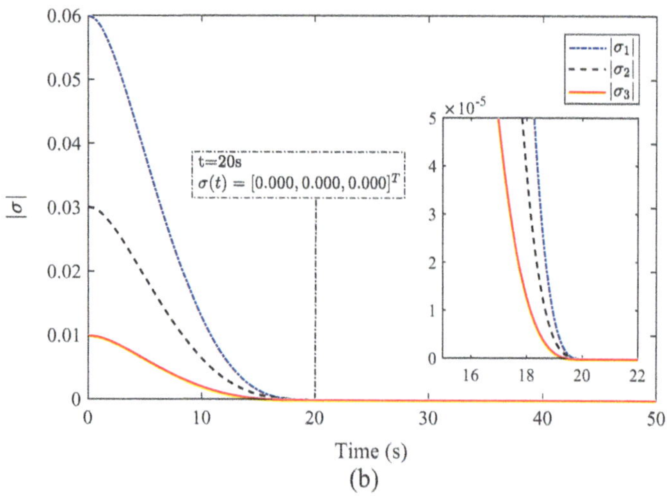

Fig. 5.5 Norm of the system state σ: **a** performance of the fixed-time control law (5.44); **b** performance of the fixed-time-synchronized control law (5.43)

the classical sign function sig_c decouples a vector into different scalar while the norm-normalized sign function sig_n incorporates the normalization of a vector with its specified norm. Due to the same reason, in Fig. 5.5b, fixed-time-synchronized stability is successfully achieved at $t = 20\,\mathrm{s}$. It is additionally noticeable that there exist overshoots in Fig. 5.5a, which inevitably degrades the system performance, while no such deterioration can be found in Fig. 5.5b.

Compared with the simulation results in the first case (see Fig. 5.2a, b), the difference between the control performances of the two controllers become more obvious in Fig. 5.5a, b. For example, the trajectories of the state elements in Fig. 5.2a, b are similar to a certain degree (although their times of arrival to the origin are different), while in Fig. 5.5a, b, the differences of trajectories are more distinct. It is conjectured empirically from numerous simulations that a decrease in p_1 and p_2 or an increase in g_1 and g_2 leads to a greater difference between the performances of the fixed-time-synchronized control law (5.43) and the fixed-time control law (5.44).

So far, we have presented two simulation cases to reveal the nature of fixed-time-synchronized stability. Moreover, compared with the fixed-time control law (5.44), the property of ratio persistence has been clearly illustrated. Note that all the current results assume that there exists no external disturbance. In real-world applications, however, disturbances are inevitable and sometimes unpredictable.

Therefore, to further evaluate the fixed-time-synchronized control law (5.43) under disturbances, additional simulations are conducted, where all the control parameters are chosen the same as shown in Table 5.1. We here consider that the system suffers from an external impulsive disturbance that acts on the state element σ_3 at $t = 10\,\mathrm{s}$. The relevant numerical results are presented in Figs. 5.6 and 5.7. It is shown in Fig. 5.6a, b that despite the external disturbance, fixed-time-synchronized stability is still achieved successfully. The rationale lies in that although σ_3 drifts significantly because of the disturbance, the fixed-time-synchronized control law (5.43) continues generating a proper control input, intending to preserve the ratio persistent property of the current system state. Thus in Fig. 5.7, after the impulsive disturbance, the ratio of each affected pair of the state elements quickly converged to a new constant.

5.2 Fixed-Time-Synchronized Control for General MIMO Systems Under Matched and Unmatched Disturbances

Similar to the proposal in Chap. 4, Sect. 4.2, in this section, we investigate fixed-time-synchronized control for general MIMO systems under matched and unmatched disturbances. Note that this section involves some similar manipulations and the logic flow behind is the same as that in Chap. 4, Sect. 4.2. We thus only keep the essential process here.

In the sequel, we consider a MIMO system with matched and unmatched disturbances (which is the same as the MIMO system (4.37) in Chap. 4, Sect. 4.2),

$$\dot{x} = Ax + Bu + d(t, x, u), \tag{5.50}$$

where $x \in \mathbb{R}^n$ is the system state, $u \in \mathbb{R}^m$ $(1 \leq m \leq n)$ the control input, $d(t, x, u)$ the general disturbances and uncertainties, and $A \in \mathbb{R}^{n \times n}$ and $B \in \mathbb{R}^{n \times m}$ the system matrices.

The following assumption on the system matrices is made.

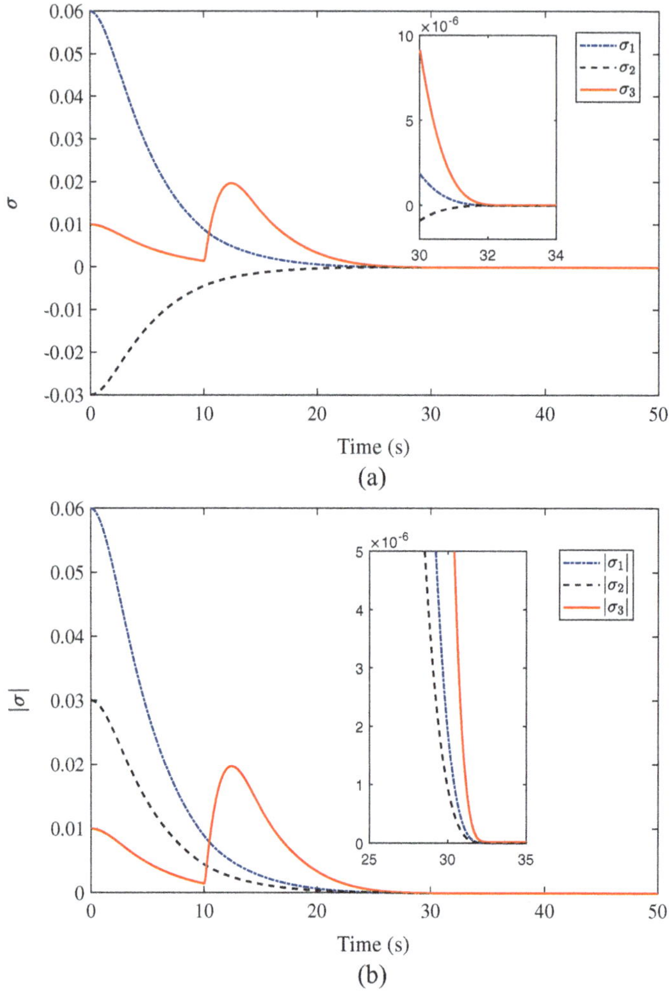

Fig. 5.6 Performance of the fixed-time-synchronized control law (5.43) with an impulsive disturbance: **a** System state σ; **b** Norm of the system state σ

Assumption 5.1 The system dynamics (5.50) contains a controllable pair (A, B). The matrix B is of full rank, and by reordering the components, it is partitioned as $B = \left[B_1^T, B_2^T\right]^T$, where $B_1 \in \mathbb{R}^{(n-m)\times m}$, $B_2 \in \mathbb{R}^{m\times m}$, and $\det(B_2) \neq 0$.

The system (5.50) is rewritten in a regular form, by the state transformation [2, 4, 6],

$$Tx = \begin{bmatrix} x_1 \\ x_2 \end{bmatrix}, \; T = \begin{bmatrix} I_{n-m} & -B_1 B_2^{-1} \\ 0 & B_2^{-1} \end{bmatrix}. \tag{5.51}$$

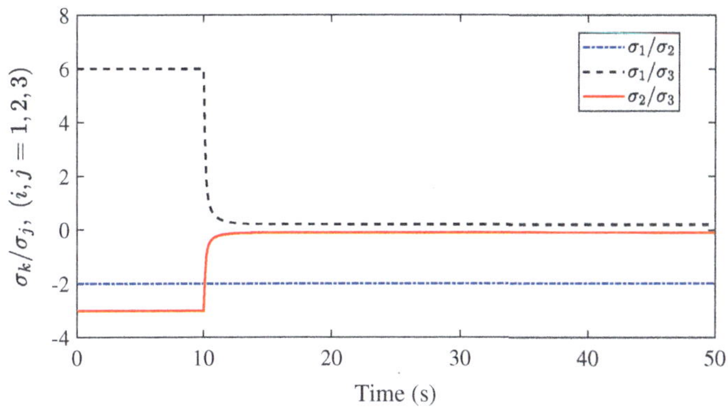

Fig. 5.7 Ratio of the state elements using fixed-time-synchronized control (5.43) under an impulsive disturbance

The regular form is provided as follows,

$$
\begin{aligned}
\dot{x}_1 &= A_{11}x_1 + A_{12}x_2 + d_u(t,x), \\
\dot{x}_2 &= A_{21}x_1 + A_{22}x_2 + u + d_m(t,x,u), \\
y &= x_1,
\end{aligned}
\tag{5.52}
$$

where $x = \left[x_1^T, x_2^T\right]^T$ denotes the system state ($x_1 \in \mathbb{R}^{n-m}$, $x_2 \in \mathbb{R}^m$), $y \in \mathbb{R}^{n-m}$ the system output, $d_u(t,x) \in \mathbb{R}^{n-m}$ and $d_m(t,x,u) \in \mathbb{R}^m$, respectively, the components of the unmatched and matched disturbances, the system matrices A_{11}, A_{12}, A_{21}, and A_{22} are of proper dimensions.

In the following subsections, we explore the fixed-time-synchronized control for input-dimension-dominant MIMO systems (introduced in Definition 4.1) and state-dimension-dominant MIMO systems (introduced in Definition 4.2), respectively.

5.2.1 *Fixed-Time-Synchronized Control for Input-Dimension-Dominant MIMO Systems*

In this subsection, fixed-time-synchronized control is investigated for the input-dimension-dominant MIMO system. The following necessary assumptions are made.

Assumption 5.2 The row echelon form of the matrix $A_{12} \in \mathbb{R}^{(n-m)\times m}$ ($m \geq \frac{n}{2}$) in the system dynamics (5.52) has pivots in each row.

Assumption 5.3 The disturbances are bounded as

$$\|d_u(t,x)\| \le \eta_u \|x\| + h_u, \tag{5.53}$$
$$\|d_m(t,x,u)\| \le \eta_m \|u\| + h_m(t,x), \tag{5.54}$$

where $\eta_u \ge 0$, $\eta_m \ge 0$, and $h_u \in [0, 1)$ are known positive constants, and $h_m(t, x)$ is a known positive function.

Assumption 5.4 The disturbances are bounded as

$$\left\|\dot{d}_u(t,x)\right\| \le \kappa_u \|x\| + \vartheta_u, \tag{5.55}$$
$$\left\|\dot{d}_m(t,x,u)\right\| \le \kappa_m \|u\| + \vartheta_m(t,x), \tag{5.56}$$

where κ_u, κ_m, and ϑ_u are known positive constants, and $\vartheta_m(t, x)$ is a known positive function.

The necessity of Assumptions 5.2, 5.3, and 5.4 have been explained in Chap. 4, Sect. 4.2.

We first define the following fixed-time-synchronized sliding mode

$$s_1 = \dot{x}_1 + \alpha_1 \mathrm{sig}_{\mathrm{n}}^{p_1}(x_1) + \beta_1 \mathrm{sig}_{\mathrm{n}}^{g_1}(x_1), \tag{5.57}$$

where α_1 and β_1 are positive constants, and p_1 and g_1 are defined as $0 < p_1 = p_1^*/p_2^* < 1$ and $g_1 = g_1^*/g_2^* > 1$, with positive odd integers p_1^*, p_2^*, g_1^* and g_2^*.

Recall Lemma 5.1, on the fixed-time-synchronized sliding surface (5.57) ($s_1 = 0$), x_1 achieves fixed-time-synchronized convergence with synchronized settling time,

$$T_{m1} \le \frac{2^{\frac{1-p_1}{2}}}{\alpha_1(1-p_1)} + \frac{2^{\frac{1-g_1}{2}}}{\beta_1(g_1-1)}. \tag{5.58}$$

Aiming at driving $s_1 \to 0$, a virtual velocity tracking error is defined as

$$z = x_2 - \phi_2, \tag{5.59}$$

where

$$\phi_2 = \phi_{2eq} + \phi_{2n}, \tag{5.60}$$
$$\phi_{2eq} = -A_{12}^{\dagger}\left(A_{11}x_1 + \alpha_1 \mathrm{sig}_{\mathrm{n}}^{p_1}(x_1) + \beta_1 \mathrm{sig}_{\mathrm{n}}^{g_1}(x_1)\right), \tag{5.61}$$
$$\dot{\phi}_{2n} = -A_{12}^{\dagger}\left(\Theta_1(x)\mathrm{sign}_{\mathrm{n}}(s_1) + \alpha_2 \mathrm{sig}_{\mathrm{n}}^{p_2}(s_1) + \beta_2 \mathrm{sig}_{\mathrm{n}}^{g_2}(s_1)\right), \tag{5.62}$$
$$\Theta_1(x) = \kappa_u \|x\| + \vartheta_u, \tag{5.63}$$

where α_2, β_2, p_2 and g_2 are positive constants, satisfying $0 < p_2 = p_3^*/p_4^* < 1$ and $g_2 = g_3^*/g_4^* > 1$, with positive odd integers p_3^*, p_4^*, g_3^* and g_4^*.

In what follows, it is shown that s_1 synchronously converges to the origin in fixed time in the case of $z = \dot{z} = 0$.

Lemma 5.5 *In the case of* $z = \dot{z} = 0$, *the sliding variable* s_1 *achieves fixed-time convergence.*

Proof Since $x_2 = z + \phi_2$, we have

$$\dot{x}_1 = A_{11}x_1 + A_{12}(\phi_2 + z) + d_u(t, x). \tag{5.64}$$

Then, s_1 becomes

$$\begin{aligned} s_1 &= A_{11}x_1 + A_{12}(\phi_2 + z) + d_u(t, x) + \alpha_1 \mathrm{sig}_\mathrm{n}^{p_1}(x_1) + \beta_1 \mathrm{sig}_\mathrm{n}^{g_1}(x_1) \\ &= A_{11}x_1 + A_{12}\phi_{2eq} + A_{12}\phi_{2n} + A_{12}z + d_u(t, x) + \alpha_1 \mathrm{sig}_\mathrm{n}^{p_1}(x_1) + \beta_1 \mathrm{sig}_\mathrm{n}^{g_1}(x_1) \\ &= A_{12}\phi_{2n} + A_{12}z + d_u(t, x). \end{aligned} \tag{5.65}$$

Construct a Lyapunov function $V_1 = \frac{1}{2}s_1^T s_1$. Its time derivative is

$$\begin{aligned} \dot{V}_1 &= s_1^T\left(A_{12}\dot{\phi}_{2n} + A_{12}\dot{z} + \dot{d}_u\right) \\ &= -s_1^T\Theta_1 \mathrm{sign}_\mathrm{n}(s_1) - \alpha_2\|s_1\|^{p_2+1} - \beta_2\|s_1\|^{g_2+1} + s_1^T\dot{d}_u + s_1^T A_{12}\dot{z} \\ &\le -\Theta_1\|s_1\| + \|s_1\|(\kappa_u\|x\| + \vartheta_u) + s_1^T A_{12}\dot{z} - \alpha_2\|s_1\|^{p_2+1} - \beta_2\|s_1\|^{g_2+1} \\ &\le -2^{\frac{1+p_2}{2}}\alpha_2 V^{\frac{1+p_2}{2}} - 2^{\frac{1+g_2}{2}}\beta_2 V^{\frac{1+g_2}{2}} + s_{21}^T A_{12}\dot{z}. \end{aligned} \tag{5.66}$$

When $z = \dot{z} = 0$, from Lemma 5.1, the system output x_1 reaches the sliding manifold $s_1 = 0$ in fixed time bounded by

$$T_{m2} \le \frac{2^{\frac{1-p_2}{2}}}{\alpha_2(1-p_2)} + \frac{2^{\frac{1-g_2}{2}}}{\beta_2(g_2-1)}. \tag{5.67}$$

The proof ends here. □

We next force $z \to 0$ synchronously in fixed time, by introducing the following fixed-time-synchronized sliding mode,

$$s_2 = \dot{z} + \alpha_3 \mathrm{sig}_\mathrm{n}^{p_3}(z) + \beta_3 \mathrm{sig}_\mathrm{n}^{g_3}(z), \tag{5.68}$$

where α_3, β_3, p_3 and g_3 are positive constants, satisfying $0 < p_3 = p_5^*/p_6^* < 1$ and $g_3 = g_5^*/g_6^* > 1$, with positive odd integers p_5^*, p_6^*, g_5^* and g_6^*. Again, based on Lemma 5.1, the auxiliary variable z achieves fixed-time-synchronized convergence with synchronized settling time bounded by

$$T_{m3} \le \frac{2^{\frac{1-p_3}{2}}}{\alpha_3(1-p_3)} + \frac{2^{\frac{1-g_3}{2}}}{\beta_3(g_3-1)}. \tag{5.69}$$

A full-order sliding mode controller is proposed based on the sliding mode (5.68),

$$u = u^* + \tau^*, \tag{5.70}$$
$$\dot{u}^* = -\Theta_2 \mathrm{sign_n}(s_2) - \alpha_4 \mathrm{sig_n^{p_4}}(s_2) - \beta_4 \mathrm{sig_n^{g_4}}(s_2), \tag{5.71}$$
$$\begin{aligned}\tau^* = &-\Big(A_{21}x_1 + A_{22}x_2 + A_{12}^{\dagger}(A_{11} + \theta(x_1))(A_{11}x_1 + A_{12}x_2) \\ &+ A_{12}^{\dagger}\big(\Theta_1(x)\mathrm{sign_n}(s_1) + \alpha_2 \mathrm{sig_n^{p_2}}(s_1) + \beta_2 \mathrm{sig_n^{g_2}}(s_1)\big) \\ &+ \alpha_2 \mathrm{sig_n^{p_2}}(s_1) + \beta_2 \mathrm{sig_n^{g_2}}(s_1)\Big),\end{aligned} \tag{5.72}$$
$$\begin{aligned}\Theta_2 = &\kappa_m \left\|\tau^*\right\| + \vartheta_m(t,x) + \left\|A_{12}^{\dagger}(A_{11} + \theta(x_1))\right\|(\kappa_u \|x\| + \vartheta_u) \\ &+ \left\|A_{12}^{\dagger}\dot{\theta}(x_1)\right\| \|\eta_u \|x\| + h_u\| + \kappa_m \left\|\int_0^t \dot{u}^* dt\right\|,\end{aligned} \tag{5.73}$$

where α_4, β_4, p_4 and g_4 are positive constants, satisfying $0 < p_4 = p_7^*/p_8^* < 1$ and $g_4 = g_7^*/g_8^* > 1$, with positive odd integers p_7^*, p_8^*, g_7^* and g_8^*, and a function $\theta(x_1) : \mathbb{R}^{n-m} \to \mathbb{R}^{n-m} \times \mathbb{R}^{n-m}$ is defined as

$$\begin{aligned}\theta(x_1) = &\alpha_1(p_1 - 1)\|x_1\|^{p_1-3}x_1x_1^T + \alpha_1\|x_1\|^{p_1-1}I_{n-m} \\ &+ \beta_1(g_1 - 1)\|x_1\|^{g_1-3}x_1x_1^T + \beta_1\|x_1\|^{g_1-1}I_{n-m}.\end{aligned} \tag{5.74}$$

The full-order controller (5.70) is quasi-continuous due to the discontinuity at $s_1 = 0$ induced by τ^*. It should be thus understood in the Filippov sense [3].

Theorem 5.5 *Considering the disturbed input-dimension-dominant MIMO system (5.52) under Assumptions 5.1, 5.2, 5.3, and 5.4, and fixed-time-synchronized control law (5.70), the system state successively reaches the sliding manifolds $s_2 = 0$ and $s_1 = 0$ in fixed time, while the closed-loop system output y is ratio persistent on the sliding surface $s_1 = 0$ and achieves fixed-time-synchronized convergence.*

Proof The sliding mode s_2 can be expended as,

$$\begin{aligned}s_2 &= \dot{x}_2 - \dot{\phi}_2 + \alpha_3 \mathrm{sig_n^{p_3}}(z) + \beta_3 \mathrm{sig_n^{g_3}}(z) \\ &= A_{21}x_1 + A_{22}x_2 + u + d_m \\ &\quad + A_{12}^{\dagger}(A_{11} + \theta(x_1))(A_{11}x_1 + A_{12}x_2 + d_u(t,x))) \\ &\quad + A_{12}^{\dagger}\big(\Theta_1(x)\mathrm{sign_n}(s_1) + \alpha_2 \mathrm{sig_n^{p_2}}(s_1) + \beta_2 \mathrm{sig_n^{g_2}}(s_1)\big) \\ &\quad + \alpha_3 \mathrm{sig_n^{p_3}}(z) + \beta_3 \mathrm{sig_n^{g_3}}(z).\end{aligned} \tag{5.75}$$

Taking the full-order controller (5.70) into (5.75), the sliding mode vector and its derivative become,

$$s_2 = u^* + A_{12}^{\dagger}(A_{11} + \theta(x_1))d_u(t,x) + d_m, \tag{5.76}$$
$$\dot{s}_2 = \dot{u}^* + A_{12}^{\dagger}(A_{11} + \theta(x_1))\dot{d}_u(t,x) + A_{12}^{\dagger}\dot{\theta}(x_1)d_u(t,x) + \dot{d}_m. \tag{5.77}$$

(The derivation in (5.77) can be obtained following the discussion in Remark 4.14.)

Consider a Lyapunov function $V_2 = \frac{1}{2} s_2^T s_2$. From (5.70) and (5.77), its time derivative takes the form,

$$
\begin{aligned}
\dot{V}_2 &= s_2^T \left(\dot{u}^* + A_{12}^{\dagger} \left(A_{11} + \theta \left(x_1 \right) \right) \dot{d}_u(t, x) + A_{12}^{\dagger} \dot{\theta} \left(x_1 \right) d_u(t, x) + \dot{d}_m \right) \\
&\leq -\Theta_2 \left\| s_2 \right\| + \left\| s_2 \right\| \left\| \dot{d}_m \right\| + \left\| s_2 \right\| \left\| A_{12}^{\dagger} \left(A_{11} + \theta \left(x_1 \right) \right) \right\| \left\| \dot{d}_u \right\| \\
&\quad + \left\| s_2 \right\| \left\| A_{12}^{\dagger} \dot{\theta} \left(x_1 \right) \right\| \left\| \eta_u \left\| x \right\| + h_u \right\| - \alpha_4 \| s_2 \|^{p_4+1} - \beta_4 \| s_2 \|^{g_4+1} \\
&\leq - \Big(\Theta_2 - \kappa_m \left\| u \right\| - \vartheta_m(t, x) - \left\| A_{12}^{\dagger} \left(A_{11} + \theta \left(x_1 \right) \right) \right\| \left(\kappa_u \left\| x \right\| + \vartheta_u \right) \\
&\quad - \left\| A_{12}^{\dagger} \dot{\theta} \left(x_1 \right) \right\| \left\| \eta_u \left\| x \right\| + h_u \right\| \Big) \left\| s_2 \right\| - \alpha_4 \| s_2 \|^{p_4+1} - \beta_4 \| s_2 \|^{g_4+1} \\
&\leq - \left\| s_2 \right\| \Bigg(\Theta_2 - \kappa_m \left\| \tau^* \right\| - \kappa_m \left\| \int_0^t \dot{u}^* dt \right\| - \vartheta_m(t, x) \\
&\quad - \left\| A_{12}^{\dagger} \left(A_{11} + \theta \left(x_1 \right) \right) \right\| \left(\kappa_u \left\| x \right\| + \vartheta_u \right) - \left\| A_{12}^{\dagger} \dot{\theta} \left(x_1 \right) \right\| \left\| \eta_u \left\| x \right\| + h_u \right\| \Bigg) \\
&\quad - \alpha_4 \| s_2 \|^{p_4+1} - \beta_4 \| s_2 \|^{g_4+1} \\
&\leq -2^{\frac{1+p_4}{2}} \alpha_4 V^{\frac{1+p_4}{2}} - 2^{\frac{1+g_4}{2}} \beta_4 V^{\frac{1+g_4}{2}}.
\end{aligned} \tag{5.78}
$$

According to Lemma 2.5, the system state reaches the sliding manifold $s_2 = 0$ in fixed time bounded by

$$
T_{m4} \leq \frac{2^{\frac{1-p_4}{2}}}{\alpha_4 \left(1 - p_4 \right)} + \frac{2^{\frac{1-g_4}{2}}}{\beta_4 \left(g_4 - 1 \right)}. \tag{5.79}
$$

We can now derive the total synchronized settling time to achieve fixed-time-synchronized convergence,

$$
T_m \leq T_{m1} + T_{m2} + T_{m3} + T_{m4}, \tag{5.80}
$$

where T_{m1} and T_{m3} are the synchronized settling time defined in (5.58) and (5.69). □

Remark 5.3 The fixed-time-synchronized controller (5.70) contains a possible singular term $\theta\left(x_1\right)$ (when $s_1 \neq 0$ and $x_1 \rightarrow 0$). To deal with the singularity, a switching-based fixed-time-synchronized sliding mode as in (5.24) and (5.25) can be applied to replace (5.57), and the time-synchronized convergence result in Theorem 5.5 still holds. Detailed singularity-avoidance analysis can be found in Lemma 5.2.

5.2.2 Semi-fixed-Time-Synchronized Control for State-Dimension-Dominant MIMO Systems

In the previous subsection, fixed-time-synchronized control for the state-dimension-dominant MIMO system (5.52) is proposed. We here investigate time-synchronized control for the state-dimension-dominant MIMO system, where the right inverse of A_{12} no longer exists.

Not all the state-dimension-dominant MIMO systems are capable of achieving fixed-time-synchronized stability. For those able to "achieve fixed-time-synchronized convergence", only a subset of the output elements (rather than all of them) reach the origin at the same time. This type of convergence property is named "*Semi-Fixed-Time-Synchronized Convergence*". Despite semi-fixed-time-synchronized convergence, there still exist d_1 ($d_1 \leq m$) elements of the system output that achieve fixed-time-synchronized convergence, where d_1 is the largest number in the controllability indices (called the controllability index) of the system $\dot{x}_1 = A_{11}x_1 + A_{12}x_2 + d_u(t, x)$, after proper state transformations by treating x_2 as the control input.

The following assumption on the unmatched disturbance is made to semi-fixed-time-synchronized control.

Assumption 5.5 The unmatched disturbance satisfies $d_u(t, x) = A_{12}\omega(t, x)$, where $\omega(t, x)$ is an unknown function characterizing the disturbances. In addition, $\omega(t, x)$ meets the following constraints,

$$\|\omega(t, x)\| \leq \overline{\eta}_u \|x\| + \overline{h}_u, \tag{5.81}$$

$$\|\dot{\omega}(t, x)\| \leq \bar{\kappa}_u \|x\| + \bar{\vartheta}_u, \tag{5.82}$$

where $\bar{\kappa}_u$ and $\bar{\vartheta}_u$ are two known positive constants.

Assumption 5.5 is necessary to stable control for the state-dimension-dominant MIMO system, otherwise, neither u (the control input) or x_2 (the virtual control input) can be used to eliminate the unmatched disturbance d_u lumped in the tunnel of the system output.

We make two state transformations (that are the same as those in Chap. 4, Sect. 4.2) for the system (5.52), given by

$$x_1 = \Gamma_1^{-1}\Gamma_2 z, \tag{5.83}$$

$$z = \Gamma_2^{-1}\Gamma_1 x_1, \tag{5.84}$$

where z is the transformed state vector, and Γ_1 and Γ_2 are transformation matrices to be defined. This process enables us to extract an "invertable part" from A_{12}. As will shown clearly in the later analysis, the invertiable part is only effective under Assumption 5.5.

We get the block controllable form by the first state transformation $x' = \Gamma_1 x_1$, where Γ_1 is a constant matrix (for details, refer to [5, 7, 8]),

$$\begin{bmatrix} \dot{x}'_r \\ \dot{x}'_{r-1} \\ \vdots \\ \dot{x}'_3 \\ \dot{x}'_2 \\ \dot{x}'_1 \end{bmatrix} = \begin{bmatrix} A'_{r,r} & B'_{r,r-1} & \cdots & 0 & 0 & 0 \\ A'_{r-1,r} & A'_{r-1,r-1} & \cdots & 0 & 0 & 0 \\ \vdots & \vdots & \ddots & \vdots & \vdots & \vdots \\ A'_{3,r} & A'_{3,r-1} & \cdots & A'_{3,3} & B'_{3,2} & 0 \\ A'_{2,r} & A'_{2,r-1} & \cdots & A'_{2,3} & A'_{2,2} & B'_{2,1} \\ A'_{1,r} & A'_{1,r-1} & \cdots & A'_{1,3} & A'_{1,2} & A'_{1,1} \end{bmatrix} \begin{bmatrix} x'_r \\ x'_{r-1} \\ \vdots \\ x'_3 \\ x'_2 \\ x'_1 \end{bmatrix} + \begin{bmatrix} 0 \\ 0 \\ \vdots \\ 0 \\ 0 \\ B_{1,0}(x_2 + \omega) \end{bmatrix}. \tag{5.85}$$

The dimension of each block has the following property,

$$d_1 = \text{rank}\left(B'_{1,0}\right) = \dim\left(x'_1\right), \tag{5.86}$$

$$d_i = \text{rank}\left(B'_{i,i-1}\right) = \dim\left(x'_i\right), \ i = 2, \cdots, r, \tag{5.87}$$

where clearly, $\sum_{i=1}^{r} d_i = n - m$. Equations (5.86) and (5.87) validate the existence of the right inverse of $B'_{i,i-1}$, $i \in \{1, \ldots, r\}$ as the dimension of the subsystem state in each block is the same as the dimension of its corresponding fictitious input. The dimensions $(d_1, \ldots, d_r)$ are the controllability indices (also called the controllability index vector). In addition, d_1 is the controllability index, since $d_1 \geq d_2 \geq \cdots \geq d_r$.

Remark 5.4 Intuitively, the ith controllability index yields the dimension of the "accessible" space by using only the ith control input, which is not "accessible" by using any previous control input. Thus for a controllable system $\in \mathbb{R}^{(n-m)}$, the sum of the controllability indices are $n - m$.

Based on the block controllable form in (5.85), the decoupled block controllable form is then constructed by a second state transformation $x' = \Gamma_2 z$ (for details, refer to [4]),

$$\begin{bmatrix} \dot{z}_r \\ \dot{z}_{r-1} \\ \vdots \\ \dot{z}_3 \\ \dot{z}_2 \\ \dot{z}_1 \end{bmatrix} = \begin{bmatrix} B_r & B'_{r,r-1} & \cdots & 0 & 0 & 0 \\ 0 & N_{r-1} & \cdots & 0 & 0 & 0 \\ \vdots & \vdots & \ddots & \vdots & \vdots & \vdots \\ 0 & 0 & \cdots & N_3 & B'_{2,2} & 0 \\ 0 & 0 & \cdots & 0 & N_2 & B'_{2,1} \\ A''_{1,r} & A''_{1,r-1} & \cdots & A''_{1,3} & A''_{1,2} & A''_{1,1} \end{bmatrix} \begin{bmatrix} z_r \\ z_{r-1} \\ \vdots \\ z_3 \\ z_2 \\ z_1 \end{bmatrix} + \begin{bmatrix} 0 \\ 0 \\ \vdots \\ 0 \\ B_{1,0}(x_2 + \omega) \end{bmatrix}, \tag{5.88}$$

where the transformation matrix Γ_2 has the following form,

$$\Gamma_2 = \begin{bmatrix} I & 0 & 0 & \cdots & 0 & 0 \\ -K_{r-1,r} & I & 0 & \cdots & 0 & 0 \\ -K_{r-2,r} & -K_{r-2,r-1} & I & \cdots & 0 & 0 \\ \vdots & \vdots & \vdots & \ddots & \vdots & \vdots \\ -K_{2,r} & -K_{2,r-1} & -K_{2,r-2} & \cdots & I & 0 \\ -K_{1,r} & -K_{1,r-1} & -K_{1,r-2} & \cdots & -K_{1,2} & I \end{bmatrix}, \tag{5.89}$$

where in addition, for $i = 1, \ldots, r - 1$,

$$K_{i,i+1} = B'^{\dagger}_{i+1,i} \left(K_{i+1,i+2} B'_{i+2,i+1} + A'_{i+1,i+1} - N_{i+1} \right), \tag{5.90}$$

for $j = i + 2, \cdots, r,\ i = 1, \cdots, r - 2,$

$$K_{i,j} = B'^{\dagger}_{i+1,i} \left(K_{i+1,j} N_j + A'_{i+1,j} + K_{i+1,j+1} B'_{j+1,j} - \sum_{k=j-1}^{i+1} A'_{i+1,k} K_{k,j} \right), \tag{5.91}$$

and

$$K_{i,j} = 0,\ \forall i \notin \{1, \cdots, r-1\} \text{ and } j \notin \{2, \cdots, r\}, \tag{5.92}$$

$$B'_{i,j} = 0,\ \forall i \notin \{1, \cdots, r\} \text{ and } j \notin \{0, \cdots, r-1\}. \tag{5.93}$$

The decoupled block control form can be reformulated as

$$\dot{z}_1 = \sum_{i=1}^{r} A''_{1,i} z_i + B_{1,0} (x_2 + \omega), \tag{5.94}$$

$$\dot{z}_i = N_i z_i + B'_{i,i-1} z_{i-1}, \quad i = 2, \ldots, r, \tag{5.95}$$

where N_i is a matrix defined as $N_i = -\ell_i I_{\pi_i}$, ℓ_i is a positive constant satisfying $0 < \ell_r < \ldots < \ell_2$, and π_i denotes the dimension of z_i.

Remark 5.5 The recursive structure in (5.94) and (5.95) explains why only semi-fixed-time-synchronized convergence rather than fixed-time-synchronized one can be achieved. After z_1 is time-synchronously driven to the origin in fixed time, $z_2, \ldots, z_r$ will successively reach the equilibrium asymptotically (rather than time-synchronously). Thus why fixed-time-synchronized convergence is achieved only in d_1 dimensions.

A fixed-time-synchronized sliding mode is constructed to force $z_1 \to 0$,

$$s_1 = \dot{z}_1 + \alpha_1 \mathrm{sig}_{\mathrm{n}}^{p_1}(z_1) + \beta_1 \mathrm{sig}_{\mathrm{n}}^{g_1}(z_1), \tag{5.96}$$

where α_1 and β_1 are positive constants, and p_1 and g_1 are defined as $0 < p_1 = p_1^*/p_2^* < 1$ and $g_1 = g_1^*/g_2^* > 1$, with positive odd integers p_1^*, p_2^*, g_1^* and g_2^*. On the sliding surface $s_1 = 0$, z_1 achieves fixed-time-synchronized convergence with synchronized settling time,

$$T_{m1} \leq \frac{2^{\frac{1-p_1}{2}}}{\alpha_1(1 - p_1)} + \frac{2^{\frac{1-g_1}{2}}}{\beta_1(g_1 - 1)}. \tag{5.97}$$

Aiming at driving $s_1 \to 0$, a virtual velocity tracking error is defined as

$$\mu = x_2 - \phi_2, \tag{5.98}$$

where ϕ_2 is designed to drive $s_1 \to 0$,

$$\phi_2 = \phi_{2eq} + \phi_{2n}, \tag{5.99}$$

$$\phi_{2eq} = -B_{1,0}^{\dagger}\left(\sum_{i=1}^{r} A''_{1,i} z_i + \alpha_1 \mathrm{sig}_{\mathrm{n}}^{p_1}(z_1) + \beta_1 \mathrm{sig}_{\mathrm{n}}^{g_1}(z_1)\right), \tag{5.100}$$

$$\dot{\phi}_{2n} = -B_{1,0}^{\dagger}\left(\Theta_1(x)\mathrm{sign}_{\mathrm{n}}(s_1) + \alpha_2 \mathrm{sig}_{\mathrm{n}}^{p_2}(s_1) + \beta_2 \mathrm{sig}_{\mathrm{n}}^{g_2}(s_1)\right), \tag{5.101}$$

$$\Theta_1(x) = \left(\bar{\kappa}_u \|x\| + \bar{\vartheta}_u\right) \left\|B_{1,0}\right\|, \tag{5.102}$$

where α_2, β_2, p_2 and g_2 are positive constants, satisfying $0 < p_2 = p_3^*/p_4^* < 1$ and $g_2 = g_3^*/g_4^* > 1$, with positive odd integers p_3^*, p_4^*, g_3^* and g_4^*, and κ_u and ϑ_u are introduced in Assumption 5.4.

Lemma 5.6 *In the case of $\mu = \dot{\mu} = 0$, the system dynamics (5.94) reaches the sliding manifold $s_1 = 0$ in fixed time.*

Proof Since $x_2 = \mu + \phi_2$, based on (5.99), we have

$$\begin{aligned}\dot{z}_1 &= \sum_{i=1}^{r} A''_{1,i} z_i + B_{1,0}\left(\mu + \phi_2 + \omega\right) \\ &= -\alpha_1 \mathrm{sig}_{\mathrm{n}}^{p_1}(z_1) - \beta_1 \mathrm{sig}_{\mathrm{n}}^{g_1}(z_1) + B_{1,0}\phi_{2n} + B_{1,0}\omega + B_{1,0}\mu.\end{aligned} \tag{5.103}$$

Taking (5.103) into (5.96), s_1 and its derivative become

$$s_1 = B_{1,0}\phi_{2n} + B_{1,0}\omega + B_{1,0}\mu \tag{5.104}$$

$$\dot{s}_1 = -\Theta_1(x)\mathrm{sign}_{\mathrm{n}}(s_1) - \alpha_2 \mathrm{sig}_{\mathrm{n}}^{p_2}(s_1) - \beta_2 \mathrm{sig}_{\mathrm{n}}^{g_2}(s_1) + B_{1,0}\dot{\omega} + B_{1,0}\dot{\mu}. \tag{5.105}$$

Consider a Lyapunov function $V_1 = \frac{1}{2}s_1^T s_1$. Applying Assumption 5.5, its time derivative is

$$\begin{aligned}\dot{V}_1 &= s_1^T\left(-\Theta_1(x)\mathrm{sign}_{\mathrm{n}}(s_1) - \alpha_2 \mathrm{sig}_{\mathrm{n}}^{p_2}(s_1) - \beta_2 \mathrm{sig}_{\mathrm{n}}^{g_2}(s_1) + B_{1,0}\dot{\omega} + B_{1,0}\dot{\mu}\right) \\ &\le -\Theta_1(x)\|s_1\| + \left\|B_{1,0}\right\|\left(\bar{\kappa}_u\|x\| + \bar{\vartheta}_u\right)\|s_1\| + s_1^T B_{1,0}\dot{\mu} - \alpha_2\|s_1\|^{p_2+1} - \beta_2\|s_1\|^{g_2+1} \\ &\le -\alpha_2\|s_1\|^{p_2+1} - \beta_2\|s_1\|^{g_2+1} + s_1^T B_{1,0}\dot{\mu}.\end{aligned} \tag{5.106}$$

When $\mu = \dot{\mu} = 0$, the inequality (5.106) becomes

$$\dot{V}_1 \le -2^{\frac{1+p2}{2}}\alpha_2 V^{\frac{1+p_2}{2}} - 2^{\frac{1+g_2}{2}}\beta_2 V^{\frac{1+g_2}{2}}. \tag{5.107}$$

From Lemma 5.1, the system output z_1 reaches the sliding manifold $s_1 = 0$ in fixed time bounded by

$$T_{m2} \leq \frac{2^{\frac{1-p_2}{2}}}{\alpha_2(1-p_2)} + \frac{2^{\frac{1-g_2}{2}}}{\beta_2(g_2-1)}. \tag{5.108}$$

□

We next force $\mu \to 0$ by using the following fixed-time-synchronized sliding mode,

$$s_2 = \dot{\mu} + \alpha_3 \mathrm{sig}_{\mathrm{n}}^{p_3}(\mu) + \beta_3 \mathrm{sig}_{\mathrm{n}}^{g_3}(\mu), \tag{5.109}$$

where α_3, β_3, p_3 and g_3 are positive constants, satisfying $0 < p_3 = p_5^*/p_6^* < 1$ and $g_3 = g_5^*/g_6^* > 1$, with positive odd integers p_5^*, p_6^*, g_5^* and g_6^*. From Lemma 5.1, μ achieves fixed-time-synchronized convergence with synchronized settling time bounded by

$$T_{m3} \leq \frac{2^{\frac{1-p_3}{2}}}{\alpha_3(1-p_3)} + \frac{2^{\frac{1-g_3}{2}}}{\beta_3(g_3-1)}. \tag{5.110}$$

Based on the sliding mode (5.109), a full-order sliding mode controller is proposed,

$$u = u^* + \tau^*, \tag{5.111}$$

$$\dot{u}^* = -\Theta_2 \mathrm{sign}_{\mathrm{n}}(s_2) - \alpha_4 \mathrm{sig}_{\mathrm{n}}^{p_4}(s_2) - \beta_4 \mathrm{sig}_{\mathrm{n}}^{g_4}(s_2), \tag{5.112}$$

$$\begin{aligned} \tau^* = -\Bigg(& A_{21}x_1 + A_{22}x_2 + B_{1,0}^{\dagger} \sum_{i=2}^{r} A_{1,i}'' \left(N_i z_i + B_{i,i-1}' z_{i-1}\right) \\ & + B_{1,0}^{\dagger}\left(A_{1,1}'' + \theta(z_1)\right)\left(\sum_{i=2}^{r} A_{1,i}'' + B_{1,0}x_2\right) \\ & + B_{1,0}^{\dagger}\left(\Theta_1(x)\mathrm{sign}_{\mathrm{n}}(s_1) + \alpha_3 \mathrm{sig}_{\mathrm{n}}^{p_3}(\mu) + \beta_3 \mathrm{sig}_{\mathrm{n}}^{g_3}(\mu)\right)\Bigg), \end{aligned} \tag{5.113}$$

$$\begin{aligned} \Theta_2 = {} & \kappa_m \left\|\tau^*\right\| + \vartheta_m(t,x) + \left\|B_{1,0}^{\dagger}\left(A_{1,1}'' + \theta(z_1)\right) B_{1,0}\right\| \left(\bar{\kappa}_u \|x\| + \bar{\vartheta}_u\right) \\ & + \left\|B_{1,0}^{\dagger}\dot{\theta}(z_1)\right\| \left\|\bar{\eta}_u \|x\| + \bar{h}_u\right\| + \kappa_m \left\|\int_0^t \dot{u}^* dt\right\|, \end{aligned} \tag{5.114}$$

where α_4, β_4, p_4 and g_4 are positive constants, satisfying $0 < p_4 = p_7^*/p_8^* < 1$ and $g_4 = g_7^*/g_8^* > 1$, with positive odd integers p_7^*, p_8^*, g_7^* and g_8^*, and a function $\theta(z_1): \mathbb{R}^{d_1} \to \mathbb{R}^{d_1} \times \mathbb{R}^{d_1}$ is defined as

$$\begin{aligned} \theta(z_1) = {} & \alpha_1(p_1-1)\|z_1\|^{p_1-3} x_1 x_1^T + \alpha_1 \|z_1\|^{p_1-1} I_{d_1} \\ & + \beta_1(g_1-1)\|z_1\|^{g_1-3} z_1 z_1^T + \beta_1 \|z_1\|^{g_1-1} I_{d_1}. \end{aligned} \tag{5.115}$$

The controller (5.111) is quasi-continuous due to the discontinuity at $s_1 = 0$ induced by τ^*. It should be thus understood in the Filippov sense [3].

Theorem 5.6 *Considering the disturbed state-dimension-dominant MIMO system (5.52) under Assumptions 5.1, 5.3 and 5.5, and the full-order sliding-mode control law (5.111), the system state reaches the fixed-time-synchronized sliding manifold* $s_2 = 0$ *and* $s_1 = 0$ *successively in fixed time. The closed-loop system output* z *achieves semi-fixed-time-synchronized convergence. Specifically,* z_1 *is ratio persistent on the sliding surface* $s_1 = 0$ *and achieves fixed-time-synchronized convergence, while* z_i*,* $i \in \{2, ..., r\}$ *achieves asymptotic convergence.*

Proof The sliding mode s_2 can be expended as,

$$\begin{aligned} s_2 &= \dot{x}_2 - \dot{\phi}_2 + \alpha_3 \text{sig}_{\text{n}}^{p_3}(\mu) + \beta_3 \text{sig}_{\text{n}}^{g_3}(\mu) \\ &= A_{21}x_1 + A_{22}x_2 + u + d_m \\ &\quad + B_{1,0}^{\dagger}\left(\sum_{i=2}^{r} A_{1,i}''\left(N_i z_i + B_{i,i-1}' z_{i-1}\right) + \left(A_{1,1}'' + \theta(z_1)\right)\left(\sum_{i=1}^{r} A_{1,i}'' z_i + B_{1,0}(x_2 + \omega)\right)\right) \\ &\quad - B_{1,0}^{\dagger}\left(\Theta_1(x)\text{sign}_{\text{n}}(s_1) + \alpha_2 \text{sig}_{\text{n}}^{p_2}(s_1) + \beta_2 \text{sig}_{\text{n}}^{g_2}(s_1)\right). \end{aligned} \tag{5.116}$$

Taking the full-order controller (5.111) into (5.116), the sliding mode vector and its derivative become,

$$s_2 = u^* + B_{1,0}^{\dagger}\left(A_{1,1}'' + \theta(z_1)\right) B_{1,0}\omega + d_m, \tag{5.117}$$

$$\dot{s}_2 = \dot{u}^* + B_{1,0}^{\dagger}\left(A_{1,1}'' + \theta(z_1)\right) B_{1,0}\dot{\omega} + B_{1,0}^{\dagger}\dot{\theta}(z_1) B_{1,0}\omega + \dot{d}_m. \tag{5.118}$$

(The derivation in (5.118) can be obtained following the discussion in Remark 4.14.)

Consider a Lyapunov function $V_2 = \frac{1}{2}s_2^T s_2$. Based on (5.111) and (5.118), the time derivative of V_2 becomes,

$$\begin{aligned} \dot{V}_2 &= s_2^T\left(\dot{u}^* + B_{1,0}^{\dagger}\left(A_{1,1}'' + \theta(z_1)\right) B_{1,0}\dot{\omega} + B_{1,0}^{\dagger}\dot{\theta}(z_1) B_{1,0}\omega + \dot{d}_m\right) \\ &\le -\Theta_2\|s_2\| + \|s_2\|\left\|\dot{d}_m\right\| + \|s_2\|\left\|B_{1,0}^{\dagger}(A_{11} + \theta(z_1)) B_{1,0}\right\|\left(\bar{\kappa}_u\|x\| + \bar{\vartheta}_u\right) \\ &\quad + \left\|B_{1,0}^{\dagger}\dot{\theta}(z_1)\right\|\left\|\bar{\eta}_u\|x\| + \bar{h}_u\right\|\|s_2\| - \alpha_4\|s_2\|^{p_4+1} - \beta_4\|s_2\|^{g_4+1} \\ &\le -\left(\Theta_2 - \kappa_m\|u\| - \vartheta_m(t,x) - \left\|B_{1,0}^{\dagger}(A_{11} + \theta(z_1)) B_{1,0}\right\|\left(\bar{\kappa}_u\|x\| + \bar{\vartheta}_u\right)\right. \\ &\quad \left. - \left\|B_{1,0}^{\dagger}\dot{\theta}(z_1)\right\|\left\|\bar{\eta}_u\|x\| + \bar{h}_u\right\|\right)\|s_2\| - \alpha_4\|s_2\|^{p_4+1} - \beta_4\|s_2\|^{g_4+1} \\ &\le -\alpha_4\|s_2\|^{p_4+1} - \beta_4\|s_2\|^{g_4+1} - \left(\Theta_2 - \kappa_m\|\tau^*\| - \kappa_m\left\|\int_0^t \dot{u}^* dt\right\| - \vartheta_m(t,x)\right. \\ &\quad \left. - \left\|B_{1,0}^{\dagger}(A_{11} + \theta(z_1)) B_{1,0}\right\|\left(\bar{\kappa}_u\|x\| + \bar{\vartheta}_u\right) - \left\|B_{1,0}^{\dagger}\dot{\theta}(z_1)\right\|\left\|\bar{\eta}_u\|x\| + \bar{h}_u\right\|\right)\|s_2\| \\ &\le -2^{\frac{1+p_4}{2}}\alpha_4 V^{\frac{1+p_4}{2}} - 2^{\frac{1+g_4}{2}}\beta_4 V^{\frac{1+g_4}{2}}. \end{aligned} \tag{5.119}$$

According to Lemma 5.1, the system state reaches the sliding manifold $s_2 = 0$ in fixed time bounded by

$$T_{m4} \leq \frac{2^{\frac{1-p_4}{2}}}{\alpha_4(1-p_4)} + \frac{2^{\frac{1-g_4}{2}}}{\beta_4(g_4-1)}. \tag{5.120}$$

We now derive the total synchronized settling time to achieve semi-fixed-time-synchronized convergence,

$$T_m \leq T_{m1} + T_{m2} + T_{m3} + T_{m4}, \tag{5.121}$$

where T_{m1} and T_{m3} are the synchronized settling time introduced in (5.97) and (5.110). The subsequent state $z_i, i \in \{2, ..., r\}$ achieves asymptotic convergence successively as discussed in Remark 5.5. □

Remark 5.6 Following the procedure in Remark 5.3, the full-order sliding-mode controller (5.111) becomes a non-singular one. The trivial details are omitted here.

5.3 Summary

Fixed-time-synchronized control has been investigated in this chapter, with the considerations of the singularity avoidance. For the affine system, the Euler-Lagrange system, and the general MIMO system, fixed-time-synchronized controllers have been proposed respectively, where the switching manifold technique and the full-order sliding mode have been employed to solve the singularity issue. In addition, the matched and unmatched disturbances that inevitable in practice, have been considered into the design of the proposed fixed-time-synchronized controllers. Simulation results have been presented to show that the proposed schemes outperform the traditional fixed-time controllers from multiple aspects.

Appendix 1: Proof of Lemma 5.2

The proof can be given by the following three cases.

Case 1: $\|q\| > \varepsilon$. It is trivial that there is no singularity problem.

Case 2: $\|q\| \leq \varepsilon$, $s^* = 0$. According to the switching law (5.24), we have

$$\dot{q} = -\alpha_1 \mathrm{sig}_{\mathrm{n}}^{p_1}(q) - \beta_1 \mathrm{sig}_{\mathrm{n}}^{g_1}(q). \tag{5.122}$$

Substituting (5.122) into the control law (5.31) yields,

$$\begin{aligned} u = & -b^{-1}(q, \dot{q}) \left(\alpha_2 \mathrm{sig}_{\mathrm{n}}^{p_2}(s) + \beta_2 \mathrm{sig}_{\mathrm{n}}^{g_2}(s) \right. \\ & \left. + Q + f(q, \dot{q})\right), \end{aligned} \tag{5.123}$$

where we define

$$Q = \rho_1 q q^T \alpha_1 \mathrm{sig}_{\mathrm{n}}^{p_1}(q) + \rho_1 q q^T \beta_1 \mathrm{sig}_{\mathrm{n}}^{g_1}(q) \\ + \rho_2 \alpha_1 \mathrm{sig}_{\mathrm{n}}^{p_1}(q) + \rho_2 \beta_1 \mathrm{sig}_{\mathrm{n}}^{g_1}(q). \tag{5.124}$$

Substituting ρ_1 and ρ_2 (see (5.17) and (5.18)) into (5.124), we have

$$\begin{aligned} Q = & \alpha_1 (p_1 - 1) \|q\|^{p_1-3} q q^T \alpha_1 \mathrm{sig}_{\mathrm{n}}^{p_1}(q) \\ & + \beta_1 (g_1 - 1) \|q\|^{g_1-3} q q^T \alpha_1 \mathrm{sig}_{\mathrm{n}}^{p_1}(q) \\ & + \alpha_1 (p_1 - 1) \|q\|^{p_1-3} q q^T \beta_1 \mathrm{sig}_{\mathrm{n}}^{g_1}(q) \\ & + \beta_1 (g_1 - 1) \|q\|^{g_1-3} q q^T \beta_1 \mathrm{sig}_{\mathrm{n}}^{g_1}(q) \\ & + \left(\alpha_1 \|q\|^{p_1-1} + \beta_1 \|q\|^{g_1-1}\right) \alpha_1 \mathrm{sig}_{\mathrm{n}}^{p_1}(q) \\ & + \left(\alpha_1 \|q\|^{p_1-1} + \beta_1 \|q\|^{g_1-1}\right) \beta_1 \mathrm{sig}_{\mathrm{n}}^{g_1}(q). \end{aligned} \tag{5.125}$$

Using Lemma 2.2, we know

$$q q^T \mathrm{sig}_{\mathrm{n}}^{p_1}(q) = \mathrm{sig}_{\mathrm{n}}^{p_1+2}(q). \tag{5.126}$$

Taking (5.126) into (5.125), it reads

$$\begin{aligned} Q = & \alpha_1^2 (p_1 - 1) \mathrm{sig}_{\mathrm{n}}^{2p_1-1}(q) + \alpha_1 \beta_1 (g_1 - 1) \mathrm{sig}_{\mathrm{n}}^{p_1+g_1-1}(q) \\ & + \alpha_1 \beta_1 (p_1 - 1) \mathrm{sig}_{\mathrm{n}}^{p_1+g_1-1}(q) + \beta_1^2 (g_1 - 1) \mathrm{sig}_{\mathrm{n}}^{2g_1-1}(q) \\ & + \alpha_1^2 \mathrm{sig}_{\mathrm{n}}^{2p_1-1}(q) + \alpha_1 \beta_1 \mathrm{sig}_{\mathrm{n}}^{p_1+g_1-1}(q) \\ & + \alpha_1 \beta_1 \mathrm{sig}_{\mathrm{n}}^{p_1+g_1-1}(q) + \beta_1^2 \mathrm{sig}_{\mathrm{n}}^{2g_1-1}(q) \\ = & \alpha_1^2 p_1 \mathrm{sig}_{\mathrm{n}}^{2p_1-1}(q) + \beta_1^2 g_1 \mathrm{sig}_{\mathrm{n}}^{2g_1-1}(q) \\ & + \alpha_1 \beta_1 (p_1 + g_1) \mathrm{sig}_{\mathrm{n}}^{p_1+g_1-1}(q). \end{aligned} \tag{5.127}$$

Since we know $\frac{1}{2} < p_1 < 1$ and $g_1 > 1$, the singularity can thus be avoided in (5.127) as $2p_1 - 1 > 0$, $2g_1 - 1 > 0$, and $p_1 + g_1 - 1 > 0$.

Case 3: $\|q\| \le \varepsilon$, $s^* \neq 0$, the sliding manifold (5.23) becomes a general one without any singularity by the switching law (5.24). Therefore, according to the above discussion, the singularity problem can be avoided. This finishes the proof.

Appendix 2: Proof of Lemma 5.3

The proof of Lemma 5.3 can then be addressed as follows.

In the case of $s^* \neq 0$, $\|q\| \le \varepsilon$, according to the switching law (5.24), we have

$$\dot{q} = -l_1 q - l_2 \mathrm{sig}_{\mathrm{n}}^4(q) + s. \tag{5.128}$$

Consider a Lyapunov function

$$V_{ISS} = \frac{1}{2} q^T q. \tag{5.129}$$

Its time derivative takes the form

$$\begin{aligned}\dot{V}_{ISS} &= -l_1 q^T q - l_2 \|q\|^3 q^T q + q^T s \\ &= -\left(l_1 + l_2 \|q\|^3\right) q^T q + q^T s.\end{aligned} \tag{5.130}$$

Denoting $L = l_1 + l_2 \|q\|^3$, based on (5.27) and (5.28), we have

$$\begin{aligned}L = &\alpha_1 \left(\frac{4}{3} - \frac{p_1}{3}\right) \|\varepsilon\|^{p_1 - 1} + \alpha_1 \left(\frac{p_1}{3} - \frac{1}{3}\right) \|\varepsilon\|^{p_1 - 4} \|q\|^3 \\ &+ \beta_1 \left(\frac{4}{3} - \frac{g_1}{3}\right) \|\varepsilon\|^{g_1 - 1} + \beta_1 \left(\frac{g_1}{3} - \frac{1}{3}\right) \|\varepsilon\|^{g_1 - 4} \|q\|^3.\end{aligned}$$

Noticing $1 < g_1 < 4$ and $\|q\| \le \varepsilon$, we thus can get

$$\begin{aligned}L \ge &\alpha_1 \left(\frac{4}{3} - \frac{p_1}{3}\right) \|\varepsilon\|^{p_1 - 1} + \alpha_1 \left(\frac{p_1}{3} - \frac{1}{3}\right) \|\varepsilon\|^{p_1 - 1} \\ &+ \beta_1 \left(\frac{4}{3} - \frac{g_1}{3}\right) \|\varepsilon\|^{g_1 - 1}.\end{aligned} \tag{5.131}$$

Define a positive constant

$$\lambda = \alpha_1 \|\varepsilon\|^{p_1 - 1} + \beta_1 \left(\frac{4}{3} - \frac{g_1}{3}\right) \|\varepsilon\|^{g_1 - 1}. \tag{5.132}$$

Therefore, (5.130) becomes

$$\begin{aligned}\dot{V}_{ISS} &\le -\lambda q^T q + q^T s \\ &= -\lambda (1 - \theta) q^T q - \left(\lambda \theta q^T q - q^T s\right),\end{aligned} \tag{5.133}$$

where $\theta \in (0, 1)$.

Then, for $\lambda \theta q^T q - q^T s > 0$, we have

$$\dot{V}_{ISS} \le -\lambda (1 - \theta) q^T q. \tag{5.134}$$

By taking

$$\kappa_1 (\delta) = \lambda (1 - \theta) \|\delta\|^2 \tag{5.135}$$

$$\kappa_2 (\delta) = \frac{\|\delta\|}{\lambda \theta}, \tag{5.136}$$

where the variable $\delta \in \mathbb{R}^n$, we obtain that

$$\dot{V}_{ISS} \leq -\kappa_1 (q), \tag{5.137}$$

if $\|q\| \geq \kappa_2 (s)$. This concludes the proof by showing that, according to Definition 2.5 and Lemma 2.4, the closed-loop system is ISS. The corresponding ISS-Lyapunov function is $V_{ISS} (q) = \frac{1}{2} q^T q$, with the Lyapunov gain $\kappa_2 (s)$.

References

1. Ahmed, J., Coppola, V.T., Bernstein, D.S.: Adaptive asymptotic tracking of spacecraft attitude motion with inertia matrix identification. Journal of Guidance, Control, and Dynamics **21**(5), 684–691 (1998)
2. Decarlo, R.A., Zak, S.H., Matthews, G.P.: Variable Structure Control of Nonlinear Multivariate Systems: A Tutorial. Proceedings of the IEEE **76**(3), 212–232 (1988). 10.1109/5.4400
3. Filippov, A.F.: Differential Equations with Discontinuous Right-Hand Side. Dordrecht, The Netherlands: Kluwer (1988)
4. Loukianov, A.G., Utkin, V.I.: Time-varying linear system decomposed control. In: Proceedings of the 1998 American Control Conference. ACC (IEEE Cat. No. 98CH36207), vol. 5, pp. 2884–2888. IEEE (1998)
5. Rao, S., Brandtstädter, H., Utkin, V., Buss, M.: Generalized Block Control Principle. IFAC Proceedings Volumes **38**(1), 229–234 (2005)
6. Utkin, V., Guldner, J., Shijun, M.: Sliding mode control in electro-mechanical systems. CRC press (1999)
7. Utkin, V.I., Chen, D.S., Chang, H.C.: Block control principle for mechanical systems. J. Dyn. Sys., Meas., Control **122**(1), 1–10 (2000)
8. Utkin, V.I., Drakunov, S.V., Izosimov, D.E., Lukyanov, A.G., Utkin, V.A.: A hierarchical principle of the control system decomposition based on motion separation. IFAC Proceedings Volumes **17**(2), 1553–1558 (1984)
9. Wang, L., Chai, T., Zhai, L.: Neural-network-based terminal sliding-mode control of robotic manipulators including actuator dynamics. IEEE Transactions on Industrial Electronics **56**(9), 3296–3304 (2009). https://doi.org/10.1109/TIE.2008.2011350

Chapter 6
Fixed-Time-Synchronized Control with Settling Time Estimation

Abstract For fixed-time-synchronized control, one critic performance specification consists in the estimation of its synchronized settling time. One possible drawback of fixed-time-synchronized control lies in that the upper bound of the synchronized settling time is usually overestimated significantly. In this chapter, fixed-time-synchronized control is further investigated, aiming to find the most proper and least conservative upper bound to estimate the synchronized settling time. The *beta function* is introduced to facilitate this enhanced estimation. Base on the beta function, the *least upper bound* of synchronized settling time is articulated and proved to be effective. Based on the established least upper bound, a series of robust and singularity-free fixed-time-synchronized controllers are proposed for different classes of multivariable disturbed systems under matched and unmatched disturbances. The techniques to be introduced in this chapter will significantly alleviate the overestimation suffered by traditional fixed-time-synchronized control results. Finally, extensions to predefined-time-synchronized control laws are addressed, where the upper bound of the synchronized setting time explicitly appears in the control parameters. Under the proposed controllers, the target system is proved to achieve strong predefined-time-synchronized convergence and true predefined-time-synchronized convergence.

6.1 Technical Preliminaries

First, The gamma and the beta functions are introduced from [5].

Definition 6.1 [5] A gamma function $\Gamma(z)$ with $\mathrm{Re}\,(z) > 0$ is defined by

$$\Gamma(z) = \int_0^{\infty} e^{-t} t^{z-1} \mathrm{d}t = \int_0^{1} (\log 1/t)^{z-1} \mathrm{d}t. \tag{6.1}$$

(There are other expressions of the gamma function, and they represent the same function. For details, please refer to [5], Chap. 1.1, *The Gamma Function*).

D. Li et al., *Time-Synchronized Control: Analysis and Design*,
https://doi.org/10.1007/978-981-16-3089-7_6

Definition 6.2 [5] A beta function $B(x, y)$ with $\mathrm{Re}(x) > 0$ and $\mathrm{Re}(y) > 0$ is defined by

$$B(x,y) = \int_0^1 t^{x-1}(1-t)^{y-1}\mathrm{d}t = \frac{\Gamma(x)\Gamma(y)}{\Gamma(x+y)}. \tag{6.2}$$

(This derivation can be found in [5], Chap. 2.5, *Hypergeometric Function*.)

Lemma 6.1 *[1] Given α, β, p, g and χ satisfying $p\chi < 1$ and $g\chi > 1$, we have*

$$\int_0^{+\infty} \frac{\mathrm{d}x}{(\alpha x^p + \beta x^q)^\chi} = \frac{B(\omega_p, \omega_g)}{\alpha^{\chi-\omega_p}\beta^{\omega_p}(g-p)}, \tag{6.3}$$

where $\omega_p = \frac{1-\chi p}{g-p}$, $\omega_g = \frac{\chi g - 1}{g-p}$.

Proof The proof is given in [1], Proposition 1. □

In the next section, we will study fixed-time-synchronized control with an enhanced synchronized settling time estimation.

6.2 Fixed-Time-Synchronized Stability with the Least Upper Bound of Synchronized Settling Time

Based on the Lyapunov condition of fixed-time stability in Lemma 2.6, we introduce the least upper bound for the (synchronized) settling time. By integrating both sides of (2.24), we have

$$T(x_0) \le \int_0^{V(x_0)} \frac{\mathrm{d}V(x)}{(\alpha V^p(x) + \beta V^g(x))^\chi}. \tag{6.4}$$

From (6.4), the least upper bound for the (synchronized) settling time can be derived by letting $\|x_0\| \to \infty$, which is defined as follows.

Definition 6.3 Under Lemma 2.6, if the system (2.18) is fixed-time (-synchronized) stable, the least upper bound for the (synchronized) settling time is defined as

$$T_{\text{least}} = \lim_{\|x_0\| \to +\infty} \int_0^{V(x_0)} \frac{\mathrm{d}V(x)}{(\alpha V^p(x) + \beta V^g(x))^\chi}. \tag{6.5}$$

Remark 6.1 Definition 6.3 can be applied to both finite-time stability and its special case, fixed-time-synchronized stability.

For convenience of later analysis, we define a function $\gamma : \mathbb{R}^5 \to \mathbb{R}$,

$$\gamma(\alpha, \beta, p, g, \chi) = \frac{B\left(\omega_p, \omega_g\right)}{\alpha^{\chi-\omega_p}\beta^{\omega_p}(g-p)}, \tag{6.6}$$

where α, β, p, g and $\chi \in \mathbb{R}$, satisfying $\chi p < 1$ and $\chi g > 1$, and ω_p and ω_g are defined as

$$\omega_p = \frac{1-\chi p}{g-p}, \quad \omega_g = \frac{\chi g - 1}{g-p}. \tag{6.7}$$

In the sequel, we provide the main result on fixed-time-synchronized stability with rigorous proof, where an enhanced settling time estimation is proposed.

Theorem 6.1 *Considering a Lyapunov function $V(x)$, the origin of the system (2.18) is fixed-time-synchronized stable, with the least upper bound for the synchronized settling time*

$$T_{\text{least}} = \gamma(\alpha, \beta, p, g, \chi), \tag{6.8}$$

if the following conditions hold:

i. *there exist positive constants α, β, p, g and χ satisfying $p < 1$ and $g > 1$, such that*

$$\dot{V}(x) \leq -(\alpha V^p(x) + \beta V^g(x))^\chi. \tag{6.9}$$

ii. *the state x is ratio persistent.*

In addition, if $\|x_0\| \to +\infty$ and $\dot{V}(x) = -(\alpha V^p(x) + \beta V^g(x))^\chi$, we have the synchronized settling time takes the exact value

$$T(x_0) = T_{\text{least}}. \tag{6.10}$$

Proof Noticing (6.9) and using Lemma 2.6, we can prove that the system (2.18) is fixed-time stable, i.e., $\forall x_0$, $x(t) = 0$ for $t > T(x_0)$. Hence by (6.9), we have

$$\frac{\mathrm{d}V(x)}{-(\alpha V^p(x) + \beta V^g(x))^\chi} \geq \mathrm{d}t, \tag{6.11}$$

$$\int_0^{V(x_0)} \frac{\mathrm{d}V(x)}{(\alpha V^p(x) + \beta V^g(x))^\chi} \geq \int_0^{T(x_0)} \mathrm{d}t. \tag{6.12}$$

From Lemma 6.1, it follows that

$$\begin{aligned} &\lim_{\|x_0\|\to+\infty} \int_0^{V(x_0)} \frac{\mathrm{d}V(x)}{(\alpha V^p(x) + \beta V^g(x))^\chi} \\ =&\frac{B\left(\omega_p, \omega_g\right)}{\alpha^{\chi-\omega_p}\beta^{\omega_p}(g-p)} \\ =&\gamma(\alpha, \beta, p, g, \chi). \end{aligned} \tag{6.13}$$

Therefore, according to Definition 6.3, the least upper bound for the synchronized settling time is $T_{least} = \gamma(\alpha, \beta, p, g, \chi)$.

Further, in the case of $\|x_0\| \to +\infty$ and $\dot{V}(x) = -(\alpha V^p(x) + \beta V^g(x))^\chi$, it reads

$$\lim_{\|x_0\| \to +\infty} T(x_0) = \sup_{x_0 \in \mathbb{R}^n} \int_0^{T(x_0)} \mathrm{d}t = \gamma(\alpha, \beta, p, g, \chi). \tag{6.14}$$

The rest of the proof is given by contradiction. Assume that at least two state elements x_i and x_j reach the equilibrium at different time instants T_i and T_j. Without loss of generality, let x_j be the latter one to arrive, i.e., $T_j > T_i$. According to the property of ratio persistence, we have

$$\lim_{t \to T_i^-} \frac{\mathrm{d}}{\mathrm{d}t} \left(\frac{x_i(t)}{x_j(t)} \right) = 0, \tag{6.15}$$

$$\lim_{t \to T_i^-} x_i(t) = c_{ij} \lim_{t \to T_i^-} x_j(t), \tag{6.16}$$

where $c_{ij} = x_i(0)/x_j(0)$. Further, since the element x_i arrives at the origin at T_i, it yields

$$\lim_{t \to T_i^-} x_i(t) = 0 \Rightarrow c_{ij} \lim_{t \to T_i^-} x_j(t) = 0. \tag{6.17}$$

Due to the continuity of the system (2.18), we have $x_j(T_i) = 0$, where the contradiction occurs because x_j reaches zero at T_j not at T_i, which completes the proof. □

Lemma 6.2 *Considering a Lyapunov function $V(x)$, the origin of the system (2.18) is fixed-time-stable, with the least upper bound for the settling time*

$$T_{\text{least}} = \gamma(\alpha, \beta, p, g, \chi), \tag{6.18}$$

if there exist positive constants α, β, p, g and χ satisfying $p < 1$ and $g > 1$, such that

$$\dot{V}(x) \leq -(\alpha V^p(x) + \beta V^g(x))^\chi. \tag{6.19}$$

Moreover, if $\|x_0\| \to +\infty$ and $\dot{V}(x) = -(\alpha V^p(x) + \beta V^g(x))^\chi$, we have the settling time takes the exact value

$$T(x_0) = T_{\text{least}}. \tag{6.20}$$

Proof Based on Theorem 6.1 , the proof is clear. □

6.3 Fixed-Time-Synchronized Control for Second-Order Systems with Enhanced Settling Time Estimation

In this section, fixed-time-synchronized stability with relevant Lyapunov theorems is proposed. A fixed-time-synchronized controller is then properly designed for a class of affine systems under unknown disturbances and uncertainties to achieve the proposed stability results.

6.3.1 Robust Fixed-Time-Synchronized Control Design

In this subsection, we consider the following affine system in $\mathbb{R}^n$,

$$\begin{aligned} \dot{x}_1 &= x_2, \\ \dot{x}_2 &= f_0(x_1, x_2) + g_0(x_1, x_2)\, u + \Delta, \end{aligned} \tag{6.21}$$

where $x_1 \in \mathbb{R}^n$ and $x_2 \in \mathbb{R}^n$, respectively, denote the general position vector and velocity vector, $f_0(x) \in \mathbb{R}^n$ and $g_0(x) \in \mathbb{R}^{n\times n}$ ($\det(g_0(x)) \neq 0$) are known portions, and $\Delta \in \mathbb{R}^n$ is the unknown disturbance and uncertainty. The dynamics (6.21) can be reformed to an Euler-Lagrange system for mechanical applications, where x_1 stands for the state vector of the Lagrangian coordinates.

A necessary assumption on the unknown system dynamics is made.

Assumption 6.1 The derivative of the disturbance $\dot{\Delta}$ is bounded by a known scalar δ, i.e., $||\dot{\Delta}|| \leq \delta$.

Assumption 6.2 The initial value of the disturbance $\Delta(0)$ is bounded by a known scalar σ, i.e., $||\Delta(0)|| \leq \sigma$.

Remark 6.2 Assumptions 6.1 and 6.2 are mild with regard to the considered disturbance, which are common in the literature, e.g., in [2, 11]. In addition, Assumption 6.2 is only made on the initial value $\Delta(0)$.

In what follows, we first introduce a disturbance observer to accurately estimate the unknown term Δ in fixed time. Based on the disturbance observer, a fixed-time-synchronized control scheme is proposed. Then, according to the proposed Lyapunov conditions in Theorem 6.1, the equilibrium of the closed-loop system (6.21) is proved to be fixed-time-synchronized stable with enhanced settling time estimation.

Inspired by the super-twisting techniques in [3, 4, 12, 14, 17], We design the following fixed-time disturbance observer,

$$\dot{z}_0 = -k_1 \mathrm{sig}_{\mathrm{c}}^{\frac{1}{2}}(\tilde{z}_0) - k_2 \mathrm{sig}_{\mathrm{c}}^{\lambda}(\tilde{z}_0) + f_0 + g_0 u + z_1, \tag{6.22}$$

$$\dot{z}_1 = -k_3 \mathrm{sign}_{\mathrm{c}}(\tilde{z}_0), \tag{6.23}$$

where z_0 and z_1 are used to approximate the state vector x_2 and the unknown term Δ, respectively, $\tilde{z}_0$ and $\tilde{z}_1$ are defined as $\tilde{z}_0 = z_0 - x_2$ and $\tilde{z}_1 = z_1 - \Delta$, k_1, k_2 and k_3 are positive observer gains, and the power term $\lambda > 1$.

We propose the following result for the observer (6.22)–(6.23).

Theorem 6.2 *Considering the system (6.22)–(6.23), under Assumptions 6.1 and 6.2, $\tilde{z}_0$ and $\tilde{z}_1$ can be derived to the origin in fixed time with an enhanced bound*

$$T_{ob} \leq \gamma\left(2k_1, 2k_2, \frac{3}{4}, \frac{1+\lambda}{2}, 1\right) + \frac{\sigma + M\gamma_{ob}}{(1 - Mh/k_1)\, m} + \frac{\sigma}{m} \tag{6.24}$$

where the parameters in (6.24) are all calculable constants regardless of initial states, h, M and m take the forms

$$h = \frac{1}{k_1} + \left(\frac{2e}{mk_1}\right)^{\frac{1}{3}}, \tag{6.25}$$

$$M = k_3 + \delta, \tag{6.26}$$

$$m = k_3 - \delta, \tag{6.27}$$

and the following inequalities hold:

$$k_3 > \delta, \tag{6.28}$$

$$k_1 > Mh. \tag{6.29}$$

Proof From (6.22) and (6.23), the error dynamics can be further derived

$$\dot{\tilde{z}}_0 = -k_1 \mathrm{sig}_c^{\frac{1}{2}}(\tilde{z}_0) - k_2 \mathrm{sig}_c^{\lambda}(\tilde{z}_0) + \tilde{z}_1, \tag{6.30}$$

$$\dot{\tilde{z}}_1 = -k_3 \mathrm{sign}_c(\tilde{z}_0) - \dot{\Delta}. \tag{6.31}$$

The enhanced bound for settling time in (6.24) is obtained by analyzing the following two cases.

Case 1: $\mathrm{sign}_c(\tilde{z}_1(0)) = 0$ or $\mathrm{sign}_c(\tilde{z}_1(0)) = -\mathrm{sign}_c(\tilde{z}_0(0))$.

Consider a Lyapunov candidate $V_{ob} = \tilde{z}_0^T \tilde{z}_0$, whose derivative has the form

$$\dot{V}_{ob} = -2k_1 \tilde{z}_0^T \mathrm{sig}_c^{\frac{1}{2}}(\tilde{z}_0) - 2k_2 \tilde{z}_0^T \mathrm{sig}_c^{\lambda}(\tilde{z}_0) + 2\tilde{z}_0^T \tilde{z}_1. \tag{6.32}$$

In this case, from (6.28) and (6.31), we know $\mathrm{sign}_n(\tilde{z}_1)$ remains opposite to $\mathrm{sign}_n(\tilde{z}_0)$ before $\tilde{z}_0$ arrives at the origin. Thus, we have $\tilde{z}_0^T \tilde{z}_1 \leq 0$. For $\forall i \in \{1, 2, \ldots, n\}$, define $V_{ob,i} = \tilde{z}_{0,i}^2$, where $\tilde{z}_{0,i}$ denotes the ith element of the vector $\tilde{z}_0$. It further leads to

$$\begin{aligned}\dot{V}_{ob,i} \leq & -2k_1\tilde{z}_{0,i}^T\mathrm{sig}_c^{\frac{1}{2}}\left(\tilde{z}_{0,i}\right) - 2k_2\tilde{z}_{0,i}^T\mathrm{sig}_c^{\lambda}\left(\tilde{z}_{0,i}\right) \\ = & -2k_1 V_{ob,i}^{\frac{3}{4}} - 2k_2 V_{ob,i}^{\frac{1+\lambda}{2}}.\end{aligned} \tag{6.33}$$

Following (6.11)–(6.14) in the proof of Theorem (6.1), we know that $\tilde{z}_{0,i}$ converges to the origin in fixed time for $\forall i \in \{1, 2, \ldots, n\}$, with the least upper bound for the settling time

$$T_{ob1} \leq \gamma_{ob} = \gamma\left(2k_1, 2k_2, \frac{3}{4}, \frac{1+\lambda}{2}, 1\right). \tag{6.34}$$

In addition, when $\|\tilde{z}_0(0)\| \to \infty$, it reads

$$T_{ob1} = \gamma_{ob}. \tag{6.35}$$

Next, when $t = T_{ob1}$, $\tilde{z}_{1,i}(T_{ob1})$ is bounded by

$$\left|\tilde{z}_{1,i}(T_{ob1})\right| \leq \sigma + MT_{ob1}. \tag{6.36}$$

Considering $t > T_{ob1}$, for $\forall i \in \{1, 2, \ldots, n\}$, the ith element of the error dynamics (6.30)–(6.31) starts at $(0, \tilde{z}_{0,i}(T_{ob1}))$, which is dominated by the ith element of the solution of the system (6.30)–(6.31) starting at $(0, \sigma + MT_{ob1})$.

From the proof of Theorem 4.5 in [12] and the analysis in [14], Section III, under the necessary condition $Mh/k_1 < 1$, we can get the ith element of the solution starting at $(0, \tilde{z}_{0,i}(\sigma + MT_{ob1}))$ has the settling time bounded by

$$T_{ob2} \leq \frac{\sigma+MT_{ob1}}{(1 - Mh/k_1)\, m}. \tag{6.37}$$

Thus, the overall settling time for Case 1 is bounded by

$$T_{ob}^{(1)} \leq T_{ob1} + T_{ob2} \leq \gamma_{ob} + \frac{\sigma + M\gamma_{ob}}{(1 - Mh/k_1)\, m}. \tag{6.38}$$

Case 2: $\exists i \in \{1, 2, ..., n\}$, $\mathrm{sign_c}(\tilde{z}_{1,i}(0)) = \mathrm{sign_c}(\tilde{z}_{0,i}(0))$.

In this case, $\tilde{z}_{0,i}$ cannot converge to the origin before $\mathrm{sign_c}(\tilde{z}_{1,i}) = -\mathrm{sign_c}(\tilde{z}_{0,i})$. $\tilde{z}_{1,i}$ will thus converge to the origin first. From (6.31), $\tilde{z}_{1,i}$ reaches zero in fixed time

$$T_{ob0} \leq \frac{\sigma}{m}. \tag{6.39}$$

Next, when $t \geq T_{ob0}$, we have $\mathrm{sign_n}(\tilde{z}_1(T_{ob0})) = 0$ or for $\forall i \in \{1, 2, ..., n\}$, $\mathrm{sign_c}(\tilde{z}_{1,i}(T_{ob0})) = -\mathrm{sign_c}(\tilde{z}_{0,i}(T_{ob0}))$, which becomes the same initial conditions as those in Case 1. Following the analysis in Case 1, the settling time is calculated as follows.

First, the original T_{ob1} in (6.34) becomes

$$T^{*}_{ob1} = T_{ob0} + T_{ob1}. \tag{6.40}$$

Next, the original T_{ob2} in (6.37) becomes

$$T^{*}_{ob2} \leq \frac{M\gamma_{ob}}{(1 - Mh/k_1)\, m}. \tag{6.41}$$

The total settling time for Case 2 is given by

$$\begin{aligned} T^{(2)}_{ob} \leq & T^{*}_{ob1} + T^{*}_{ob2} \\ = & \left(1 + \frac{1}{(1/M - h/k_1)\, m}\right)\gamma_{ob} + \frac{\sigma}{m}. \end{aligned} \tag{6.42}$$

Finally, from $T^{(1)}_{ob}$ and $T^{(2)}_{ob}$, we derive the enhanced settling time

$$T_{ob} \leq \gamma_{ob} + \frac{\sigma}{m} + \frac{\sigma + M\gamma_{ob}}{(1 - Mh/k_1)\, m}. \tag{6.43}$$

□

Remark 6.3 Compared with the super-twisting works in the literature [3, 12, 14, 17], Theorem 6.2 proposes an enhanced settling time estimation. In addition, some results, e.g., in [17], $\tilde{z}_1(0)$ is assumed to be 0, while Theorem 6.2 allows $\tilde{z}_1(0)$ to be any bounded value.

Remark 6.4 Theorem 6.24 requires Assumption 6.2 to be valid. When the bound of the initial value of the disturbance $\Delta(0)$ is not available in advance, one can use an additional fixed-time observer to estimate the bound of $\Delta(0)$ as in [4]. In this case, Assumption 6.2 is no longer necessary. This extension is omitted here.

Based on the proposed observer (6.22) and (6.23), we will next design a robust fixed-time-synchronized control law the system (6.21). Before moving forward, we first construct a fixed-time-synchronized sliding variable

We design a sliding-mode manifold for the system (6.21),

$$s = x_2 + \alpha_1 \mathrm{sig}_{\mathrm{n}}^{p_1}(x_1) + \beta_1 \mathrm{sig}_{\mathrm{n}}^{g_1}(x_1), \tag{6.44}$$

where α_1 and β_1 are positive constants, and p_1 and g_1 are defined as $0 < p_1 = p_1^*/p_2^* < 1$ and $g_1 = g_1^*/g_2^* > 1$, with positive odd integers p_1^*, p_2^*, g_1^* and g_2^*. In accordance with the sliding-mode manifold (6.44), we have the following result.

Lemma 6.3 *For the sliding variable (6.44), when $s = 0$, the system state x_1 is ratio persistent and the closed-loop system is fixed-time-synchronized stable with the least upper bound for the synchronized settling time*

$$T_{\text{least}} = \gamma\left(2\alpha_1, 2\beta_1, \frac{1+p_1}{2}, \frac{1+g_1}{2}, 1\right). \tag{6.45}$$

Proof In the case of $s = 0$, it leads to

$$x_2 = -\alpha_1 \mathrm{sig}_{\mathrm{n}}^{p_1}(x_1) - \beta_1 \mathrm{sig}_{\mathrm{n}}^{g_1}(x_1). \tag{6.46}$$

By considering a Lyapunov function $V = x_1^T x_1$, it is clear that

$$\begin{aligned}\dot{V} =& -2\alpha_1 x_1^T \mathrm{sig}_{\mathrm{n}}^{p_1}(x_1) - 2\beta_1 x_1^T \mathrm{sig}_{\mathrm{n}}^{g_1}(x_1)\\ =& -2\alpha_1 V^{\frac{1+p_1}{2}} - 2\beta_1 V^{\frac{1+g_1}{2}}.\end{aligned} \tag{6.47}$$

From (6.46) and according to Definition 2.11, we know x_1 is ratio persistent.

Based on Theorem 6.1, the system (6.46) is fixed-time-synchronized stable, with the least upper bound of the synchronized settling time

$$T_{\text{least}} = \gamma\left(2\alpha_1, 2\beta_1, \frac{1+p_1}{2}, \frac{1+g_1}{2}, 1\right). \tag{6.48}$$

The proof finishes here. □

Designing a fixed-time-synchronized control algorithm directly based on the sliding variable (6.44), however, may result in a singular control input when $x_1, x_2 \to 0$. Usually, it is a non-trivial task to obtain a singularity-free controller.

Here, inspired by the proposed fixed-time-synchronized sliding variable (6.44) and with the aid of the fixed-time observer (6.22)–(6.23), we design a robust fixed-time-synchronized control law with a switching fixed-time-synchronized sliding variable for the system (6.21) as follows,

$$u = -g_0^{-1}\left(\alpha_2 \mathrm{sig}_{\mathrm{n}}^{p_2}(s) + \beta_2 \mathrm{sig}_{\mathrm{n}}^{g_2}(s) + \dot{s}_s + f_0 + z_1\right), \tag{6.49}$$

where the corresponding powers are defined as $0 < p_2 = p_3^*/p_4^* < 1$ and $g_2 = p_3^*/p_4^* > 1$ with positive odd integers p_3^*, p_4^*, g_3^* and g_4^*.

In the control law (6.49), switching fixed-time-synchronized sliding variable is defined as

$$s = x_2 + s_s. \tag{6.50}$$

The switching law s_s and its derivative have the forms as in (6.51) and (6.52),

$$s_s = \begin{cases} \alpha_1 \mathrm{sig}_{\mathrm{n}}^{p_1}(x_1) + \beta_1 \mathrm{sig}_{\mathrm{n}}^{g_1}(x_1), & \text{if } s^* = 0 \text{ or } s^* \neq 0,\ \|x_1\| > \varepsilon,\\ l_1 x_1 + l_2 \mathrm{sig}_{\mathrm{n}}^{4}(x_1), & \text{if } s^* \neq 0,\ \|x_1\| \leq \varepsilon,\end{cases} \tag{6.51}$$

$$\dot{s}_s = \begin{cases} \rho_1 x_1 x_1^T x_2 + \rho_2 x_2, & \text{if } s^* = 0 \text{ or } s^* \neq 0,\ \|x_1\| > \varepsilon,\\ l_1 x_2 + 3 l_2 \|x_1\| x_1 x_1^T x_2 + l_2 \|x_1\|^3 x_2, & \text{if } s^* \neq 0,\ \|x_1\| \leq \varepsilon,\end{cases} \tag{6.52}$$

where ε is a small positive constant, the trigger sliding-mode variable is defined as

$$s^* = x_2 + \alpha_1 \mathrm{sig}_{\mathrm{n}}^{p_1}(x_1) + \beta_1 \mathrm{sig}_{\mathrm{n}}^{g_1}(x_1), \tag{6.53}$$

variables ρ_1 and ρ_2 are defined as

$$\rho_1 = \alpha_1 (p_1 - 1) \|x_1\|^{p_1 - 3} + \beta_1 (g_1 - 1) \|x_1\|^{g_1 - 3}, \tag{6.54}$$

$$\rho_2 = \alpha_1 \|x_1\|^{p_1 - 1} + \beta_1 \|x_1\|^{g_1 - 1}, \tag{6.55}$$

and constants l_1 and l_2 take the following forms

$$l_1 = \alpha_1 \left(\frac{4}{3} - \frac{p_1}{3}\right) \|\varepsilon\|^{p_1 - 1} + \beta_1 \left(\frac{4}{3} - \frac{g_1}{3}\right) \|\varepsilon\|^{g_1 - 1}, \tag{6.56}$$

$$l_2 = \alpha_1 \left(\frac{p_1}{3} - \frac{1}{3}\right) \|\varepsilon\|^{p_1 - 4} + \beta_1 \left(\frac{g_1}{3} - \frac{1}{3}\right) \|\varepsilon\|^{g_1 - 4}, \tag{6.57}$$

where the corresponding powers are redefined as

$$\frac{1}{2} < p_1 = p_1^*/p_2^* < 1, \tag{6.58}$$

$$1 < g_1 = p_1^*/p_2^* < 4, \tag{6.59}$$

with positive odd integers p_1^*, p_2^*, g_1^* and g_2^*. The rationale of these parameter ranges are addressed later in Theorem 6.4.

We can verify that the design of l_1 and l_2 guarantee the continuity of the switching law s_s (6.51) as well as its time derivative $\dot{s}_s$ (6.52).

We are now ready to present the main result for the robust fixed-time-synchronized control law (6.49).

Theorem 6.3 *For the system (6.21), if Assumptions (6.1) and (6.2) hold, under the robust fixed-time-synchronized control law (6.49) and the disturbance observer (6.22)–(6.23), the sliding-mode variable s is ratio persistent after the observer error* $\tilde{z}_1$ *converges to zeron, the system state* x_1 *is ratio persistent on the sliding surface* $s = 0$*, and the closed-loop system is fixed-time-synchronized stable with enhanced settling time estimation*

$$T_c \leq T_{ob} + \gamma_{c1} + \gamma_{c2}, \tag{6.60}$$

where T_{ob} *is the enhanced settling time of the observer as shown in (6.24), and* γ_{c1} *and* γ_{c2} *are* γ *functions defined as*

$$\gamma_{c1} = \gamma\left(2\alpha_1, 2\beta_1, \frac{1 + p_1}{2}, \frac{1 + g_1}{2}, 1\right), \tag{6.61}$$

$$\gamma_{c2} = \gamma\left(2\alpha_2, 2\beta_2, \frac{1 + p_2}{2}, \frac{1 + g_2}{2}, 1\right). \tag{6.62}$$

Proof Consider a Lyapunov function $V_{c2} = s^T s$. Under the control law (6.49), the derivative of V_{c2} can be written as

$$\dot{V}_{c2} = 2s^T \left(-\alpha_2 \mathrm{sig}_{\mathrm{n}}^{p_2}(s) - \beta_2 \mathrm{sig}_{\mathrm{n}}^{g_2}(s) - z_1 + \Delta\right). \tag{6.63}$$

When $t \geq T_{ob}$, according to Theorem 6.2, we know $z_1 = \Delta$. Then, (6.63) becomes

$$\begin{aligned} \dot{V}_{c2} &= -2\alpha_2 s^T \mathrm{sig}_{\mathrm{n}}^{p_2}(s) - 2\beta_2 s^T \mathrm{sig}_{\mathrm{n}}^{g_2}(s) \\ &= -2\alpha_2 V_{c2}^{\frac{1+p_2}{2}} - 2\beta_2 V_{c2}^{\frac{1+g_2}{2}}. \end{aligned} \tag{6.64}$$

Here, we have $\dot{s} = -\alpha_2 \mathrm{sig}_{\mathrm{n}}^{p_2}(s) - \beta_2 \mathrm{sig}_{\mathrm{n}}^{g_2}(s)$. From Definition 2.11, it yields that s is ratio persistent when $t \geq T_{ob}$.

Based on Theorem 6.1, the sliding variable s is fixed-time-synchronized stable, and its corresponding least upper bound of the synchronized settling time takes the form

$$T_{c2} \leq \gamma_{c2} = \gamma\left(2\alpha_2, 2\beta_2, \frac{1+p_2}{2}, \frac{1+g_2}{2}, 1\right). \tag{6.65}$$

The system state q reaches the sliding manifold $s = 0$ when $t \geq T_{ob} + T_{c2}$. According to the sliding-mode variable (6.50) and (6.51), we can obtain

$$x_2 = -\alpha_1 \mathrm{sig}_{\mathrm{n}}^{p_1}(x_1) - \beta_1 \mathrm{sig}_{\mathrm{n}}^{g_1}(x_1), \tag{6.66}$$

which leads to the ratio persistent state x_1 according to Definition 2.11.

By using Lemma 6.3, the system state x_1 is fixed-time-synchronized stable with the enhanced synchronized settling time

$$T_{c1} \leq \gamma_{c1} = \gamma\left(2\alpha_1, 2\beta_1, \frac{1+p_1}{2}, \frac{1+g_1}{2}, 1\right). \tag{6.67}$$

Thus, for the system (6.21) under disturbance, the total synchronized setting time of the closed-loop system state x_1 can be derived

$$T_c \leq T_{ob} + T_{c1} + T_{c2}, \tag{6.68}$$

which concludes the proof. □

Remark 6.5 In (6.60), $T_{ob} + T_{c1} + T_{c2}$ is not the least upper bound of the synchronized settling time T_c due to the fact that T_{ob} does not directly meet Definition 6.3. Therefore, it is more accurate to refer it as an *enhanced* synchronized settling time estimation rather than the least upper bound of the synchronized settling time. Specifically, it contains three parts T_{ob}, T_{c1} and T_{c2} which are derived based on their least upper bounds. Therefore, (6.60) is less conservative than those in classical fixed-time control results. This will be analyzed and compared in the simulation.

We next discuss the singularity-avoidance property of the controller (6.49).

Theorem 6.4 *Using the proposed control law (6.49) with the switching sliding manifold (6.50), the singularity problem can be avoided.*

Proof The proof is given by three cases: 1) $\|x_1\| > \varepsilon$; 2) $\|x_1\| \leq \varepsilon$, $s^* = 0$; and 3) $\|x_1\| \leq \varepsilon, s^* \neq 0$. When $\|x_1\| > \varepsilon$, obviously, no singularity exists. When $\|x_1\| \leq \varepsilon$, $s^* = 0$, from the control law (6.49) and the switching law (6.51), it yields

$$u = - g_0^{-1} \left(\alpha_2 \text{sig}_{\text{n}}^{p_2}(s) + \beta_2 \text{sig}_{\text{n}}^{g_2}(s) + f_0 + J(x_1) \right), \tag{6.69}$$

where $J(x_1)$ is defined by

$$\begin{aligned} J(x_1) = & \ \alpha_1 \rho_1 x_1 x_1^T \text{sig}_{\text{n}}^{p_1}(x_1) + \beta_1 \rho_1 x_1 x_1^T \text{sig}_{\text{n}}^{g_1}(x_1) \\ & + \alpha_1 \rho_2 \text{sig}_{\text{n}}^{p_1}(x_1) + \beta_1 \rho_2 \text{sig}_{\text{n}}^{g_1}(x_1). \end{aligned} \tag{6.70}$$

Substituting ρ_1 and ρ_2 (introduced in (6.54)–(6.55)) into (6.70) and noticing $x_1 x_1^T \text{sig}_{\text{n}}^{p_1}(x_1) = \text{sig}_{\text{n}}^{p_1+2}(x_1)$, we have

$$\begin{aligned} J(x_1) = & \ \alpha_1^2 p_1 \text{sig}_{\text{n}}^{2p_1-1}(x_1) + \beta_1^2 g_1 \text{sig}_{\text{n}}^{2g_1-1}(x_1) \\ & + \alpha_1 \beta_1 (p_1 + g_1) \text{sig}_{\text{n}}^{p_1+g_1-1}(x_1). \end{aligned} \tag{6.71}$$

Since $\frac{1}{2} < p_1 < 1$ and $g_1 > 1$, the singularity is avoided in (6.71). Next, according to the switching law (6.51), when $\|x_1\| \leq \varepsilon$, $s^* \neq 0$, the sliding manifold (6.50) becomes a general one and no singularity occurs. □

Remark 6.6 For the switching law (6.51), the general sliding manifold $s_s = l_1 x_1 + l_2 \text{sig}_{\text{n}}^4(x_1)$ only takes place when $s^* \neq 0$, $\|x_1\| \leq \varepsilon$. By selecting a sufficiently small ε, such that the trajectory of x_1 is governed by $s_s = \alpha_1 \text{sig}_{\text{n}}^{p_1}(x_1) + \beta_1 \text{sig}_{\text{n}}^{g_1}(x_1)$ rather than $s_s = l_1 x_1 + l_2 \text{sig}_{\text{n}}^4(x_1)$. Despite the fact that the general one does not arises when ε is small enough, the closed-loop convergence property in this case ($s^* \neq 0$, $\|x_1\| \leq \varepsilon$) can still be explored. One may choose an input-to-state stable (ISS) Lyapunov function $V_{ISS} = x_1^T x_1/2$ associated with a Lyapunov gain

$$\kappa(\delta) = \frac{\|\delta\|}{\theta \alpha_1 \|\varepsilon\|^{p_1-1} + \theta \beta_1 \left(\frac{4}{3} - \frac{g_1}{3}\right) \|\varepsilon\|^{g_1-1}}, \tag{6.72}$$

where $\delta \in \mathbb{R}^n$ and $\theta \in (0, 1)$. This directly proves input-to-state stability with the state x_1 and the input s, since a system is ISS iff there exists an ISS Lyapunov function for it [13].

Remark 6.7 The singularity problem widely exists in the literature of fixed/finite-time control design, and some effective techniques have been proposed in the existing results. Actually, this is not the first time a switching sliding-mode manifold is proposed. The switching variable (6.50) is inspired by the finite-time control result [16], where a switching law is designed to avoid the singularity in the controller. Similar techniques have also been applied in [8, 10] for spacecraft attitude tracking. However, the switching manifold in [8, 10, 16] is designed for the classical sign function $\mathrm{sign_c}(\cdot)$, which is based on each element $|x_{1,i}|$ in the function $\mathrm{sig_c}(x_1)$, and is thus not applicable to $\mathrm{sig_n}(\cdot)$. Compared with the results in [8, 10, 16], the proposed switching law (6.51) is designed based on the vector x_1 of the norm-normalized sign function $\mathrm{sign_n}(x_1)$. In summary, this design extends the former scalar ones to a generalized multivariable case, which completely changes the relevant design procedure and the analysis due to the compatibility with the norm-normalized sign function. Therefore, the new designed switching law (6.51) takes a different form, and contributes not only to the avoidance of the singularity point but also to the achievement of fixed-time-synchronized stability.

6.3.2 Numerical Results

In this section, we examine the proposed synchronized controller (6.49) to show the properties of the fixed-time-synchronized convergence with enhanced settling time estimation. In what follows, to yield a comprehensive comparison, we first rewrite the fixed-time-synchronized control law (6.49) to a non-synchronized one.

Lemma 6.4 *Considering the system governed by (6.21), under the following fixed-time control law integrated with the disturbance observer (6.22)–(6.23),*

$$\bar{u} = -g_0^{-1}\left(\alpha_2 \mathrm{sig_c}^{p_2}(\bar{s}) + \beta_2 \mathrm{sig_c}^{g_2}(\bar{s}) + \dot{\bar{s}}_s + f_0 + z_1\right), \tag{6.73}$$

where the sliding manifold $\bar{s} = x_2 + \bar{s}_s$, *the trigger sliding-mode variable* $\bar{s}^* = x_2 + \alpha_1 \mathrm{sig_c}^{p_1}(x_1) + \beta_1 \mathrm{sig_c}^{g_1}(x_1)$, *and the switching law take the forms as shown in (6.74)–(6.75),*

$$\bar{s}_{s,i} = \begin{cases} \alpha_1 \mathrm{sig_c}(x_{1,i})^{p_1} + \beta_1 \mathrm{sig_c}(x_{1,i})^{g_1}, & \text{if } \bar{s}^* = 0, \text{ or } \bar{s}^* \neq 0, \ |x_{1,i}| > \varepsilon, \\ l_3 x_{1,i} + l_4 \mathrm{sig_c}(x_{1,i})^2, & \text{if } \bar{s}^* \neq 0, \ |x_{1,i}| \leq \varepsilon, \end{cases} \tag{6.74}$$

$$\dot{\bar{s}}_{s,i} = \begin{cases} \alpha_1 p_1 |x_{1,i}|^{p_1-1} x_{2,i} + \beta_1 g_1 |x_{1,i}|^{g_1-1} x_{2,i}, & \text{if } \bar{s}^* = 0, \text{ or } \bar{s}^* \neq 0, \ |x_{1,i}| > \varepsilon, \\ l_3 x_{2,i} + l_4 |x_{1,i}| x_{2,i}, & \text{if } \bar{s}^* \neq 0, \ |x_{1,i}| \leq \varepsilon, \end{cases} \tag{6.75}$$

where $x_{1,i}$, $x_{2,i}$, $\bar{s}_{s,i}$ *and* $\dot{\bar{s}}_{s,i}$ *are respectively the* i*th element of* x_1, x_2, $\bar{s}_s$ *and* $\dot{\bar{s}}_s$, $i = 1, 2, \ldots, n$, *and constant parameters* l_3 *and* l_4 *are then designed as follows to maintain the continuity of* $\bar{s}_s$ *and* $\dot{\bar{s}}_s$

$$l_3 = (2 - p_1)\alpha_1 \varepsilon^{p_1-1} + (2 - g_1)\beta_1 \varepsilon^{g_1-1}, \tag{6.76}$$

$$l_4 = (p_1 - 1)\alpha_1 \varepsilon^{p_1-2} + (g_1 - 1)\beta_1 \varepsilon^{g_1-2}, \tag{6.77}$$

Table 6.1 Parameters for simulation

Parameters	Values	Parameters	Values
f_0	$5x$	g_0	$2I_3$
k_1	10	k_2	50
k_3	20	λ	11/9
α_1	1	α_2	5
β_1	0.2	β_2	0.2
p_1	7/13	p_2	7/13
g_1	15/9	g_2	15/9
δ	0.657	θ	0.0001
Δ	$[2, 3\sin(0.02\pi t), 4\sin(0.05\pi t + \pi/2)]^T$		

the closed-loop system state converges to the origin in fixed time.

Proof The proof can be obtained by following the proof of Theorem 6.3. □

Intuitively, it is clear that the controller (6.73) cannot yield a time-synchronized behavior as it decouples the system state into n individual scalars, i.e.,

$$\bar{u}_i(x_1, x_2) = \bar{u}_i\left(x_{1,i}, x_{2,i}\right). \tag{6.78}$$

Thus, the synchronized convergence cannot be expected from the controller (6.73). We here add this fixed-time control design into the comparison to further verify and highlight the time-synchronized features attained.

Remark 6.8 The switching law $\bar{s}_s$ in Lemma 6.4 basically takes the same form as the ones in [8, 10, 16], where the relevant proof and more detailed analysis are addressed. As shown in Lemma 6.4, the switching law $\bar{s}_{s,i}$ is designed based on the element $x_{1,i}$ of the classical sign function $\mathrm{sig}_c(x_{1,i})$, while the proposed switching law (6.51) herein is designed based on vector x_1 of the norm-normalized sign function $\mathrm{sig}_n(x_1)$, making the design of $\bar{s}_s$ and s_s (or $\dot{\bar{s}}_s$ and $\dot{s}_s$), the forms of l_i, $i = 1, 2, 3, 4$, and the corresponding triggering conditions of the switching laws differ significantly. This is in line with the discussion in Remark 6.7 that fundamental differences lie in the forms as well as the performance of these two switching laws.

Simulation parameters are shown in Table 6.1, where the fixed-time-synchronized controller (6.49) and the fixed-time controller (6.73) share the same control and observer parameters to produce a fair comparison.

As presented in Figs. 6.1 and 6.2, under the fixed-time-synchronized control law (6.49), all the system state elements reach the equilibrium at the same time in the presence of disturbances, while under the fixed-time law (6.73), three state elements arrive at the equilibrium separately. This could be observed more clearly in Figs. 6.3 and 6.4, where the norms of the state elements arrive at the origin time-synchronously at $t = 6.3$s under the proposed controller (6.49), and arrive at the origin one after

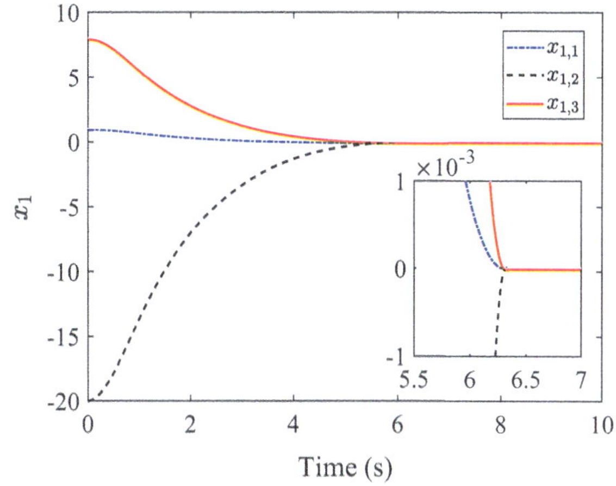

Fig. 6.1 System state x under the fixed-time-synchronized control law (6.49)

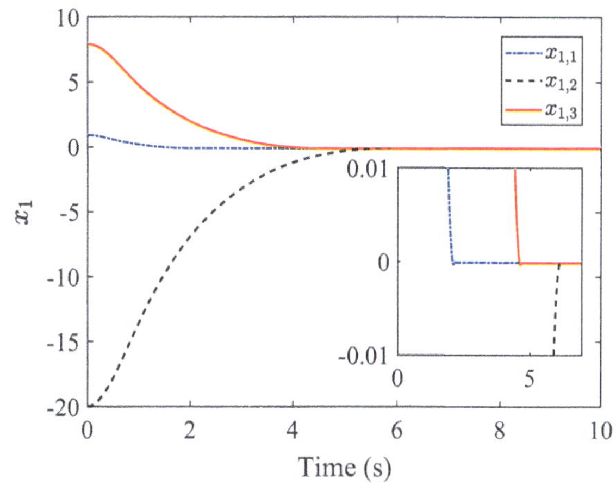

Fig. 6.2 System state x under the fixed-time control law (6.73)

another at $t = 2.1$s, $t = 4.5$s, and $t = 6.3$s under the fixed-time controller (6.73). Moreover, in Fig. 6.3, the state elements converge time-synchronously even when we zoom in at the level of $\times 10^{-4}$. Next, the relevant control inputs of both controllers are shown in Fig. 6.5.

Ratio persistence is verified in Fig. 6.6. It evidently approves declared fixed-time-synchronized stability under the controller (6.49), where after T_1, the ratio of each pair of the state elements is time-invariant. Specific subfigures are also included to showcase how the ratio converges to a constant.

The 3-dimensional state trajectories under the fixed-time-synchronized control law (6.49) and the finite-time law (6.73) are presented in Fig. 6.7. The trajectory under the fixed-time-synchronized control law (6.49) (the blue curve) is divided into two parts: the trajectory s.t. $t < T_{ob}$ and the trajectory s.t. $t \geq T_{ob}$. It is noted that the trajectory s.t. $t < T_{ob}$ is a curve since the estimation from the disturbance observer

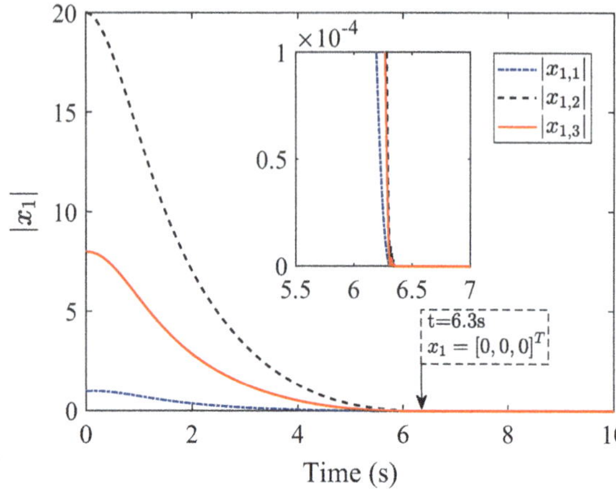

Fig. 6.3 Norms of the system state elements under the fixed-time-synchronized control law (6.49)

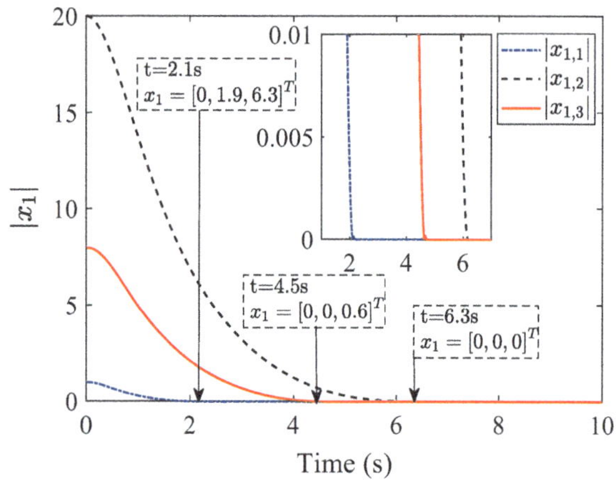

Fig. 6.4 Norms of the system state elements under the fixed-time control law (6.73)

(6.22)–(6.23) is not accurate, i.e., $\tilde{z}_1 \neq 0$. Given a precise disturbance estimation, it suggests that fixed-time-synchronized stability helps the system generate a smoother and shorter path. As according to Theorem 6.3, when $\tilde{z}_1 = 0$, $t \geq T_{ob}$, the output trajectory of the closed-loop system under the fixed-time-synchronized control law (6.49) is ratio persistent, being a straight line in the 3-dimensional space (optimal in light of the travel length). On the contrary, the trajectory governed by the fixed-time law (6.73) (the red curve) is much longer and unpredictable in advance.

We will next verify the enhanced synchronized settling time estimation, and explore its performance compared with existing settling time estimation methods.

Given the parameters in Table 6.1, the enhanced settling time estimation under is derived from (6.68),

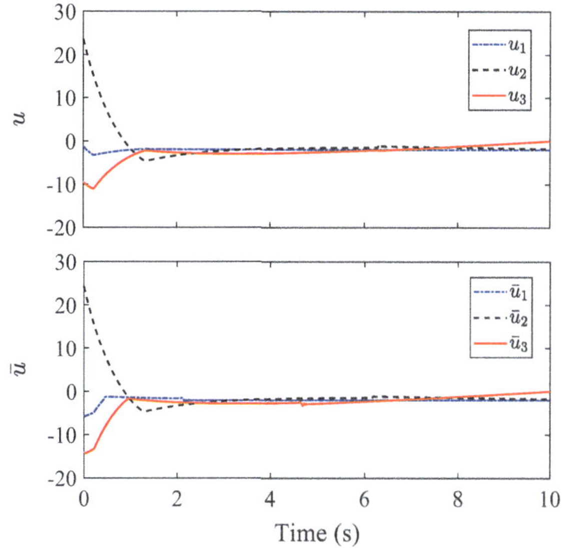

Fig. 6.5 Control inputs of the fixed-time-synchronized control law (6.49) u and the fixed-time control law $\bar{u}$ (6.73)

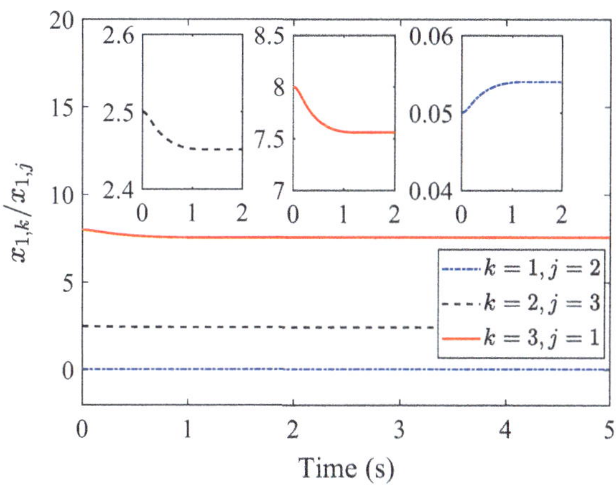

Fig. 6.6 Ratio of each pair of the state elements

$$
\begin{aligned}
T_c \leq & T_{ob} + T_{c1} + T_{c2} \\
= & \gamma\left(2k_1, 2k_2, \frac{3}{4}, \frac{1+\lambda}{2}, 1\right) + \frac{\sigma + M\gamma_{ob}}{(1 - Mh/k_1)\, m} + \frac{\sigma}{m} \\
& + \gamma\left(2\alpha_1, 2\beta_1, \frac{1+p_1}{2}, \frac{1+g_1}{2}, 1\right) \\
& + \gamma\left(2\alpha_2, 2\beta_2, \frac{1+p_2}{2}, \frac{1+g_2}{2}, 1\right) \\
= & 9.3019\text{s}.
\end{aligned}
\tag{6.79}
$$

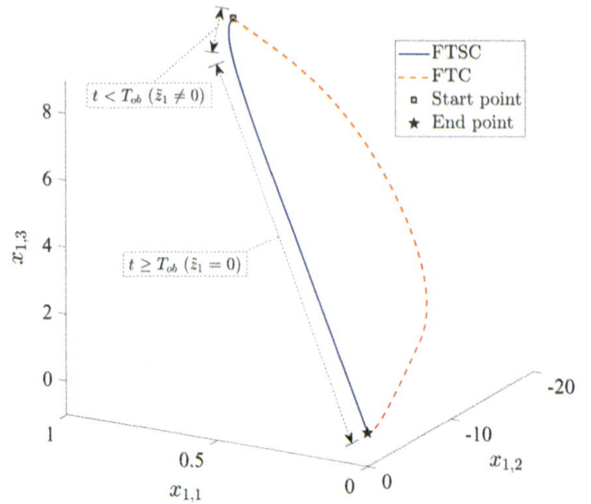

Fig. 6.7 Trajectories of x under the fixed-time-synchronized controller (FTSC) and under the fixed-time controller (FTC)

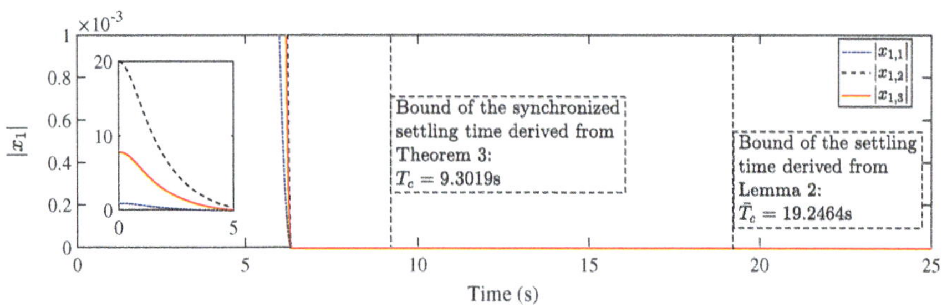

Fig. 6.8 Norms of the system state x under the fixed-time-synchronized control law (6.49), with the enhanced synchronized settling time estimation T_c and the settling time estimation $\bar{T}_c$ derived from Lemma 2.6

By using the settling time estimation method (2.25) from Lemma 2.6, the bound of the settling time is given,

$$\begin{aligned}
\bar{T}_c \leq & \bar{T}_{ob} + \bar{T}_{c1} + \bar{T}_{c2} \\
= & \frac{1}{k_1(1-3/4)} + \frac{2}{k_2(\lambda - 1)} + \frac{\sigma + M\gamma_{ob}}{(1 - Mh/k_1)\, m} + \frac{\sigma}{m} \\
& + \frac{1}{\alpha_1(1-p_1)} + \frac{1}{\beta_1(g_1 - 1)} \\
& + \frac{1}{\alpha_2(1-p_2)} + \frac{1}{\beta_2(g_2 - 1)} \\
= & 19.2464\text{s}.
\end{aligned} \tag{6.80}$$

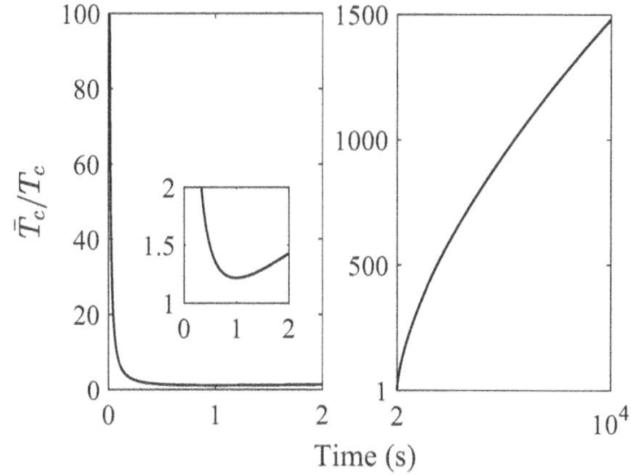

Fig. 6.9 Comparison between the estimation of the synchronized settling time T_c and that of the settling time $\bar{T}_c$

The enhanced estimation (6.79) and the traditional estimation (6.80) are applicable to both the synchronized settling time under the fixed-time-synchronized controller (6.49) and the settling time under the fixed-time controller (6.73). Obviously, the bound of the settling time acquired by the proposed algorithm is less conservative than those by classical methods. In Fig. 6.8, it is clear that, by using the proposed estimation, the slack between the real settling time and the proposed estimation T_c is substantially lower compared with that between the real settling time and the estimation $\bar{T}_c$ from Lemma 2.6. The reason is that the proposed estimation is composed of several least upper bounds for the synchronized settling time. This makes our approach less over-engineered. We here explore more details on the comparison between T_c and $\bar{T}_c$. Define a positive variable ϖ. Set $\alpha_1 = \alpha_2 = \varpi$ and $\beta_1 = \beta_2 = 1/\varpi$. Let the other parameters remain the same. Fig. 6.9 shows the value of $\bar{T}_c(\varpi)/\ T_c(\varpi)$ over ϖ. It shows that for $\forall \varpi > 0$, $\bar{T}_c/T_c > 1$. Specifically, $\bar{T}_c/T_c \rightarrow \infty$ when $\varpi \rightarrow \infty$ or $\varpi \rightarrow 0$ (the left subfigure in Fig. 6.9 only shows a max value 100 for greater clarity). This validates the effectiveness of the enhanced synchronized settling time estimation.

6.4 Fixed-Time-Synchronized Control for General MIMO Systems Under Matched and Unmatched Disturbances with Enhanced Settling Time Estimation

In this section, fixed-time-synchronized control is studied for general MIMO systems under matched and unmatched disturbances, with the consideration of enhanced settling time estimation. The control design herein is mainly based on the formulation in Chap. 5, Sect. 5.2, where some overlapped content is omitted.

The following MIMO system is considered with matched and unmatched disturbances

$$\dot{x} = Ax + Bu + d(t, x, u), \tag{6.81}$$

where $x \in \mathbb{R}^n$ is the system state, $u \in \mathbb{R}^m$ $(1 \leq m \leq n)$ the control input, $d(t, x, u)$ the general disturbances and uncertainties, and $A \in \mathbb{R}^{n\times n}$ and $B \in \mathbb{R}^{n\times m}$ system matrices.

The following assumption on controllability is made.

Assumption 6.3 The system dynamics (6.81) contains a controllable pair (A, B). The matrix B is of full rank, and by reordering the components, it is partitioned as $B = \left[B_1^T, B_2^T\right]^T$, where $B_1 \in \mathbb{R}^{(n-m)\times m}$, $B_2 \in \mathbb{R}^{m\times m}$, and $\det(B_2) \neq 0$.

We rewrite the system (6.81) in a regular form, by a proper state transformation [6, 9, 15],

$$Tx = \begin{bmatrix} x_1 \\ x_2 \end{bmatrix}, \; T = \begin{bmatrix} I_{n-m} & -B_1 B_2^{-1} \\ 0 & B_2^{-1} \end{bmatrix}. \tag{6.82}$$

The regular form can be expressed as,

$$\begin{aligned} \dot{x}_1 &= A_{11}x_1 + A_{12}x_2 + d_u(t, x), \\ \dot{x}_2 &= A_{21}x_1 + A_{22}x_2 + u + d_m(t, x, u), \\ y &= x_1, \end{aligned} \tag{6.83}$$

where $x = \left[x_1^T, x_2^T\right]^T$ denotes the system state ($x_1 \in \mathbb{R}^{n-m}$, $x_2 \in \mathbb{R}^m$), $y \in \mathbb{R}^{n-m}$ the system output , $d_u(t, x) \in \mathbb{R}^{n-m}$ and $d_m(t, x, u) \in \mathbb{R}^m$, respectively, the components of the unmatched and matched disturbances, the system matrices A_{11}, A_{12}, A_{21}, and A_{22} are of proper dimensions.

We will propose fixed-time-synchronized control for input-dimension-dominant MIMO systems (introduced in Definition 4.1) and state-dimension-dominant MIMO systems (introduced in Definition 4.2), respectively.

6.4.1 Control Design for Input-Dimension-Dominant MIMO Systems

We first focus on the input-dimension-dominant MIMO system. Before moving forward, the following necessary assumptions are made.

Assumption 6.4 The row echelon form of the matrix $A_{12} \in \mathbb{R}^{(n-m)\times m}$ $(m \geq \frac{n}{2})$ in the system dynamics (6.83) has pivots in each row.

Assumption 6.5 The disturbances are bounded as

$$\|d_u(t, x)\| \leq \eta_u \|x\| + h_u, \tag{6.84}$$

$$\|d_m(t, x, u)\| \leq \eta_m \|u\| + h_m(t, x), \tag{6.85}$$

where $\eta_u \geq 0$, $\eta_m \geq 0$, and $h_u \in [0, 1)$ are known positive constants, and $h_m(t, x)$ is a known positive function.

Assumption 6.6 The disturbances are bounded as

$$\left\|\dot{d}_u(t, x)\right\| \leq \kappa_u \|x\| + \vartheta_u, \tag{6.86}$$

$$\left\|\dot{d}_m(t, x, u)\right\| \leq \kappa_m \|u\| + \vartheta_m(t, x), \tag{6.87}$$

where κ_u, κ_m, and ϑ_u are known positive constants, and $\vartheta_m(t, x)$ is a known positive function.

The necessity of Assumptions 6.4, 6.5, and 6.6 have been explained in Chap. 4, Sect. 4.2.

We first define the following fixed-time-synchronized sliding mode

$$s_1 = \dot{x}_1 + \alpha_1 \mathrm{sig}_{\mathrm{n}}^{p_1}(x_1) + \beta_1 \mathrm{sig}_{\mathrm{n}}^{g_1}(x_1), \tag{6.88}$$

where α_1 and β_1 are positive constants, and p_1 and g_1 are defined as $0 < p_1 = p_1^*/p_2^* < 1$ and $g_1 = g_1^*/g_2^* > 1$, with positive odd integers p_1^*, p_2^*, g_1^* and g_2^*.

Based on Theorem 6.1, on the fixed-time-synchronized sliding surface (6.88) ($s_1 = 0$), x_1 is fixed-time-synchronized stable, with the least upper bound for the synchronized settling time,

$$T_{m1} \leq T_{\mathrm{least},1} = \gamma(2^{\frac{1+p_1}{2}}\alpha_1, 2^{\frac{1+g_1}{2}}\beta_1, \frac{1+p_1}{2}, \frac{1+g_1}{2}, 1), \tag{6.89}$$

where in the case of $\|x_1(0)\| \to +\infty$, we have $T_{m1} = T_{\mathrm{least},1}$.

Aiming at driving $s_1 \to 0$, a virtual velocity tracking error is defined as

$$z = x_2 - \phi_2, \tag{6.90}$$

where

$$\phi_2 = \phi_{2eq} + \phi_{2n}, \tag{6.91}$$

$$\phi_{2eq} = -A_{12}^{\dagger}\left(A_{11}x_1 + \alpha_1 \mathrm{sig}_{\mathrm{n}}^{p_1}(x_1) + \beta_1 \mathrm{sig}_{\mathrm{n}}^{g_1}(x_1)\right), \tag{6.92}$$

$$\dot{\phi}_{2n} = -A_{12}^{\dagger}\left(\Theta_1(x)\mathrm{sign}_{\mathrm{n}}(s_1) + \alpha_2 \mathrm{sig}_{\mathrm{n}}^{p_2}(s_1) + \beta_2 \mathrm{sig}_{\mathrm{n}}^{g_2}(s_1)\right), \tag{6.93}$$

$$\Theta_1(x) = \kappa_u \|x\| + \vartheta_u, \tag{6.94}$$

where α_2, β_2, p_2 and g_2 are positive constants, satisfying $0 < p_2 = p_3^*/p_4^* < 1$ and $g_2 = g_3^*/g_4^* > 1$, with positive odd integers p_3^*, p_4^*, g_3^* and g_4^*.

In what follows, it is shown that s_1 synchronously converges to the origin in fixed time in the case of $z = \dot{z} = 0$.

Lemma 6.5 *In the case of $z = \dot{z} = 0$, the sliding variable s_1 achieves fixed-time convergence, with the least upper bound for the settling time,*

$$T_{m2} \le T_{\text{least},2} = \gamma(2^{\frac{1+p_2}{2}}\alpha_2, 2^{\frac{1+g_2}{2}}\beta_2, \frac{1+p_2}{2}, \frac{1+g_2}{2}, 1). \tag{6.95}$$

In the case of $\|s_1(0)\| \to +\infty$, $T_{m2} = T_{\text{least},2}$.

Proof From Lemmas 5.5 and 6.2, the system output x_1 reaches the sliding manifold $s_1 = 0$ in fixed time bounded by

$$T_{m2} \le T_{\text{least},2} = \gamma(2^{\frac{1+p_2}{2}}\alpha_2, 2^{\frac{1+g_2}{2}}\beta_2, \frac{1+p_2}{2}, \frac{1+g_2}{2}, 1), \tag{6.96}$$

where in the case of $\|s_1(0)\| \to +\infty$, we have $T_{m2} = T_{\text{least},2}$. The proof ends here. □

We next force $z \to 0$ synchronously in fixed time, by introducing the following fixed-time-synchronized sliding mode,

$$s_2 = \dot{z} + \alpha_3 \text{sig}_{\text{n}}^{p_3}(z) + \beta_3 \text{sig}_{\text{n}}^{g_3}(z), \tag{6.97}$$

where α_3, β_3, p_3 and g_3 are positive constants, satisfying $0 < p_3 = p_5^*/p_6^* < 1$ and $g_3 = g_5^*/g_6^* > 1$, with positive odd integers p_5^*, p_6^*, g_5^* and g_6^*. Again, based on Lemma 5.1, the auxiliary variable z achieves fixed-time-synchronized convergence, with the least upper bound for the synchronized settling time,

$$T_{m3} \le T_{\text{least},3} = \gamma(2^{\frac{1+p_3}{2}}\alpha_3, 2^{\frac{1+g_3}{2}}\beta_3, \frac{1+p_3}{2}, \frac{1+g_3}{2}, 1), \tag{6.98}$$

where in the case of $\|z(0)\| \to +\infty$, $T_{m3} = T_{\text{least},3}$.

A full-order sliding mode controller is proposed based on the sliding mode (6.97),

$$u = u^* + \tau^*, \tag{6.99}$$

$$\dot{u}^* = -\Theta_2 \text{sign}_{\text{n}}(s_2) - \alpha_4 \text{sig}_{\text{n}}^{p_4}(s_2) - \beta_4 \text{sig}_{\text{n}}^{g_4}(s_2), \tag{6.100}$$

$$\begin{aligned} \tau^* = &-\Big(A_{21}x_1 + A_{22}x_2 + A_{12}^{\dagger}(A_{11} + \theta(x_1))(A_{11}x_1 + A_{12}x_2) \\ &+ A_{12}^{\dagger}\left(\Theta_1(x)\text{sign}_{\text{n}}(s_1) + \alpha_2 \text{sig}_{\text{n}}^{p_2}(s_1) + \beta_2 \text{sig}_{\text{n}}^{g_2}(s_1)\right) \\ &+ \alpha_2 \text{sig}_{\text{n}}^{p_2}(s_1) + \beta_2 \text{sig}_{\text{n}}^{g_2}(s_1)\Big), \end{aligned} \tag{6.101}$$

$$\begin{aligned} \Theta_2 = &\kappa_m \left\|\tau^*\right\| + \vartheta_m(t, x) + \left\|A_{12}^{\dagger}(A_{11} + \theta(x_1))\right\| (\kappa_u \|x\| + \vartheta_u) \\ &+ \left\|A_{12}^{\dagger}\dot{\theta}(x_1)\right\| \|\eta_u \|x\| + h_u\| + \kappa_m \left\| \int_0^t \dot{u}^* dt \right\|, \end{aligned} \tag{6.102}$$

where α_4, β_4, p_4 and g_4 are positive constants, satisfying $0 < p_4 = p_7^*/p_8^* < 1$ and $g_4 = g_7^*/g_8^* > 1$, with positive odd integers p_7^*, p_8^*, g_7^* and g_8^*, and a function $\theta(x_1)$: $\mathbb{R}^{n-m} \to \mathbb{R}^{n-m} \times \mathbb{R}^{n-m}$ is defined as

$$\theta(x_1) = \alpha_1 (p_1 - 1) \|x_1\|^{p_1-3} x_1 x_1^T + \alpha_1 \|x_1\|^{p_1-1} I_{n-m}$$
$$+ \beta_1 (g_1 - 1) \|x_1\|^{g_1-3} x_1 x_1^T + \beta_1 \|x_1\|^{g_1-1} I_{n-m}. \tag{6.103}$$

The full-order controller (6.99) is quasi-continuous due to the discontinuity at $s_1 = 0$ induced by τ^*. It should be thus understood in the Filippov sense [7].

Theorem 6.5 *Considering the disturbed input-dimension-dominant MIMO system (6.83) under Assumptions 6.3, 6.4, 6.5, and 6.6, and fixed-time-synchronized control law (6.99), the system state successively reaches the sliding manifolds $s_2 = 0$ and $s_1 = 0$ in fixed time, while the closed-loop system output y is ratio persistent on the sliding surface $s_1 = 0$ and achieves fixed-time-synchronized convergence, with the least upper bound for the total synchronized settling time,*

$$T_m \le T_{\text{least},1} + T_{\text{least},2} + T_{\text{least},3} + T_{\text{least},4}, \tag{6.104}$$

where $T_{\text{least},4} = \gamma(2^{\frac{1+p_4}{2}}\alpha_4, 2^{\frac{1+g_4}{2}}\beta_4, \frac{1+p_4}{2}, \frac{1+g_4}{2}, 1)$.

Proof According to Theorem 5.5 and Lemma 6.2, the system state reaches the sliding manifold $s_2 = 0$ in fixed time, with the least upper bound for the synchronized settling time

$$T_{m4} \le T_{\text{least},4} = \gamma(2^{\frac{1+p_4}{2}}\alpha_4, 2^{\frac{1+g_4}{2}}\beta_4, \frac{1+p_4}{2}, \frac{1+g_4}{2}, 1), \tag{6.105}$$

where in the case of $\|s_2(0)\| \to +\infty$, $T_{m4} = T_{\text{least},4}$.

We can now derive the least upper bound for the total synchronized settling time to achieve fixed-time-synchronized convergence,

$$T_m \le T_{\text{least},1} + T_{\text{least},2} + T_{\text{least},3} + T_{\text{least},4}, \tag{6.106}$$

where T_{m1} and T_{m3} are the synchronized settling time defined in (6.89) and (6.98). □

Remark 6.9 The fixed-time-synchronized controller (6.99) contains a possible singular term $\theta(x_1)$ (when $s_1 \neq 0$ and $x_1 \to 0$). To deal with the singularity, a switching-based fixed-time-synchronized sliding mode as in (5.24) and (5.25) can be applied to replace (6.88), and the time-synchronized convergence result in Theorem 6.5 still holds. Detailed singularity-avoidance analysis can be found in Lemma 5.2.

6.4.2 *Control Design for State-Dimension-Dominant MIMO Systems*

In the previous subsection, fixed-time-synchronized control for the state-dimension-dominant MIMO system (6.83) is proposed. We here investigate fixed-time-synchronized control for the state-dimension-dominant MIMO system, where the right inverse of A_{12} no longer exists.

Not all the state-dimension-dominant MIMO systems are capable of achieving fixed-time-synchronized stability. For those able to "achieve fixed-time-synchronized convergence", only a subset of the output elements (rather than all of them) reach the origin at the same time. This type of convergence property is named "*Semi-Fixed-Time-Synchronized Convergence*". Despite semi-fixed-time-synchronized convergence, there still exist d_1 ($d_1 \leq m$) elements of the system output that achieve fixed-time-synchronized convergence, where d_1 is the largest number in the controllability indices (called the controllability index) of the system $\dot{x}_1 = A_{11}x_1 + A_{12}x_2 + d_u(t, x)$, after proper state transformations by treating x_2 as the control input.

The following assumption on the unmatched disturbance is made to semi-fixed-time-synchronized control.

Assumption 6.7 The unmatched disturbance satisfies $d_u(t, x) = A_{12}\omega(t, x)$, where $\omega(t, x)$ is an unknown function characterizing the disturbances. In addition, $\omega(t, x)$ meets the following constraints,

$$\|\omega(t, x)\| \leq \overline{\eta}_u \|x\| + \overline{h}_u, \tag{6.107}$$

$$\|\dot{\omega}(t, x)\| \leq \bar{\kappa}_u \|x\| + \bar{\vartheta}_u, \tag{6.108}$$

where $\bar{\kappa}_u$ and $\bar{\vartheta}_u$ are two known positive constants.

We make two state transformations, given by

$$x_1 = \Gamma_1^{-1}\Gamma_2 z, \tag{6.109}$$

$$z = \Gamma_2^{-1}\Gamma_1 x_1, \tag{6.110}$$

where z is the transformed state vector, and Γ_1 and Γ_2 are transformation matrices defined in Chap. 4, Sect. 4.2.

The transformed state vector $z = [z_1, ..., z_r]^T$ can be reformulated as

$$\dot{z}_1 = \sum_{i=1}^{r} A''_{1,i} z_i + B_{1,0}(x_2 + \omega), \tag{6.111}$$

$$\dot{z}_i = N_i z_i + B'_{i,i-1} z_{i-1}, \quad i = 2, \ldots, r, \tag{6.112}$$

where N_i is a matrix defined as $N_i = -\ell_i I_{\pi_i}$, ℓ_i is a positive constant satisfying $0 < \ell_r < ... < \ell_2$, and π_i denotes the dimension of z_i.

Remark 6.10 The recursive structure in (6.111) and (6.112) explains why only semi-fixed-time-synchronized convergence rather than fixed-time-synchronized one can be achieved. After z_1 is time-synchronously driven to the origin in fixed time, $z_2, ..., z_r$ will successively reach the equilibrium asymptotically (rather than time-synchronously). Thus why fixed-time-synchronized convergence is achieved only in d_1 dimensions.

A fixed-time-synchronized sliding mode is constructed to force $z_1 \to 0$,

$$s_1 = \dot{z}_1 + \alpha_1 \mathrm{sig}_{\mathrm{n}}^{p_1}(z_1) + \beta_1 \mathrm{sig}_{\mathrm{n}}^{g_1}(z_1), \tag{6.113}$$

where α_1 and β_1 are positive constants, and p_1 and g_1 are defined as $0 < p_1 = p_1^*/p_2^* < 1$ and $g_1 = g_1^*/g_2^* > 1$, with positive odd integers p_1^*, p_2^*, g_1^* and g_2^*. From Theorem 6.1, on the sliding surface $s_1 = 0$, z_1 achieves fixed-time-synchronized convergence, with the least upper bound for the synchronized settling time,

$$T_{m1} \le T_{\mathrm{least},1} = \gamma(2^{\frac{1+p_1}{2}}\alpha_1, 2^{\frac{1+g_1}{2}}\beta_1, \frac{1+p_1}{2}, \frac{1+g_1}{2}, 1), \tag{6.114}$$

where in the case of $\|z_1(0)\| \to +\infty$, $T_{m1} = T_{\mathrm{least},1}$.

Aiming at driving $s_1 \to 0$, a virtual velocity tracking error is defined as

$$\mu = x_2 - \phi_2, \tag{6.115}$$

where ϕ_2 is designed to drive $s_1 \to 0$,

$$\phi_2 = \phi_{2eq} + \phi_{2n}, \tag{6.116}$$

$$\phi_{2eq} = -B_{1,0}^{\dagger}\left(\sum_{i=1}^{r} A_{1,i}'' z_i + \alpha_1 \mathrm{sig}_{\mathrm{n}}^{p_1}(z_1) + \beta_1 \mathrm{sig}_{\mathrm{n}}^{g_1}(z_1)\right), \tag{6.117}$$

$$\dot{\phi}_{2n} = -B_{1,0}^{\dagger}\left(\Theta_1(x)\mathrm{sign}_{\mathrm{n}}(s_1) + \alpha_2 \mathrm{sig}_{\mathrm{n}}^{p_2}(s_1) + \beta_2 \mathrm{sig}_{\mathrm{n}}^{g_2}(s_1)\right), \tag{6.118}$$

$$\Theta_1(x) = \left(\bar{\kappa}_u \|x\| + \bar{\vartheta}_u\right)\left\|B_{1,0}\right\|, \tag{6.119}$$

where α_2, β_2, p_2 and g_2 are positive constants, satisfying $0 < p_2 = p_3^*/p_4^* < 1$ and $g_2 = g_3^*/g_4^* > 1$, with positive odd integers p_3^*, p_4^*, g_3^* and g_4^*, and κ_u and ϑ_u are introduced in Assumption 6.6.

Lemma 6.6 *In the case of $\mu = \dot{\mu} = 0$, the system dynamics (6.111) reaches the sliding manifold $s_1 = 0$ in fixed time, with the least upper bound for the settling time,*

$$T_{m2} \le T_{\mathrm{least},2} = \gamma(2^{\frac{1+p_2}{2}}\alpha_2, 2^{\frac{1+g_2}{2}}\beta_2, \frac{1+p_2}{2}, \frac{1+g_2}{2}, 1). \tag{6.120}$$

In the case of $\|s_1(0)\| \to +\infty$, $T_{m2} = T_{\mathrm{least},2}$.

Proof From Lemmas 5.6 and 6.2, the system output z_1 reaches the sliding manifold $s_1 = 0$ in fixed time, with the least upper bound for the synchronized settling time

$$T_{m2} \le T_{\mathrm{least},2} = \gamma(2^{\frac{1+p_2}{2}}\alpha_2, 2^{\frac{1+g_2}{2}}\beta_2, \frac{1+p_2}{2}, \frac{1+g_2}{2}, 1), \tag{6.121}$$

where in the case of $\|s_1(0)\| \to +\infty$, $T_{m2} = T_{\mathrm{least},2}$. □

We next force $\mu \to 0$ by using the following fixed-time-synchronized sliding mode,

$$s_2 = \dot{\mu} + \alpha_3 \mathrm{sig}_{\mathrm{n}}^{p_3}(\mu) + \beta_3 \mathrm{sig}_{\mathrm{n}}^{g_3}(\mu), \tag{6.122}$$

where α_3, β_3, p_3 and g_3 are positive constants, satisfying $0 < p_3 = p_5^*/p_6^* < 1$ and $g_3 = g_5^*/g_6^* > 1$, with positive odd integers p_5^*, p_6^*, g_5^* and g_6^*. From Lemma 6.1, μ achieves fixed-time-synchronized convergence, with the least upper bound for the synchronized settling time

$$T_{m3} \le T_{\mathrm{least},3} = \gamma(2^{\frac{1+p_3}{2}}\alpha_3, 2^{\frac{1+g_3}{2}}\beta_3, \frac{1+p_3}{2}, \frac{1+g_3}{2}, 1), \tag{6.123}$$

where in the case of $\|\mu(0)\| \to +\infty$, $T_{m3} = T_{\mathrm{least},3}$.

Based on the sliding mode (6.122), a full-order sliding mode controller is proposed,

$$\begin{aligned}
u &= u^* + \tau^*, && (6.124)\\
\dot{u}^* &= -\Theta_2 \mathrm{sign}_{\mathrm{n}}(s_2) - \alpha_4 \mathrm{sig}_{\mathrm{n}}^{p_4}(s_2) - \beta_4 \mathrm{sig}_{\mathrm{n}}^{g_4}(s_2), && (6.125)\\
\tau^* &= -\left(A_{21}x_1 + A_{22}x_2 + B_{1,0}^{\dagger} \sum_{i=2}^{r} A_{1,i}'' \left(N_i z_i + B_{i,i-1}' z_{i-1}\right) \right.\\
&\quad + B_{1,0}^{\dagger}\left(A_{1,1}'' + \theta(z_1)\right)\left(\sum_{i=2}^{r} A_{1,i}'' + B_{1,0}x_2\right)\\
&\quad \left. + B_{1,0}^{\dagger}\left(\Theta_1(x)\mathrm{sign}_{\mathrm{n}}(s_1) + \alpha_3 \mathrm{sig}_{\mathrm{n}}^{p_3}(\mu) + \beta_3 \mathrm{sig}_{\mathrm{n}}^{g_3}(\mu)\right)\right), && (6.126)\\
\Theta_2 &= \kappa_m \left\|\tau^*\right\| + \vartheta_m(t,x) + \left\|B_{1,0}^{\dagger}\left(A_{1,1}'' + \theta(z_1)\right)B_{1,0}\right\| \left(\bar{\kappa}_u \|x\| + \bar{\vartheta}_u\right)\\
&\quad + \left\|B_{1,0}^{\dagger}\dot{\theta}(z_1)\right\| \left\|\bar{\eta}_u \|x\| + \bar{h}_u\right\| + \kappa_m \left\|\int_0^t \dot{u}^* dt\right\|, && (6.127)
\end{aligned}$$

where α_4, β_4, p_4 and g_4 are positive constants, satisfying $0 < p_4 = p_7^*/p_8^* < 1$ and $g_4 = g_7^*/g_8^* > 1$, with positive odd integers p_7^*, p_8^*, g_7^* and g_8^*, and a function $\theta(z_1) : \mathbb{R}^{d_1} \to \mathbb{R}^{d_1} \times \mathbb{R}^{d_1}$ is defined as

$$\begin{aligned}
\theta(z_1) &= \alpha_1(p_1 - 1)\|z_1\|^{p_1-3} x_1 x_1^T + \alpha_1 \|z_1\|^{p_1-1} I_{d_1}\\
&\quad + \beta_1(g_1 - 1)\|z_1\|^{g_1-3} z_1 z_1^T + \beta_1 \|z_1\|^{g_1-1} I_{d_1}.
\end{aligned} \tag{6.128}$$

The controller (6.124) is quasi-continuous due to the discontinuity at $s_1 = 0$ induced by τ^*. It should be thus understood in the Filippov sense [7].

Theorem 6.6 *Considering the disturbed state-dimension-dominant MIMO system (6.83) under Assumptions 6.3, 6.5 and 6.7, and the full-order sliding-mode control law (6.124), the system state reaches the fixed-time-synchronized sliding manifold $s_2 = 0$ and $s_1 = 0$ successively in fixed time. The closed-loop system output*

z achieves semi-fixed-time-synchronized convergence. Specifically, z_i, $i \in \{2, ..., r\}$ achieves asymptotic convergence, and z_1 is ratio persistent on the sliding surface $s_1 = 0$ and achieves fixed-time-synchronized convergence, with the least upper bound for the total synchronized settling time,

$$T_m \leq T_{\text{least},1} + T_{\text{least},2} + T_{\text{least},3} + T_{\text{least},4}, \tag{6.129}$$

where $T_{\text{least},4} = \gamma(2^{\frac{1+p_4}{2}}\alpha_4, 2^{\frac{1+g_4}{2}}\beta_4, \frac{1+p_4}{2}, \frac{1+g_4}{2}, 1)$.

Proof According to Theorem 5.5 and Lemma 6.2, the system state reaches the sliding manifold $s_2 = 0$ in fixed time, with the least upper bound for the synchronized settling time

$$T_{m4} \leq T_{\text{least},4} = \gamma(2^{\frac{1+p_4}{2}}\alpha_4, 2^{\frac{1+g_4}{2}}\beta_4, \frac{1+p_4}{2}, \frac{1+g_4}{2}, 1), \tag{6.130}$$

where in the case of $\|s_2(0)\| \to +\infty$, $T_{m4} = T_{\text{least},4}$.

The least upper bound for the total synchronized settling time to achieve semi-fixed-time-synchronized convergence is given by,

$$T_m \leq T_{\text{least},1} + T_{\text{least},2} + T_{\text{least},3} + T_{\text{least},4}, \tag{6.131}$$

where T_{m1} and T_{m3} are the synchronized settling time introduced in (6.114) and (6.123). The subsequent state z_i, $i \in \{2, ..., r\}$ achieves asymptotic convergence successively as discussed in Remark 6.10. □

Remark 6.11 Following the procedure in Remark 6.9, the full-order sliding-mode controller (6.124) becomes a non-singular one. The trivial details are omitted here.

6.5 An Extension to Predefined-Time-Synchronized Control

In the previous sections, we derive a least upper bound of the synchronized settling time, which is less conservative compared with standard estimation techniques. In line with Definition 2.14, in this section, we continue to investigate how to integrate the desired predefined synchronized settling time into fixed-time-synchronized controllers and to suitably utilize relevant Lyapunov-like theorem to analyze such a control system.

6.6 Strong Predefined-Time-Synchronized Control

Based on Theorem 6.1, we first propose the following theorem to articulate a strong predefined-time-synchronized stable system.

Theorem 6.7 *Considering a Lyapunov function $V(x)$, the system (2.112) is strong predefined-time-synchronized stable with strong predefined synchronized settling time T_{pre} that is a tunable system parameter consisting in ϕ, if the following conditions hold:*

i. *there exist positive constants α, β, p, g and χ satisfying $p < 1$ and $g > 1$, such that*

$$\dot{V}(x) \leq -\frac{1}{T_{pre}}\gamma(\alpha, \beta, p, g, \chi)(\alpha V^{p}(x) + \beta V^{g}(x))^{\chi}, \tag{6.132}$$

where function γ is defined in Remark 6.1.

ii. *the state x is ratio persistent.*

In addition, if $\|x_0\| \rightarrow +\infty$ and (6.132) is an equality, we have the synchronized settling time takes the exact value

$$T(x_0) = T_{pre}, \tag{6.133}$$

that is $\sup_{x_0 \in \mathbb{R}^n} T(x_0) = T_{pre}$ in accordance with strong predefined-synchronized stability as in Definition 2.14.

Proof For convenience, denote $\gamma_{pre} = \gamma(\alpha, \beta, p, g, \chi)$. We rewrite (6.132) as

$$\dot{V}(x) \leq -(\alpha\left(\frac{\gamma_{pre}}{T_{pre}}\right)^{\frac{1}{\chi}} V^{p}(x) + \beta\left(\frac{\gamma_{pre}}{T_{pre}}\right)^{\frac{1}{\chi}} V^{g}(x))^{\chi}. \tag{6.134}$$

From (6.134), according to Theorem 6.1, the system (2.18) is fixed-time-synchronized stable with the least upper bound of the synchronized settling time,

$$T_{\text{least}} = \gamma\left(\alpha\left(\frac{\gamma_{pre}}{T_{pre}}\right)^{\frac{1}{\chi}}, \beta\left(\frac{\gamma_{pre}}{T_{pre}}\right)^{\frac{1}{\chi}}, p, g, \chi\right). \tag{6.135}$$

With the following manipulation, strong predefined-time-synchronized stability is proved,

$$\begin{aligned} T_{\text{least}} &= \gamma\left(\alpha\left(\frac{\gamma_{pre}}{T_{pre}}\right)^{\frac{1}{\chi}}, \beta\left(\frac{\gamma_{pre}}{T_{pre}}\right)^{\frac{1}{\chi}}, p, g, \chi\right) \\ &= \frac{B\left(\omega_p, \omega_g\right)}{\alpha^{\chi-\omega_p}\left(\frac{\gamma_{pre}}{T_{pre}}\right)^{\frac{\chi-\omega_p}{\chi}} \beta^{\omega_p}\left(\frac{\gamma_{pre}}{T_{pre}}\right)^{\frac{\omega_p}{\chi}}(g-p)} \\ &= \frac{T_{pre}}{\gamma_{pre}}\frac{B\left(\omega_p, \omega_g\right)}{\alpha^{\chi-\omega_p}\beta^{\omega_p}(g-p)} \\ &= T_{pre}. \end{aligned} \tag{6.136}$$

According to the property of the least upper bound as in Definition 6.3, we have $\sup_{x_0 \in \mathbb{R}^n} T(x_0) = T_{pre}$. The proof ends here. □

In the following, we provide some simple examples to showcase how to construct a strong/true predefined-time-synchronized stable system based on Theorem 6.7 and Lemma 2.17, respectively.

Considering a single-integrator system

$$\dot{x} = u \in \mathbb{R}^n, \tag{6.137}$$

we design the following strong predefined-time-synchronized controller,

$$u = -\frac{\gamma_{pre}}{T_{pre}} \left(\alpha \mathrm{sig}_{\mathrm{n}}^{p}(x) + \beta \mathrm{sig}_{\mathrm{n}}^{g}(x)\right), \tag{6.138}$$

where T_{pre} is the strong predefined synchronized settling time, and the other constant parameters are defined as $\gamma_{pre} = \gamma(2\alpha, 2\beta, \frac{p+1}{2}, \frac{g+1}{2}, 1)$, $\alpha, \beta > 0$, $p \in (0, 1)$ and $g \in (1, \infty)$.

Theorem 6.8 *Under the controller (6.138), the system (6.137) is strong predefined-time-synchronized stable with predefined time T_{pre}.*

Proof Consider a Lyapunov function candidate $V(x) = x^T x$. Its time derivative takes the form,

$$\begin{aligned} \dot{V} =& -2x^T \frac{\gamma_{pre}}{T_{pre}} \left(\alpha \mathrm{sig}_{\mathrm{n}}^{p}(x) + \beta \mathrm{sig}_{\mathrm{n}}^{g}(x)\right) \\ =& -\frac{\gamma_{pre}}{T_{pre}} \left(2\alpha V^{\frac{p+1}{2}} + 2\beta V^{\frac{g+1}{2}}\right). \end{aligned} \tag{6.139}$$

Based on Theorem 6.7, the closed-loop system is strong predefined-time-synchronized stable with predefined synchronized setting time T_{pre}. □

6.7 True Predefined-Time-Synchronized Control

Similarly, for the system (6.137), we provide a true predefined-time-synchronized controller,

$$u = \begin{cases} \frac{-k\left(1-e^{-\|x\|}\right)}{T_{pre}-t} \mathrm{sign}_{\mathrm{n}}(\mathrm{x}), & \text{if } 0 \le t < T_{pre}, \\ 0, & \text{otherwise}, \end{cases} \tag{6.140}$$

where k is a positive control gain and T_{pre} the desired setting time.

Given the controller (6.140), we have the following result.

Theorem 6.9 *Under the controller (6.140), the system (6.137) is true predefined-time-synchronized stable with predefined time T_{pre}.*

Proof Consider a Lyapunov function candidate $V(x) = x^T x$. Its time derivative takes the form,

$$\begin{aligned}\dot{V} &= \frac{-2k\left(1 - e^{-\|x\|}\right)}{T_{pre} - t}\|x\| \\ &= \frac{-2kV^{\frac{1}{2}}\left(e^{\|x\|} - 1\right)}{e^{\|x\|}\left(T_{pre} - t\right)}.\end{aligned} \quad (6.141)$$

Denote $Q = V^{\frac{1}{2}}$. Taking the controller (6.140) into the time-derivative of Q, we can get

$$\dot{Q} = \frac{1}{2}V^{-\frac{1}{2}}\dot{V} = \frac{-k\left(e^{Q} - 1\right)}{e^{Q}\left(T_{pre} - t\right)}. \quad (6.142)$$

By directly using Lemma 2.17 twice, one knows true predefined-time-synchronize stability (as defined in Definition 2.14) is achieved for x. The true predefined-synchronized setting time is T_{pre}. □

6.8 Summary

Fixed-time-synchronized control problem with the least upper bound of the synchronized setting time has been for the first time defined and solved in this chapter. The upper bound of the synchronized settling time is composed of a series of least upper bounds. It has significantly alleviated the overestimation suffered by existing fixed-time control results. Moreover, the important matters considered in the previous chapters, such as control under matched and unmatched disturbances and control with the consideration of singularity avoidance, have been likewise taken care of in this chapter. Extensions to predefined-time-synchronized control have been revealed to realize strong predefined-time-synchronized convergence and true predefined-time-synchronized convergence. This chapter has shown an emerging deeper perception of the established fixed-time-synchronized stability.

References

1. Aldana-López, R., Gómez-Gutiérrez, D., Jiménez-Rodríguez, E., Sánchez-Torres, J.D., Defoort, M.: Enhancing the settling time estimation of a class of fixed-time stable systems. International Journal of Robust and Nonlinear Control **29**(12), 4135–4148 (2019)
2. Basin, M.: Finite- and fixed-time convergent algorithms: Design and convergence time estimation. Annual Reviews in Control **48**, 209–221 (2019). https://doi.org/10.1016/j.arcontrol.2019.05.007

3. Basin, M., Panathula, C.B., Shtessel, Y.: Multivariable continuous fixed-time second-order sliding mode control: Design and convergence time estimation. IET Control Theory and Applications **11**(8), 1104–1111 (2017)
4. Basin, M., Rodriguez-Ramirez, P., Garza-Alonso, A.: Continuous Fixed-Time Convergent Super-Twisting Algorithm in Case of Unknown State and Disturbance Initial Conditions. Asian Journal of Control **21**(1), 323–338 (2019)
5. Bateman, H.: Higher Transcendental Functions [Volumes I-III], vol. 1. McGraw-Hill Book Company (1953)
6. Decarlo, R.A., Zak, S.H., Matthews, G.P.: Variable Structure Control of Nonlinear Multivariate Systems: A Tutorial. Proceedings of the IEEE **76**(3), 212–232 (1988). 10.1109/5.4400
7. Filippov, A.F.: Differential Equations with Discontinuous Right-Hand Side. Dordrecht, The Netherlands: Kluwer (1988)
8. Jiang, B., Chen, Z., Hu, Q., Qi, J.: Fixed-time attitude control for rigid spacecraft with actuator saturation and faults. IEEE Transactions on Control Systems Technology **24**(5), 4375–4380 (2014)
9. Loukianov, A.G., Utkin, V.I.: Time-varying linear system decomposed control. In: Proceedings of the 1998 American Control Conference. ACC (IEEE Cat. No. 98CH36207), vol. 5, pp. 2884–2888. IEEE (1998)
10. Lu, K., Xia, Y.: Adaptive attitude tracking control for rigid spacecraft with finite-time convergence. Automatica **49**(12), 3591–3599 (2013)
11. Nagesh, I., Edwards, C.: A multivariable super-twisting sliding mode approach. Automatica **50**(3), 984–988 (2014). https://doi.org/10.1016/j.automatica.2013.12.032
12. Shtessel, Y., Edwards, C., Fridman, L., Levant, A.: Sliding mode control and observation. Springer (2014)
13. Sontag, E.D., Wang, Y.: On characterizations of the input-to-state stability property. Systems & Control Letters **24**(5), 351–359 (1995)
14. Utkin, V.: On convergence time and disturbance rejection of super-twisting control. IEEE Transactions on Automatic Control **58**(8) (2013)
15. Utkin, V., Guldner, J., Shijun, M.: Sliding mode control in electro-mechanical systems. CRC press (1999)
16. Wang, L., Chai, T., Zhai, L.: Neural-network-based terminal sliding-mode control of robotic manipulators including actuator dynamics. IEEE Transactions on Industrial Electronics **56**(9), 3296–3304 (2009). https://doi.org/10.1109/TIE.2008.2011350
17. Wang, X., Guo, J., Tang, S., Qi, S.: Fixed-time disturbance observer based fixed-time backstepping control for an air-breathing hypersonic vehicle. ISA Transactions **88**, 233–245 (2019)

Chapter 7
Time-Synchronized Consensus of Network Systems

Abstract In this chapter, the problem of multi-agent time-synchronized consensus is introduced, where all the state elements of all the agents reach the consensus (a comment value) at the same time. If the synchronized settling time is independent of the initial state of the multi-agent system, fixed-time-synchronized consensus is additionally achieved. The sufficient graphical topology is discussed and formulated to lay a solid foundation for the subsequent control design. Relevant (fixed-) time-synchronized consensus controllers are then proposed for multi-agent systems under undirected and direct graphs. Further, based on the rationality of time-synchronized consensus, time-synchronized distributed observers are proposed to accurately acquire the non-global state of a target that only accessible to a small subset of the agents via information flow.

7.1 Problem Formulations

In this section, the concept of *time-synchronized consensus* is explored.

Considering an N-agent network, we utilize a communication graph $\mathcal{G}_c = (\mathcal{V}_c, \mathcal{E}_c)$ to characterize the underlying information flow among agents, where $\mathcal{E}_c \subseteq \mathcal{V}_c \times \mathcal{V}_c$ and $\mathcal{V}_c = \{1, 2, \ldots, N\}$ denote the arc set and the vertex set, respectively. $\mathcal{G}_c$ is undirected if an edge $(i, j) \in \mathcal{E}_c$ represents that agents i and j can receive information from one another. Denote $\mathcal{N}_i$ as the in-neighbour set of node i, where $\mathcal{N}_i = \{j : (j, i) \in \mathcal{E}_c\}$. The adjacency matrix $\mathcal{A}_c = [a_{ij}] \in \mathbb{R}^{N \times N}$ is defined as $a_{ij} > 0$ if $(j, i) \in \mathcal{E}_c$, and $a_{ij} = 0$ otherwise, where $a_{ii} = 0$. The Laplacian matrix is defined as $\mathcal{L}_c = [l_{ij}] \in \mathbb{R}^{N \times N}$, where $l_{ii} = \sum_{j=1, j \neq i}^{k} a_{ij}$ and $l_{ij} = -a_{ij}, i \neq j$.

Definition 7.1 (*Time-Synchronized Consensus*). A group of networked agent systems (2.18) achieve time-synchronized consensus if and only if all the agents reach consensus synchronously, i.e., we have

D. Li et al., *Time-Synchronized Control: Analysis and Design*,
https://doi.org/10.1007/978-981-16-3089-7_7

$$\lim_{t \to T} \sum_{i,j \in \mathcal{V}_c, i \neq j} \left\| x_i(t) - x_j(t) \right\| = 0, \tag{7.1}$$

with a positive time instant T, while for any time instants t_1 and t_2 satisfying $0 \leq t_1 < t_2 < T$ and any $i \in \mathcal{V}_c$, we have

$$\left\| x_{i,k}(t) - \chi_k \right\| \neq 0, \ \forall t_1 \leq t \leq t_2, \tag{7.2}$$

where $\chi = x_i(T)$, $i \in \mathcal{V}_c$ denotes the final value of the consensus, and $x_{i,k}(t)$ and χ_k denote the kth element of $x_i(t)$ and χ, respectively ($k = 1, 2, \ldots, n$).

Different from the traditional consensus problem, the time-synchronized consensus stated in Definition 7.1 drives all the elements of all the agent states to consensus at the same time, i.e., the time-synchronized consensus is achieved in all dimensions.

Definition 7.2 (*Fixed-Time-Synchronized Consensus*). A group of networked agent systems (2.18) achieve fixed-time-synchronized consensus if and only if they achieve time-synchronized consensus with bounded settling-time T, where the bound is independent of any initial multi-agent system states, i.e., for $\forall x_i(0)$, $i \in \mathcal{V}_c$, we have $T < T_m$, where T_m is a positive constant.

7.2 Time-Synchronized Consensus Control

Suppose that agent i is described as

$$\dot{x}_i = u_i, \tag{7.3}$$

where $x_i = [x_{i1}, x_{i2}, \ldots, x_{in}]^T \in \mathbb{R}^n$.

An assumption on the graphical condition is made.

Assumption 7.1 The undirected graph $\mathcal{G}_c$ is connected.

Assumption 7.1 indicates that there always exists a path between two nodes in $\mathcal{G}_c$.

Lemma 7.1 [13] *Under Assumption 7.1, the Laplacian matrix $\mathcal{L}_c$ has a simple eigenvalue 0 associated with an eigenvector $\mathbf{1}_N$ while the other eigenvalues are all positive.*

Definitions on the consensus trajectory and the consensus equilibrium persistence are introduced.

Definition 7.3 (*Consensus Trajectory*). Over a time interval $\mathcal{D} \subseteq \mathbb{R}_+$ before the consensus is finally achieved, a consensus trajectory is the trajectory of $x(t)$ defined as

$$\mathcal{T}(x, \mathcal{D}) := \left\{ x | x(t) = \left[x_1^T, \ldots, x_N^T \right]^T \in \mathbb{R}^{Nn}, t \in \mathcal{D} \subseteq \mathbb{R}_+ \right\}.$$

Definition 7.4 (*Consensus Equilibrium Persistence*). A group of agents described by the framework $\mathcal{G}_c(x)$ is consensus equilibrium persistent if there exist at most isolated time instants when for an agent $i \in \mathcal{V}_c$, we have

$$\sum_{j\in N_i} a_{ij}\left(x_i - x_j\right) = 0 \text{ and } x_i \neq x_j, \tag{7.4}$$

which means that

$$\Omega_{t^*} := \{t^* | \sum_{j\in N_i} a_{ij}\left(x_i\left(t^*\right) - x_j\left(t^*\right)\right) = 0, \\ x_i\left(t^*\right) \neq x_j\left(t^*\right), i \in \mathcal{V}_c\} \tag{7.5}$$

has measure zero (a discrete set if not empty).

We assume that the networked N-agent system satisfies the following assumption.

Assumption 7.2 The framework $\mathcal{G}_c(x)$ is consensus equilibrium persistent along the consensus trajectory.

Under Assumption 7.2, before achieving consensus, an agent will not stay on fake equilibrium points which satisfy $\sum_{j\in N_i} a_{ij}\left(x_i - x_j\right) = 0$ while $x_i \neq x_j$, $i \in \mathcal{V}_c$. Moreover, in accordance with a considerable amount of simulations with different initial states and graphical conditions, we conjecture that Assumption 7.2 can always be met.

We propose the following control laws for agent i

$$u_{1i} = -\beta_i(k\text{sign}_\text{c}(z_i) - \sum\nolimits_{j\in N_i} a_{ij}\dot{x}_j), \tag{7.6}$$

$$u_{2i} = -\beta_i(k\text{sign}_\text{n}(z_i) - \sum\nolimits_{j\in N_i} a_{ij}\dot{x}_j), \tag{7.7}$$

where the consensus error is defined as $z_i = \sum_{j\in N_i} a_{ij}\left(x_i - x_j\right)$, the control gain as k, and the constant as $\beta_i = 1/\sum_{j\in N_i} a_{ij}$.

The following theorems hold for the proposed controllers.

Theorem 7.1 *Considering multi-agent systems governed by (7.3), if Assumptions 7.1 and 7.2 hold, under the control law (7.6), the closed-loop multi-agent systems achieve consensus in finite time; and under the control law (7.7), they achieve SATO consensus.*

Proof We first analyze finite-time stability with the control law (7.6). Taking (7.6) into the derivative of z_i, we have $\dot{z}_i = -k\text{sign}_\text{c}(z_i)$. Based on Theorem 3.1, the consensus error z_i has the finite-time convergence property. Define $z = [z_1, \ldots, z_N]^T$ and $x = [x_1, \ldots, x_N]^T$. From Lemma 7.1, it is clear that $z = k\mathcal{L}_c x$ and $z = 0$ if and only if $x_1 = x_2 = \cdots = x_N$. Therefore, the consensus is achieved here in finite time, where the settling time of consensus is bounded by $T_{m1} \leq \max_i (|z_i(0)|_1)/k$. Therefore, finite time consensus is achieved.

Next, we discuss time-synchronized consensus with the control law (7.7). Taking (7.6) into the derivative of z_i, we have $\dot{z}_i = -k\text{sign}_\text{n}(z_i)$. Based on Theorem 3.1, the consensus error z_i of each single agent i, $i \in \mathcal{V}_c$ achieves time-synchronized convergence. Noticing $z = k\mathcal{L}_c x$, we know the consensus is achieved, and the settling time of the consensus is bounded by $T_{m2} \leq \max_i (\|z_i(0)\|)/k$.

We will next prove that the state elements for all the agents reach the equilibrium synchronously. The proof is given by contradiction. Assume that agent h reaches the equilibrium first at the time instant t_c and stays on it, while agent k does not. According to Assumption 7.1, the communication graph is connected, which correspondingly means there exists a path from agent k to agent h. Without loss of generality, suppose that there are l agents (labeled as $l_1, l_2, \ldots, l_l$) among the path from agent k to agent h. Based on Assumption 7.2, the position $x^+_{l_1}(t_c)$ of agent l_1 which is directly connected to agent k must not be at the equilibrium point, since $\sum_{j\in\mathcal{N}_{l_1}} a_{l_1 j}\left(x_{l_1} - x_j\right) = 0, x_{l_1} \neq x_j$ exists at most on isolated time instants. Accordingly, we have the state of $x^+_{l_2}(t_c), \ldots, x^+_{l_l}(t_c)$ must not be at the equilibrium point either. Again, based on Assumption 7.2, we know $\sum_{j\in\mathcal{N}_h} a_{hj}\left(x_h - x_j\right) = 0, x_h \neq x_j$ exists at most on isolated time instants. Thus, $x^+_h(t_c)$ will not stay at the equilibrium as the state $x^+_{l_l}(t_c)$ of its neighbour agent l_l is not at the equilibrium. This contradicts with the fact that agent h reaches the equilibrium first at time instant t_c and stay on it, which completes the proof according to Definition 7.1. □

We define $x = [x_1, x_2, \ldots, x_N]^T \in \mathbb{R}^{Nn}$. Considering 4 agents in $\mathbb{R}^3$ modeled by (7.3), simulation examples are given in Figs. 7.1 and 7.2. The initial value is given as $x_i = -i(-1)^i[4, -6, 8]^T$, and the corresponding adjacency matrix is chosen based on Assumption 7.1 to involve a connected graph.

As shown in Figs. 7.1 and 7.2, both control laws (7.6) and (7.7) can guarantee that the agent state x_i converges to the origin in finite time. Under the control law (7.7), time-synchronized consensus is achieved in Fig. 7.2.

The trajectories of agents in 3-dimensional space are demonstrated in Figs. 7.3 and 7.4.

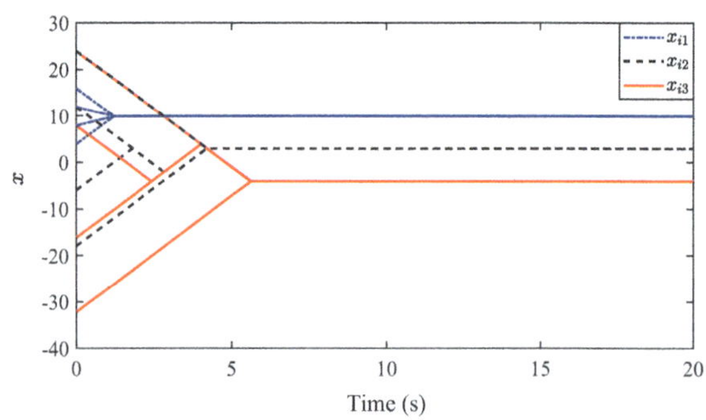

Fig. 7.1 Performance of the control law (7.6)

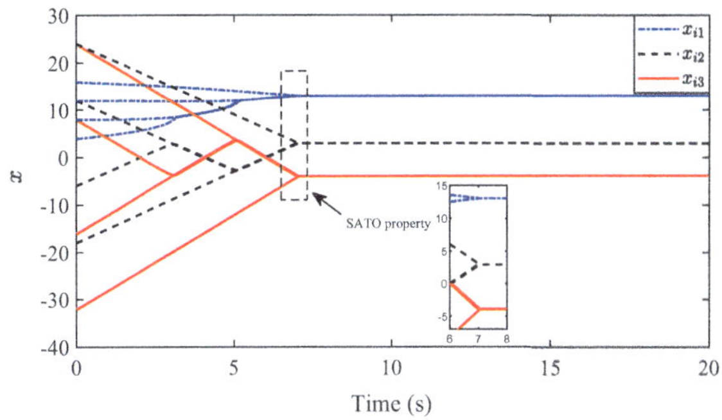

Fig. 7.2 Performance of the control law (7.7)

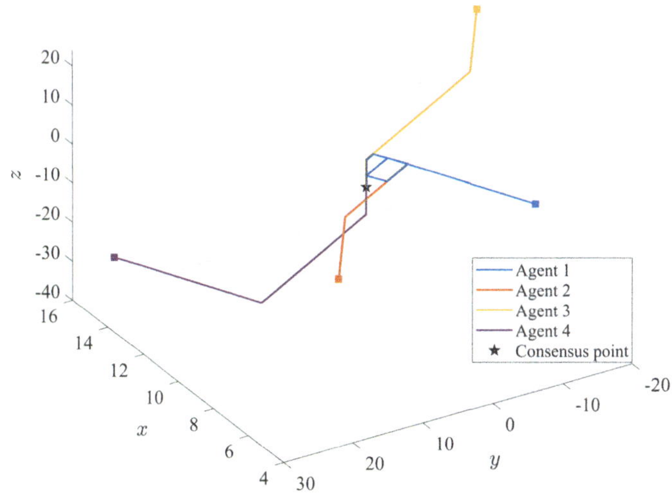

Fig. 7.3 Trajectories of the agents using the control law (7.6)

Remark 7.1 Note that the controller (7.7) contains an algebraic loop. If the controller (7.7) is simplified as $u_{2i} = -k\text{sign}_\text{n}(z_i)$, more obvious time-synchronized consensus performance can be expected in simulation results while the algebraic loop vanishes. The rigorous proof of time-synchronized consensus under this simplified control law remains open.

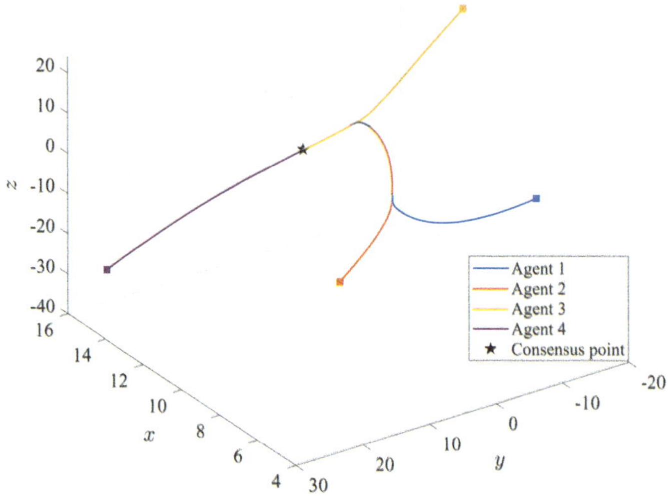

Fig. 7.4 Trajectories of the agents using the control law (7.7)

7.3 Fixed-Time-Synchronized Consensus Control

This section studies fixed-time-synchronized consensus control design based on fixed-time-synchronized stability. Consider an N-agent network, where agent i is modeled as

$$\begin{aligned} \dot{x}_i &= v_i, \\ \dot{v}_i &= f_i(x_i, v_i) + b_i(x_i, v_i)\, u_i, \end{aligned} \tag{7.8}$$

where $x_i \in \mathbb{R}^n$ and $v_i \in \mathbb{R}^n$, $i \in \{1, 2, \ldots, N\}$, denote the general position vector and velocity vector, respectively, $u_i \in \mathbb{R}^n$ is the control input, and $f_i(x_i, v_i) \in \mathbb{R}^n$ and $b_i(x_i, v_i) \in \mathbb{R}^{n\times n}$ $(\det(b_i(x_i, v_i)) \neq 0)$ are the known parts.

The following necessary assumptions are made.

Assumption 7.3 The undirected graph $\mathcal{G}_c$ is connected.

Assumption 7.4 The framework $\mathcal{G}_c(x)$ is consensus equilibrium persistent along the consensus trajectory.

We propose the following auxiliary variables

$$q_{ri} = \frac{1}{\varpi_i} \sum\nolimits_{j\in N_i} a_{ij}\left(x_i - x_j\right), \tag{7.9}$$

$$z_{ri} = \frac{1}{\varpi_i} \sum\nolimits_{j\in N_i} a_{ij} v_j, \tag{7.10}$$

where $\varpi_i = \sum_{j \in \mathcal{N}_i} a_{ij}$.

Based on the introduced norm-normalized sign function (2.6), we design a switching terminal sliding mode

$$s_i = \dot{q}_{ri} + s_{si}, \tag{7.11}$$

$$s_{si} = \begin{cases} \alpha_1 \mathrm{sig}_{\mathrm{n}}^{p_1}(q_{ri}) + \beta_1 \mathrm{sig}_{\mathrm{n}}^{g_1}(q_{ri}), & \text{if } s_i^* = 0, \text{ or } s_i^* \neq 0, \|q_{ri}\| > \varepsilon, \\ l_1 q_{ri} + l_2 \mathrm{sig}_{\mathrm{n}}^{4}(q_{ri}), & \text{if } s_i^* \neq 0, \|q_{ri}\| \leq \varepsilon, \end{cases} \tag{7.12}$$

where the trigger sliding-mode variable is given as

$$s_i^* = \dot{q}_{ri} + \alpha_1 \mathrm{sig}_{\mathrm{n}}^{p_1}(q_{ri}) + \beta_1 \mathrm{sig}_{\mathrm{n}}^{g_1}(q_{ri}), \tag{7.13}$$

α_1, β_1, g_1 and p_1 are positive constants satisfying $g_1 > 1$ and $p_1 = \ell_1/\ell_2 \in (0.5, 1)$ with positive odd integers $\ell_1$and ℓ_2, l_1 and l_2 are defined as

$$l_1 = \alpha_1 \left(\frac{4}{3} - \frac{p_1}{3}\right) \varepsilon^{p_1 - 1} + \beta_1 \left(\frac{4}{3} - \frac{g_1}{3}\right) \varepsilon^{g_1 - 1}, \tag{7.14}$$

$$l_2 = \alpha_1 \left(\frac{p_1}{3} - \frac{1}{3}\right) \varepsilon^{p_1 - 4} + \beta_1 \left(\frac{g_1}{3} - \frac{1}{3}\right) \varepsilon^{g_1 - 4}, \tag{7.15}$$

and ε is a small constant.

We can verify that the design of l_1 and l_2 guarantee the continuity of the switching law s_{si} (7.11) as well as its time derivative $\dot{s}_{si}$ (7.12).

Using the sliding mode (7.11), we propose the fixed-time-synchronized consensus control law for agent i

$$u_i = -b_i^{-1}(x_i, v_i) \left(\dot{s}_{si} - \dot{z}_{ri} + \alpha_2 \mathrm{sig}_{\mathrm{n}}^{p_2}(s_i) + \beta_2 \mathrm{sig}_{\mathrm{n}}^{g_2}(s_i) + f_i(x_i, v_i)\right), \tag{7.16}$$

$$\dot{s}_{si} = \begin{cases} \rho_{1i} q_{ri} q_{ri}^T \dot{q}_{ri} + \rho_{2i} \dot{q}_{ri}, & \text{if } s_i^* = 0 \text{ or } s_i^* \neq 0, \|q_{ri}\| > \varepsilon, \\ l_1 \dot{q}_{ri} + 3 l_2 \|q_{ri}\| q_{ri} q_{ri}^T \dot{q}_{ri} + l_2 \|q_{ri}\|^3 \dot{q}_{ri}, & \text{if } s_i^* \neq 0, \|q_{ri}\| \leq \varepsilon, \end{cases} \tag{7.17}$$

where α_2, β_2, p_2 and g_2 are positive constants satisfying $p_2 < 1$ and $g_2 > 1$, and auxiliary variables ρ_{1i} and ρ_{2i} take the forms

$$\rho_{1i} = \alpha_1 (p_1 - 1) \|q_{ri}\|^{p_1 - 3} + \beta_1 (g_1 - 1) \|q_{ri}\|^{g_1 - 3}, \tag{7.18}$$

$$\rho_{2i} = \alpha_1 \|q_{ri}\|^{p_1 - 1} + \beta_1 \|q_{ri}\|^{g_1 - 1}. \tag{7.19}$$

We can verify that no singularity problem occurs in the following cases:

(i) In the case of $s_i^* = 0$ and $q_{ri} \to 0$, we have

$$\dot{q}_{ri} = -\alpha_1 \mathrm{sig}_{\mathrm{n}}^{p_1}(q_{ri}) - \beta_1 \mathrm{sig}_{\mathrm{n}}^{g_1}(q_{ri}), \tag{7.20}$$

Taking (7.20) into the controller (7.16) yields,

$$\dot{s}_{si} = -\alpha_1^2 p_1 \mathrm{sig}_{\mathrm{n}}^{2p_1-1}(q_{ri}) - \beta_1^2 g_1 \mathrm{sig}_{\mathrm{n}}^{2g_1-1}(q_{ri})$$
$$-\alpha_1\beta_1 (p_1+g_1) \mathrm{sig}_{\mathrm{n}}^{p_1+g_1-1}(q_{ri}). \tag{7.21}$$

(ii) In the case of $s_i^* \neq 0$ and $q_{ri} \to 0$, we have

$$\dot{s}_{si} = -l_1\dot{q}_{ri} - 3l_2 \|q_{ri}\| q_{ri}q_{ri}^T\dot{q}_{ri} - l_2\|q_{ri}\|^3\dot{q}_{ri}. \tag{7.22}$$

In both cases (i) and (ii), there is no singularity as $p_1 > 1/2$ and $g_1 > 1$.

The main theorem is formally proposed.

Theorem 7.2 *Considering multi-agent systems governed by (7.8), if Assumptions 7.3 and 7.4 hold, under the control law (7.16), the closed-loop multi-agent system achieves fixed-time-synchronized consensus.*

Proof We will prove that: (i) the sliding variable s_i achieves fixed-time-synchronized convergence; (ii) the consensus error q_{ri} of each agent achieves fixed-time-synchronized convergence; and (iii) all the elements of all the agent states come to an agreement synchronously.

Consider a Lyapunov function $V_{m1} = \sum_{i\in\mathcal{V}_c} s_i^T s_i$. Substituting the control law (7.16) into the derivative of V_{m1}, we have

$$\begin{aligned}\dot{V}_{m1} &= -2s_i^T\alpha_2 \sum_{i\in\mathcal{V}_c} \mathrm{sig}_{\mathrm{n}}^{p_2}(s_i) - 2s_i^T\beta_2 \sum_{i\in\mathcal{V}_c} \mathrm{sig}_{\mathrm{n}}^{g_2}(s_i)\\ &= -2\alpha_2 \sum_{i\in\mathcal{V}_c} \|s_i\|^{p_2+1} - 2\beta_2 \sum_{i\in\mathcal{V}_c} \|s_i\|^{g_2+1}\\ &= -2\alpha_2 V_{m1}^{\frac{p_2+1}{2}} - 2\beta_2 V_{m1}^{\frac{g_2+1}{2}}.\end{aligned} \tag{7.23}$$

Based on Lemma 2.6, we know that the agent states will reach the surface $s_i = 0$ in a fixed-time instant bounded by

$$T_{m1} \le \frac{1}{\alpha_2(1-p_2)} + \frac{1}{\beta_2(g_2-1)}. \tag{7.24}$$

On the sliding manifold $s_i = 0$, we have

$$\dot{q}_{ri} = -\alpha_1 \mathrm{sig}_{\mathrm{n}}^{p_1}(q_{ri}) - \beta_1 \mathrm{sig}_{\mathrm{n}}^{g_1}(q_{ri}). \tag{7.25}$$

It follows from (7.25) and Lemma 5.1, that the consensus error q_{ri} of each agent achieves fixed-time-synchronized stabilization.

Define $x = [x_1, \ldots, x_N]^T$ and $q_r = \left[q_{r1}^T, q_{r2}^T, \ldots, q_{rN}^T\right]^T$ and we have $q_r = \mathcal{L}_c x$. From Lemma 7.1, it is clear that $x_1 = x_2 = \cdots = x_N$ if and only if $q_r = 0$. Therefore, since the consensus error q_r converges in fixed time (note that here, q_r does not necessarily converge time-synchronously), the consensus is achieved in fixed time.

We will next prove that the state elements for all the agents reach the equilibrium synchronously. The proof is given by contradiction. Assume that agent h reaches the equilibrium first at the time instant t_c and stays on it, while agent k does not. According to Assumption 7.3, the communication graph is connected, which correspondingly means there exists a path from agent k to agent h. Without loss of generality, suppose that there are l agents (labeled as $l_1, l_2, \ldots, l_l$) among the path from agent k to agent h. Based on Assumption 7.4, the position $x_{l_1}^+(t_c)$ of agent l_1 which is directly connected to agent k must not be at the equilibrium point, since $\sum_{j\in\mathcal{N}_{l_1}} a_{l_1 j}\left(x_{l_1} - x_j\right) = 0, x_{l_1} \neq x_j$ exists at most on isolated time instants. Accordingly, we have the state of $x_{l_2}^+(t_c), \ldots, x_{l_l}^+(t_c)$ must not be at the equilibrium point either. Again, based on Assumption 7.4, we know $\sum_{j\in\mathcal{N}_h} a_{hj}\left(x_h - x_j\right) = 0, x_h \neq x_j$ exists at most on isolated time instants. Thus, $x_h^+(t_c)$ will not stay at the equilibrium as the state $x_{l_l}^+(t_c)$ of its neighbour agent l_l is not at the equilibrium. This contradicts with the fact that agent h reaches the equilibrium first at time instant t_c and stay on it, which completes the proof according to Definition 7.2. □

Theorem 7.2 is valid under an undirected graph. Interestingly though, we are able to find several special cases of directed graphs, under which (fixed-) time-synchronized can be achieved. Unfortunately, the matter of how to extract their mutual topographical features and propose a proper and rigorous sufficient graphical condition is still challenging.

Remark 7.2 Based on Theorem 6.1, the least upper bound for the synchronized settling time in Theorem 7.2 is acquired,

$$T_m = T_{m1} + T_{m2} \leq T_{\text{least},1} + T_{\text{least},2}, \tag{7.26}$$

where

$$T_{\text{least},1} = \gamma\left(2\alpha_2, 2\beta_2, \frac{1+p_2}{2}, \frac{1+g_2}{2}, 1\right), \tag{7.27}$$

$$T_{\text{least},2} = \gamma\left(2\alpha_1, 2\beta_1, \frac{1+p_1}{2}, \frac{1+g_1}{2}, 1\right). \tag{7.28}$$

This estimation of the synchronized settling time substantially alleviates the overestimation problem as analyzed and verified in Chap. 6.

Remark 7.3 The proposed switching manifold (7.12) is actually inspired by the finite-time control result integrated with the switching law first proposed in [14]. Similar switching laws have also been applied in [5, 9] for spacecraft attitude tracking. However, the switching manifold in [5, 9, 14] is designed for the classical sign function $\text{sign}_\text{c}(\cdot)$, which is based on each element $|x_i|$ in the function $\text{sig}_\text{c}^\alpha(x)$, and is thus not applicable to $\text{sig}_\text{n}^\alpha(\cdot)$. Compared with the result in [5, 9, 14], the proposed switching law (7.12) is designed based on the vector q_{ri} of the norm-normalized sign function $\text{sign}_\text{n}(q_{ri})$, which completely changes the design procedure and the analysis. The new designed switching law (7.12), therefore, takes a different form, and contributes not only to the avoidance of the singularity but also to the achievement of fixed-time-synchronized stability.

Remark 7.4 The acceleration information $\dot{z}_{ri}$ is required in the fixed-time-synchronized consensus controller (7.16). In practice, in the case without acceleration sensors, this acceleration can be indirectly calculated by numerical differentiation of the velocities [3]. In terms of the possible algebraic loops induced by the usage of acceleration information, we can analyze the stability with delayed acceleration information [8] to solve this problem.

To better showcase the performance of the fixed-time-synchronized consensus controller, another fixed-time consensus control law is proposed for comparison studies.

Lemma 7.2 *Considering multi-agent systems governed by (7.8), if Assumptions 7.3 and 7.4 hold, under the following control law*

$$\begin{aligned}\bar{u}_i = &- b_i^{-1}(x_i, v_i)\left(\dot{\bar{s}}_{si} - \dot{z}_{ri} + \alpha_2 \mathrm{sig}_{\mathrm{c}}^{p_2}(s_i)\right.\\ &\left.+ \beta_2 \mathrm{sig}_{\mathrm{c}}^{g_2}(s_i) + f_i(x_i, v_i)\right), \end{aligned} \tag{7.29}$$

with the sliding manifold $\bar{s}_i = \dot{q}_{ri} + \bar{s}_{si}$*, the trigger sliding-mode variable* $\bar{s}_i^* = \dot{q}_{ri} + \alpha_1 \mathrm{sig}_{\mathrm{c}}^{p_1}(q_{ri}) + \beta_1 \mathrm{sig}_{\mathrm{c}}^{g_1}(q_{ri})$*, and the switching law* $\bar{s}_{si}$

$$\bar{s}_{si,k} = \begin{cases} \alpha_1 \mathrm{sig}_{\mathrm{c}}^{p_1}(q_{ri,k}) + \beta_1 \mathrm{sig}_{\mathrm{c}}^{g_1}(q_{ri,k}), & \text{if } \bar{s}_i^* = 0, \\ & \text{or } \bar{s}_i^* \neq 0,\ |q_{ri}| > \varepsilon, \\ l_3 q_{ri,k} + l_4 \mathrm{sig}_{\mathrm{c}}^{2}(q_{ri,k}), & \text{if } \bar{s}_i^* \neq 0,\ |q_{ri}| \leq \varepsilon, \end{cases} \tag{7.30}$$

$$\dot{\bar{s}}_{si,k} = \begin{cases} \alpha_1 p_1 \left|q_{ri,k}\right|^{p_1-1} \dot{q}_{ri,k} + \beta_1 g_1 \left|q_{ri,k}\right|^{g_1-1} \dot{q}_{ri,k}, & \text{if } \bar{s}^* = 0, \\ & \text{or } \bar{s}^* \neq 0,\ \left|q_{ri,k}\right| > \varepsilon, \\ l_3 \dot{q}_{ri,k} + l_4 \left|q_{ri,k}\right| \dot{q}_{ri,k}, & \text{if } \bar{s}^* \neq 0,\ \left|q_{ri,k}\right| \leq \varepsilon, \end{cases} \tag{7.31}$$

where $q_{ri,k}$, $\bar{s}_{si,k}$ *and* $\dot{\bar{s}}_{si,k}$ *are respectively the kth element of* q_{ri}, $\bar{s}_{si}$ *and* $\dot{\bar{s}}_{si}$, $k = 1, 2, \ldots, n$*, and constant parameters* l_3 *and* l_4 *are designed as follows to maintain the continuity of* $\bar{s}_{si}$ *and* $\dot{\bar{s}}_{si}$*:*

$$l_3 = (2 - p_1)\alpha_1 \varepsilon^{p_1 - 1} + (2 - g_1)\beta_1 \varepsilon^{g_1 - 1}, \tag{7.32}$$

$$l_4 = (p_1 - 1)\alpha_1 \varepsilon^{p_1 - 2} + (g_1 - 1)\beta_1 \varepsilon^{g_1 - 2}, \tag{7.33}$$

the closed-loop multi-agent systems achieve consensus in fixed time.

Proof The proof follows from that of Theorem 7.2. □

Lemma 7.2 implies that we can try to apply the norm-normalized sign function to existing control laws using the classical sign function, contributing an additional

(F)TSS for a single system (or (F)TSC property for multi-agent systems), as the norm-normalized sign function possesses most of the properties of the classical sign function. However, as we have shown in this chapter, certain efforts should be taken to implement the norm-normalized sign function, as there are substantial differences between the two sign functions in terms of the control design and the stability analysis.

Remark 7.5 The switching law $\bar{s}_{si}$ in Lemma 7.2 is largely inspired by the ones in [5, 9, 14], where the relevant proof and more detailed analysis are also addressed. As shown in Lemma 7.2, the switching law $\bar{s}_{si,k}$ is designed element-wisely, based on the state element $q_{ri,k}$ and the classical sign function $\text{sign}_{\text{c}}(q_{ri})$ (or $\text{sig}_{\text{c}}^{\alpha}(q_{ri})$), while the proposed switching law (7.12) is designed based on the vector q_{ri} of the norm-normalized sign function $\text{sig}_{\text{n}}^{\alpha}(q_{ri})$, making the design of $\bar{s}_{si}$ and s_{si} (or $\dot{\bar{s}}_{si}$ and $\dot{s}_{si}$), the forms of l_i, $i = 1, 2, 3, 4$, and the corresponding triggering conditions of the switching laws differ substantially. This is in line with the discussion in Remark 7.3 that fundamental differences lie in the forms as well as the performance of these two control algorithms.

Remark 7.6 Note that the controller (7.29) for the system (7.8) is in $\mathbb{R}^n$. Although in Lemma 7.2, the fixed-time stability can be achieved, it is clear that the control law (7.29) with the classical sign function generates n decoupled scalar systems from each single agent, where the achievement of (F)TSC cannot be expected.

To demonstrate the full power of the proposed fixed-time-synchronized consensus control law, we consider a group of 4 networked agents labeled from 1 to 4. The simulation parameters are given in Table 7.1.

Under the fixed-time-synchronized consensus control law (7.16), the trajectories and the closed-loop system states of the agents are given in Figs. 7.5 and 7.6. The control inputs are given in Fig. 7.7. It is shown that the consensus is achieved and the states converge to the equilibrium in fixed time. Unfortunately, we cannot clearly observe the nature of the fixed-time-synchronized consensus property from Fig. 7.6 alone, even though all the elements of all the agents appear to converge synchronously in Fig. 7.6. Thus, compared with the fixed-time consensus control law (7.29), additional simulations are conducted to find out if all the elements of all the

Table 7.1 Parameters for simulation: case 1

Parameters	Values	Parameters	Values
α_1	0.5	p_1	0.7
α_2	1	p_2	0.7
β_1	0.1	g_1	1.2
β_2	0.05	g_2	1.2
f_i	0	b_i	I_3
q_i	$-i(-1)^i[4, -6, 8]^T$	ε	0.001
$\mathcal{A}_c$	$[0, 1, 0, 0; 1, 0, 1, 0; 0, 1, 0, 1; 0, 0, 1, 0]$		

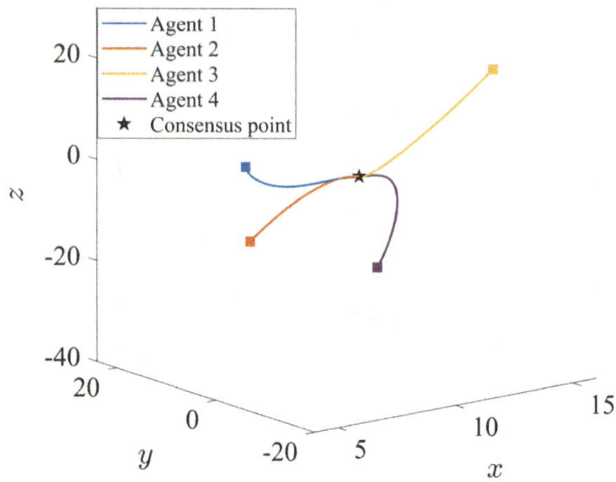

Fig. 7.5 Trajectories of 4 agents under the fixed-time-synchronized consensus control law (7.16)

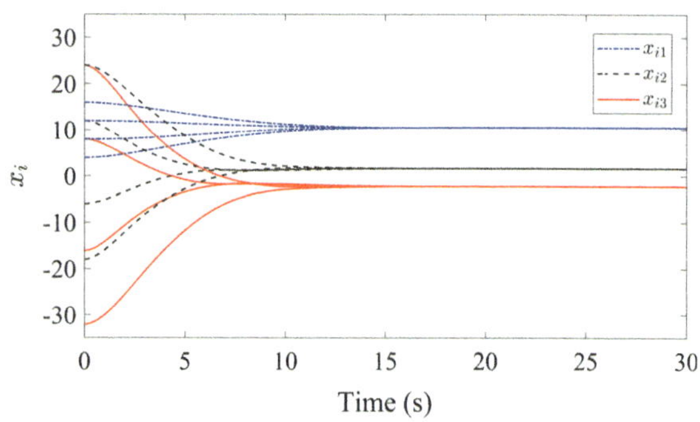

Fig. 7.6 The closed-loop system states under the fixed-time-synchronized consensus control law (7.16)

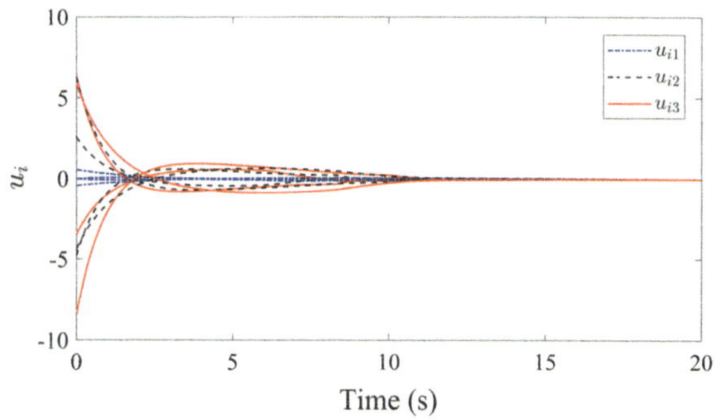

Fig. 7.7 Control inputs of the fixed-time-synchronized consensus control law (7.16)

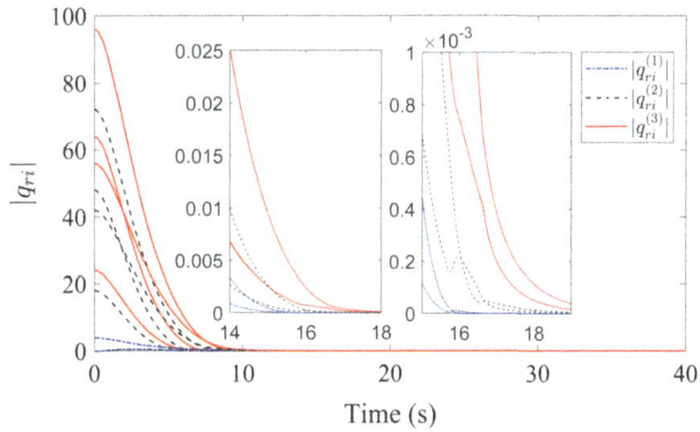

Fig. 7.8 Norms of the consensus errors under the fixed-time control law (7.29)

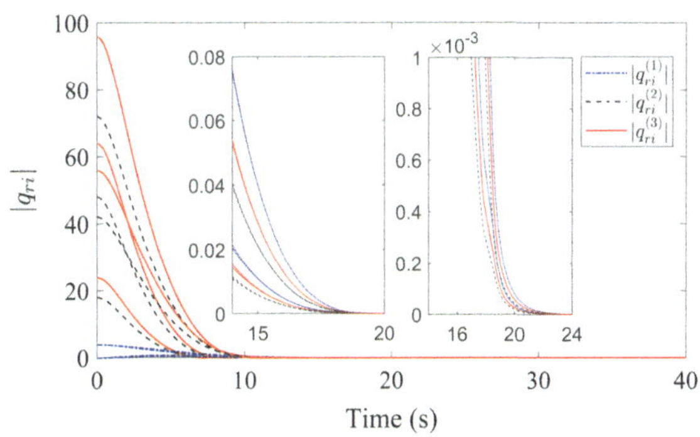

Fig. 7.9 Norms of the consensus errors under the fixed-time-synchronized consensus control law (7.16)

agents reach the equilibrium synchronously. The simulation results in Figs. 7.8 and 7.9 approve the declared fixed-time-synchronized consensus property. Clearly, in Fig. 7.8, under the fixed-time consensus control law, all the state elements of multi-agent systems converge to the equilibrium separately, while in Fig. 7.9, under the fixed-time-synchronized consensus controller, they achieve consensus time-synchronously even when we zoom in at the level of $\times 10^{-3}$. Moreover, as claimed in Step 3, the proof of Theorem 7.2, after the sliding-mode surface is reached, the proportion of the elements of the consensus error $q_{ri}(t)$ is time-invariant and preserved by the fixed-time-synchronized consensus control law (7.16). This property has also been shown in Fig. 7.10, which further validates that fixed-time-synchronized consensus is achieved.

Next, another simulation case is considered to illustrate how the proposed controller performs with a larger number of agents. Here, 20 agents are selected whose initial state elements $x_{i,k}(0), i = 1, 2, \ldots, 20, k = 1, 2, 3$ are random values belong to $[-40, 40]$. The control gains are $\alpha_1 = 0.5$, $\alpha_2 = 0.5$, $\beta_1 = 5$, $\beta_1 = 0.2$, while the other parameters remain the same as in Table 7.1. The communication topology is

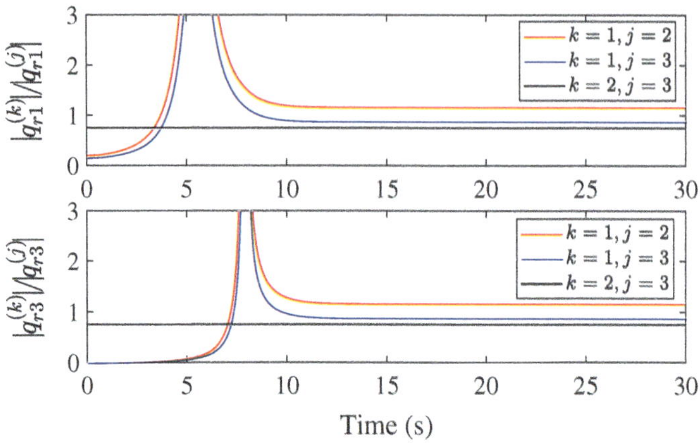

Fig. 7.10 Ratio of the elements of the consensus error under the fixed-time-synchronized consensus control law (7.16)

Fig. 7.11 Trajectories of 20 agents under the fixed-time-synchronized consensus control law (7.16)

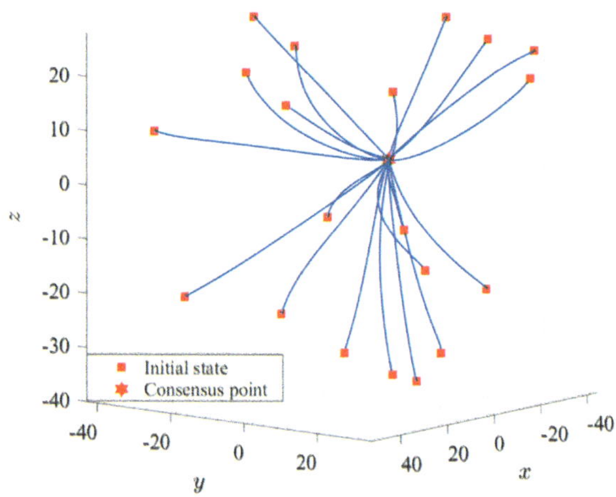

an arbitrary graph satisfying Assumption 7.3. Under the fixed-time-synchronized consensus control law (7.16), the agent trajectories and the norms of the consensus errors are shown in Figs. 7.11 and 7.12. From Fig. 7.12, it is clear that the consensus is achieved time-synchronously, whose convergence is similar to that in the case of 4 agents (see Fig. 7.9). This approves that the effectiveness of the proposed algorithm is not influenced by the number of agents.

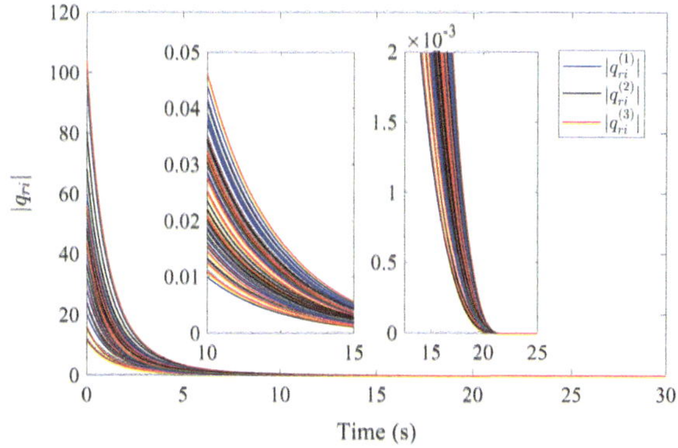

Fig. 7.12 Norms of the consensus errors under the fixed-time-synchronized consensus control law (7.16)

7.4 Fixed-Time-Synchronized Consensus Under Directed Graph

In this section, we extend the fixed-time-synchronized consensus control proposed in Sect. 7.3 to the case of directed graph. Consider N agents modeled by (7.8), communicating among a directed graph $\mathcal{G}_c$. The following definitions are provided to facilitate analysis under a directed graph.

Definition 7.5 (*Strongly Connected*) The directed graph $\mathcal{G}_c$ is strongly connected if for any two nodes, there exists a path between them.

Lemma 7.3 [12, 17, 18] *Zero is an eigenvalue of* $\mathcal{L}_c^T$ *if* $\mathcal{G}_c$ *is strongly connected, i.e.,* $\mathcal{L}_c^T p = 0$, *where in addition,*

$$P = \text{diag}\left(p_i\right), \tag{7.34}$$

$$H = P\mathcal{L}_c + \mathcal{L}_c^T P, \tag{7.35}$$

are positive definite matrices.

We choose a normalized eigenvector p such that $\mathcal{L}_c^T p = 0$ and $p^T \mathbf{1}_N = 1$. Define

$$z_{1i} = x_i - \sum_{j=1}^{N} p_j x_j, \tag{7.36}$$

$$z_{2i} = v_i - \sum_{j=1}^{N} p_j v_j, \tag{7.37}$$

which denote the position and velocity with respect to the weighted average of the multi-agent system. The transformed system is rewritten as

$$\dot{z}_{1i} = z_{2i}, \tag{7.38}$$
$$\dot{z}_{2i} = f_i(x_i, v_i) + b_i(x_i, v_i)u_i - \dot{\omega}_i, \tag{7.39}$$

where $\omega_i = \sum_{j=1}^{N} p_j v_j$.

The following column-wise concatenations are defined

$$z_1 = \left[z_{11}^T, \ldots, z_{1N}^T\right]^T, \tag{7.40}$$
$$z_2 = \left[z_{21}^T, \ldots, z_{2N}^T\right]^T, \tag{7.41}$$
$$x = \left[x_1^T, \ldots, x_N^T\right]^T, \tag{7.42}$$
$$v = \left[v_1^T, \ldots, v_N^T\right]^T. \tag{7.43}$$

Clearly, based on the above, we have

$$z_1 = (Q \otimes I_n)\, x, \tag{7.44}$$
$$z_2 = (Q \otimes I_n)\, v, \tag{7.45}$$

where $Q = I_N - \mathbf{1}p^T$. Moreover, the matrix Q has a simple eigenvalue 0 associated with the right eigenvector $\mathbf{1}_N$. Therefore, $z_1 = 0$ and $z_2 = 0$ if and only if $x_1 = x_2 = \cdots = x_N$ and $v_1 = v_2 = \cdots = v_N$.

Define a non-singular fixed-time-synchronized sliding mode

$$s_i = z_{2i} + s_{si}, \tag{7.46}$$
$$s_{si} = \begin{cases} \alpha_1 \mathrm{sig}_{\mathrm{n}}^{p_1}(z_{1i}) + \beta_1 \mathrm{sig}_{\mathrm{n}}^{g_1}(z_{1i}), & \text{if } s_i^* = 0, \text{ or} s_i^* \neq 0, \ \|z_{1i}\| > \varepsilon, \\ l_1 z_{1i} + l_2 \mathrm{sig}_{\mathrm{n}}^{4}(z_{1i}), & \text{if} s_i^* \neq 0, \ \|z_{1i}\| \leq \varepsilon, \end{cases} \tag{7.47}$$

where α_1, β_1, p_1, g_1, l_1, l_2, and ε are the same as those in (7.12), and the trigger sliding mode variable is defined as

$$s_i^* = z_{2i} + \alpha_1 \mathrm{sig}_{\mathrm{n}}^{p_1}(z_{1i}) + \beta_1 \mathrm{sig}_{\mathrm{n}}^{g_1}(z_{1i}). \tag{7.48}$$

A fixed-synchronized consensus controller is proposed,

$$u_i = -b_i^{-1}(x_i, v_i)\left(\dot{s}_{si} - \dot{\omega}_i + \alpha_2 \mathrm{sig}_{\mathrm{n}}^{p_2}(s_i) + \beta_2 \mathrm{sig}_{\mathrm{n}}^{g_2}(s_i) + f_i(x_i, v_i)\right), \tag{7.49}$$
$$\dot{s}_{si} = \begin{cases} \rho_{1i} z_{1i} z_{1i}^T z_{2i} + \rho_{2i} z_{2i}, & \text{if } s_i^* = 0 \text{ or } s_i^* \neq 0, \ \|z_{1i}\| > \varepsilon, \\ l_1 z_{2i} + 3 l_2 \|z_{1i}\| z_{1i} z_{1i}^T z_{2i} + l_2 \|z_{1i}\|^3 z_{2i}, & \text{if } s_i^* \neq 0, \ \|z_{1i}\| \leq \varepsilon, \end{cases} \tag{7.50}$$

where α_2, β_2, p_2, and g_2 are defined in (7.16), and ρ_{1i} and ρ_{2i} take the forms

$$\rho_{1i} = \alpha_1 (p_1 - 1) \|z_{1i}\|^{p_1 - 3} + \beta_1 (g_1 - 1) \|z_{1i}\|^{g_1 - 3}, \tag{7.51}$$

$$\rho_{2i} = \alpha_1 \|z_{1i}\|^{p_1 - 1} + \beta_1 \|z_{1i}\|^{g_1 - 1}. \tag{7.52}$$

The main result associated with the control law (7.49) is proposed.

Theorem 7.3 *Considering multi-agent systems governed by (7.8), if $\mathcal{G}_c$ is strongly connected, under the control law (7.49), the closed-loop multi-agent system achieves fixed-time-synchronized consensus.*

Proof The proof can be given following the three steps in the proof of Theorem (7.2). That is omitted here. □

Remark 7.7 To achieve fixed-time-synchronized consensus, the controller (7.49) for the ith agent have to utilize the control inputs of its neighbors. This drawback can be alleviated by the methods in Remark 7.4. Actually, under a finite-time controller (without the usage of neighbors' accelerations)

$$\begin{aligned} u_i = -b_i^{-1} \Bigg(& \alpha_1 \left[\sum_{j=1}^{N} l_{ij} x_j + \operatorname{sig}_{\mathrm{c}}^{p_1} \left(\sum_{j=1}^{N} l_{ij} x_j \right) \right] \\ & + \beta_1 \left[\sum_{j=1}^{N} l_{ij} v_j + \operatorname{sig}_{\mathrm{c}}^{p_1} \left(\sum_{j=1}^{N} l_{ij} v_j \right) \right] + f_i \Bigg), \end{aligned} \tag{7.53}$$

the transformed multi-agent dynamics can be rewritten in an elegant form,

$$\dot{z}_1 = z_2, \tag{7.54}$$

$$\begin{aligned} \dot{z}_2 = & -\alpha_1 \left((Q \otimes I_n) \left((\mathcal{L}_c \otimes I_n) z_1 + \operatorname{sig}_{\mathrm{c}}^{p_1} ((\mathcal{L}_c \otimes I_n) z_1) \right) \right) \\ & - \beta_1 \left((Q \otimes I_n) \left((\mathcal{L}_c \otimes I_n) z_2 + \operatorname{sig}_{\mathrm{c}}^{p_1} ((\mathcal{L}_c \otimes I_n) z_2) \right) \right). \end{aligned} \tag{7.55}$$

From $Q\mathcal{L}_c = \mathcal{L}_c$, it follows

$$\dot{z}_1 = z_2, \tag{7.56}$$

$$\begin{aligned} \dot{z}_2 = & -\alpha_1 \left((\mathcal{L}_c \otimes I_n) z_1 + \operatorname{sig}_{\mathrm{c}}^{p_1} [(\mathcal{L}_c \otimes I_n) z_1) \right] \\ & - \beta_1 \left((\mathcal{L}_c \otimes I_n) z_2 + \operatorname{sig}_{\mathrm{c}}^{p_1} ((\mathcal{L}_c \otimes I_n) z_2) \right). \end{aligned} \tag{7.57}$$

We can then construct a proper Lyapunov function based on the positive definite matrix H (introduced in Lemma 7.3) and prove finite-time consensus. However, in this case, (fixed-) time-synchronized consensus cannot be expected.

For the purpose of suitable extensions, we herein introduce some interesting concepts and lemmas to be possibly useful in (fixed-) time-synchronized multi-agent control under different considerations.

Definition 7.6 (*Weakly connected*) For the directed graph $\mathcal{G}_c$, if its underlying undirected graph is connected, then $\mathcal{G}_c$ is weakly connected.

Definition 7.7 (*In-degree and out-degree*) For node i in a graph, its in-degree is the sum of all entries in the ith column of the adjacency matrix $\mathcal{A}$ of the graph, and its out-degree the sum of all entries in the ith row of $\mathcal{A}$.

Definition 7.8 (*(Weight-) balanced graphs*) A node in $\mathcal{G}_c$ is (weight-) balanced, if its in-degree is equal to its out-degree. If all the nodes in $\mathcal{G}_c$ are balanced, then $\mathcal{G}_c$ is balanced. An undirected graph is always balanced.

Lemma 7.4 *Zero is a simple eigenvalue of the symmetric positive semi-definite matrix $\mathcal{L}_c^T + \mathcal{L}_c$ if $\mathcal{G}_c$ is balanced and strongly connected.*

Proof Since $\mathcal{L}^T + \mathcal{L}$ is the Laplacian matrix of the underlying undirected graph $\mathcal{G}_c'$ associated with $\mathcal{G}_c$, according to Lemma 7.1, $\mathcal{G}_c'$ has a simple eigenvalue 0 together with an eigenvector $\mathbf{1}_N$, and $\mathcal{L}^T + \mathcal{L}$ is symmetric positive semi-definite. □

A balanced and strongly connected graph is widely applied to solve the "average consensus" problem as in the well-established results, e.g., [11].

There is another graphical concept related to "Balanced Graphs" called "Detail-Balanced Graphs".

Definition 7.9 (*Detail-balanced*) The directed graph $\mathcal{G}_c$ is detail-balanced, if there exist positive scalars θ_i satisfying $\theta_i a_{ij} = \theta_j a_{ji}$, $i, j \in \{1, 2, \ldots, N\}$.

Detail-balanced graphical conditions are usually applied to construct a proper Lyapunov function candidate to ease the analysis of convergence of multi-agent systems [7, 15, 19], e.g.,

$$V_0 = \sum_{i=1}^{N} \sum_{j=1}^{N} \theta_i a_{ij} \left(x_i - x_j\right)^2. \tag{7.58}$$

More techniques on the design of Lyapunov functions for multi-agent systems are provided in [18] (and the references therein).

7.5 Time-Synchronized Distributed Observers

Another interesting multi-agent control topic is distributed observers [1, 2, 4, 6, 6, 10, 16]. A distributed observer is designed to obtain the non-global state of the target whose information is only accessible to a subset of the agents by distributed communication. The proposal in the chapter can be analogously sketched and extend to the case of distributed observers, where for the estimation error of the target state, time-synchronized convergence or fixed-time-synchronized stabilization is achieved via a distributed manner.

Assume that there exist M leaders and N followers in a group of agents. We herein design a distributed time-synchronized observer for each follower to acquire the velocity of the leader (if $m = 1$) or the weighted average of the velocities of the leaders (if $m > 1$).

Remark 7.8 The weighted average of leaders' velocities is often utilized in leader-follower containment control, driving followers into a specific area spanned by leaders.

The system dynamics to be investigated takes following form,

$$\begin{aligned} \dot{x}_i &= v_i, \\ \dot{v}_i &= f_i\left(x_i, v_i\right) + b_i\left(x_i, v_i\right) u_i, \end{aligned} \tag{7.59}$$

where $x_i \in \mathbb{R}^n$ and $v_i \in \mathbb{R}^n$, $i \in \{1, 2, \ldots, N + M\}$, denote the general position vector and velocity vector, respectively, $u_i \in \mathbb{R}^n$ is the control input, and $f_i\left(x_i, v_i\right) \in \mathbb{R}^n$ and $b_i\left(x_i, v_i\right) \in \mathbb{R}^{n\times n}$ $\left(\det\left(b_i\left(x_i, v_i\right)\right) \neq 0\right)$ are the known parts.

We utilize a communication graph $\mathcal{G} = (\mathcal{V}, \mathcal{E})$ to characterize the underlying information flow among agents, where $\mathcal{E} \subseteq \mathcal{V} \times \mathcal{V}$ and $\mathcal{V} = \{1, 2, \ldots, N + M\}$ denote the arc set and the vertex set, respectively. Denote $\mathcal{N}_i$ as the in-neighbour set of node i, where $\mathcal{N}_i = \{j : (j, i) \in \mathcal{E}\}$. The adjacency matrix $\mathcal{A} = \left[a_{ij}\right] \in \mathbb{R}^{(N+M)\times(N+M)}$ is defined as $a_{ij} > 0$ if $(j, i) \in \mathcal{E}$, and $a_{ij} = 0$ otherwise, where $a_{ii} = 0$. The Laplacian matrix is defined as $\mathcal{L} = \left[l_{ij}\right] \in \mathbb{R}^{(N+M)\times(N+M)}$, where $l_{ii} = \sum_{j=1, j\neq i}^{k} a_{ij}$ and $l_{ij} = -a_{ij}$, $i \neq j$. Let $\mathcal{G}_l = (\mathcal{V}_l, \mathcal{E}_l)$ and $\mathcal{G}_f = \left(\mathcal{V}_f, \mathcal{E}_f\right)$ characterize the communication graphs among the leaders and the followers, respectively. The following assumptions on graphical conditions are made.

Assumption 7.5 The leaders receive no information from the followers.

Assumption 7.6 For each follower in the graph $\mathcal{G}$, there exists at least one leader that has a path to this follower, and the graph $\mathcal{G}_f$ among the followers is undirected and connected.

The above assumptions are natural and well-utilized in the literature on multi-agent control. Based on Assumption 7.6, the following lemma holds.

Lemma 7.5 [10] *The Laplacian matrix $\mathcal{L}_{f1}$ is positive definite, the sum of each row in $-\mathcal{L}_{f1}^{-1}\mathcal{L}_{f2}$ in equal to 1, and each entry in $-\mathcal{L}_{f1}^{-1}\mathcal{L}_{f2}$ is non-negative.*

Problem 7.1 Design a distributed estimator, yielding a velocity estimation $\hat{v}_i$ for each follower, such that the estimation error $\tilde{v}_f = \hat{v}_f - v_d$ achieves time-synchronized stabilization, where the following column-wise concatenations are defined for the ease of analysis,

$$v_l = \left[v_{N+1}^T, \ldots, v_{N+M}^T\right]^T, \tag{7.60}$$

$$\tilde{v}_f = \left[\tilde{v}_1^T, \ldots, \tilde{v}_N^T\right]^T, \tag{7.61}$$

$$\hat{v}_f = \left[\hat{v}_1^T, \ldots, \hat{v}_N^T\right]^T, \tag{7.62}$$

$$v_d = -\left(\mathcal{L}_{f1}^{-1}\mathcal{L}_{f2} \otimes I_n\right) v_l. \tag{7.63}$$

The Laplacian matrix $\mathcal{L}$ is given by

$$\mathcal{L}(l_{ij}) = \begin{bmatrix} \mathcal{L}_{f1} & \mathcal{L}_{f2} \\ 0_{M\times N} & \mathcal{L}_{l1} \end{bmatrix}, \tag{7.64}$$

where $i, j \in \{1, \ldots, N+M\}$, $\mathcal{L}_{f1} \in \mathbb{R}^{N\times N}$, $\mathcal{L}_{f2} \in \mathbb{R}^{N\times M}$, and $\mathcal{L}_{l1} \in \mathbb{R}^{M\times M}$.

Define a virtual estimation error for the ith follower as

$$z_i = \sum_{j\in N_i} a_{ij}\left(\hat{v}_i - \hat{v}_j\right), \quad i \in \mathcal{V}_f, \tag{7.65}$$

and $z = [z_1, \ldots, z_N]^T$. The following distributed time-synchronized observer is designed for follower i, $i \in \mathcal{V}_f$,

$$\dot{\hat{v}}_i = \frac{1}{\varpi_i}\sum_{j\in N_i} a_{ij}\dot{\hat{v}}_j - \frac{k_1}{\varpi_i}\mathrm{sig}_{\mathrm{n}}^{\alpha}(z_i) - \frac{k_2}{\varpi_i}\mathrm{sig}_{\mathrm{n}}^{\beta}(z_i), \tag{7.66}$$

where k_1 and k_2 are positive constant gains, α and β are defined as $0 < \alpha = \alpha_1/\alpha_2 < 1$ and $\beta = \beta_1/\beta_2 > 1$, α_1, α_2, β_1 and β_2 are positive odd integers, $\hat{v}_j$ is the velocity estimation satisfying $\hat{v}_j = v_j$ for $j \in \mathcal{V}_l$, and ϖ_i is a constant $\varpi_i = \sum_{j\in N_i} a_{ij}$.

Taking the distributed fixed-time-synchronized observer (7.66) into the time derivative of estimation error (7.65), we obtain

$$\dot{z}_i = \sum_{j\in N_i} a_{ij}\left(\dot{\hat{v}}_i - \dot{\hat{v}}_j\right) \tag{7.67}$$

$$= \varpi_i \dot{\hat{v}}_{di} - \sum_{j\in N_i} a_{ij}\dot{\hat{v}}_j \tag{7.68}$$

$$= \varpi_i\left(\frac{1}{\omega_i}\sum_{j\in N_i} a_{ij}\dot{\hat{v}}_j - \frac{k_1}{\varpi_i}\mathrm{sig}_{\mathrm{n}}^{\alpha}(z_i) - \frac{k_2}{\varpi_i}\mathrm{sig}_{\mathrm{n}}^{\beta}(z_i)\right) - \sum_{j\in N_i} a_{ij}\dot{\hat{v}}_j \tag{7.69}$$

$$= -k_1\mathrm{sig}_{\mathrm{n}}^{\alpha}(z_i) - k_2\mathrm{sig}_{\mathrm{n}}^{\beta}(z_i), \tag{7.70}$$

which further yields fixed-time-synchronized convergence of z_i according to Lemma 5.1, with the synchronized settling time

$$T_m \le \frac{2^{\frac{1-\alpha}{2}}}{k_1(1-\alpha)} + \frac{2^{\frac{1-\beta}{2}}}{k_1(\beta-1)}. \tag{7.71}$$

It is proven that each virtual estimation error z_i achieves fixed-time-synchronized stabilization. Due to the connectivity of the communication topology, we can further prove that all overall virtual error z likewise achieves fixed-time-synchronized stabilization by following the contradiction in Step 3 of the proof of Theorem 7.2.

From (7.65)

$$z_i = \sum_{j\in \mathcal{N}_i} l_{ij}\hat{v}_{dj} = \sum_{j\in \mathcal{V}_f} l_{ij}\tilde{v}_j + \sum_{j\in \mathcal{V}_f} l_{ij}v_{dj} + \sum_{j\in \mathcal{V}_l} l_{ij}v_j, \tag{7.72}$$

Define $\mathcal{H} = -\mathcal{L}_{f1}^{-1}\mathcal{L}_{f2}$, and thus we have $v_d = (H \otimes I_n)\, v_l$. Further,

$$\sum_{j\in \mathcal{V}_f} l_{ij}v_{dj} = \sum_{j\in \mathcal{V}_f} l_{ij} \sum_{k\in \mathcal{V}_l} h_{jk}v_{k+N} = \sum_{k\in \mathcal{V}_l}\left(\sum_{j\in \mathcal{V}_f} l_{ij}h_{jk}\right)v_{k+N}. \tag{7.73}$$

Based on $\mathcal{L}_{f1}H = \mathcal{L}_{f1}\left(-\mathcal{L}_{f1}^{-1}\mathcal{L}_{f2}\right) = -\mathcal{L}_{f2}$, the equation (7.73) becomes

$$\sum_{j\in \mathcal{V}_f} l_{ij}v_{dj} = \sum_{k\in \mathcal{V}_l}\left(\sum_{j\in \mathcal{V}_f} l_{ij}h_{jk}\right)v_{k+N} = -\sum_{j\in \mathcal{V}_l} l_{ij}v_j. \tag{7.74}$$

Taking (7.74) into (7.72), we obtain

$$z_i = \sum_{j\in \mathcal{V}_f} l_{ij}\tilde{v}_j, \tag{7.75}$$

$$z = \left(\mathcal{L}_{f1} \otimes I_n\right)\tilde{v}. \tag{7.76}$$

Therefore, noticing that $\mathcal{L}_{f1}$ is positive definite, the estimation error $\tilde{v}$ achieves fixed-time-synchronized convergence as z does.

Remark 7.9 According to Theorem 6.1, we can obtain the least upper bound for the synchronized settling time. Therefore, the estimation is enhance by

$$T_m \le T_{\text{least}} = \gamma\left(2k_1, 2k_2, \frac{1+\alpha}{2}, \frac{1+\beta}{2}, 1\right), \tag{7.77}$$

where in the case of $\|(0)\| \to +\infty$, $T_m = T_{\text{least}}$. Please note by choosing different Lyapunov functions, the least upper bound T_{least} in (7.77) may have different forms.

Remark 7.10 In fact, given $z = \left(\mathcal{L}_{f1} \otimes I_n\right)\tilde{v}$, some elements of the estimation error $\tilde{v}$ may happen to cross zero before the synchronized settling time T_m but will not stay at zero.

Remark 7.11 Indeed, there exist other complex and important distributed estimation problems, with the consideration of time delays, uncertainties, communication constraints, etc. Interested readers are suggested to investigate distributed time-synchronized or distributed fixed-time-synchronized observers in these cases, following the design procedure in this section. The main ideas behind them are the same.

7.6 Summary

Time-synchronized consensus has been defined and solved in this chapter, which ensures simultaneous convergence of all the state elements of all the agents to the equilibrium time-synchronously. The concepts of the consensus trajectory and the consensus equilibrium persistence have been proposed and formulated to facilitate the time-synchronized consensus control design. Various graphical conditions have been discussed and analyzed, among which, the sufficient communication topology has been proposed to achieve time-synchronized consensus. As a byproduct, the time-synchronized distributed observer has also been proposed, which is useful in the distributed sensing, multi-agent exploration and network system control.

References

1. Cai, H., Huang, J.: The leader-following consensus for multiple uncertain Euler-Lagrange systems with an adaptive distributed observer. IEEE Transactions on Automatic Control **61**(10), 3152–3157 (2015)
2. Cai, H., Lewis, F.L., Hu, G., Huang, J.: The adaptive distributed observer approach to the cooperative output regulation of linear multi-agent systems. Automatica **75**, 299–305 (2017)
3. Chen, G., Lewis, F.L.: Distributed adaptive controller design for the unknown networked Lagrangian systems. In: IEEE Conference on Decision and Control (CDC), pp. 6698–6703 (2010). https://doi.org/10.1109/CDC.2010.5717728
4. Han, W., Trentelman, H.L., Wang, Z., Shen, Y.: A simple approach to distributed observer design for linear systems. IEEE Transactions on Automatic Control **64**(1), 329–336 (2018)
5. Jiang, B., Chen, Z., Hu, Q., Qi, J.: Fixed-time attitude control for rigid spacecraft with actuator saturation and faults. IEEE Transactions on Control Systems Technology **24**(5), 4375–4380 (2014)
6. Liang, H., Zhou, Y., Ma, H., Zhou, Q.: Adaptive distributed observer approach for cooperative containment control of nonidentical networks. IEEE Transactions on Systems, Man, and Cybernetics: Systems **49**(2), 299–307 (2018)
7. Lin, X., Zheng, Y.: Finite-time consensus of switched multiagent systems. IEEE Transactions on Systems, Man, and Cybernetics: Systems **47**(7), 1535–1545 (2017). https://doi.org/10.1109/TSMC.2016.2631659
8. Liu, S., Xie, L., Lewis, F.L.: Synchronization of multi-agent systems with delayed control input information from neighbors. Automatica **47**(10), 2152–2164 (2011)
9. Lu, K., Xia, Y.: Adaptive attitude tracking control for rigid spacecraft with finite-time convergence. Automatica **49**(12), 3591–3599 (2013)

10. Meng, Z., Ren, W., You, Z.: Distributed finite-time attitude containment control for multiple rigid bodies. Automatica **46**(12), 2092–2099 (2010). https://doi.org/10.1016/j.automatica.2010.09.005
11. Olfati-Saber, R., Murray, R.M.: Consensus problems in networks of agents with switching topology and time-delays. IEEE Transactions on Automatic Control **49**(9), 1520–1533 (2004). https://doi.org/10.1109/TAC.2004.834113
12. Qu, Z.: Cooperative control of dynamical systems: applications to autonomous vehicles. Springer Science & Business Media (2009)
13. Ren, W., Beard, R.W.: Consensus seeking in multiagent systems under dynamically changing interaction topologies. IEEE Transactions on Automatic Control **50**(5), 655–661 (2005)
14. Wang, L., Chai, T., Zhai, L.: Neural-network-based terminal sliding-mode control of robotic manipulators including actuator dynamics. IEEE Transactions on Industrial Electronics **56**(9), 3296–3304 (2009). https://doi.org/10.1109/TIE.2008.2011350
15. Wang, L., Xiao, F.: Finite-time consensus problems for networks of dynamic agents. IEEE Transactions on Automatic Control **55**(4), 950–955 (2010)
16. Wang, X., Li, S., Shi, P.: Distributed Finite-Time Containment Control for Double-Integrator Multiagent Systems. IEEE Transactions on Cybernetics **44**(9), 1518–1528 (2014)
17. Zhang, H., Lewis, F.L., Qu, Z.: Lyapunov, adaptive, and optimal design techniques for cooperative systems on directed communication graphs. IEEE transactions on industrial electronics **59**(7), 3026–3041 (2011)
18. Zhang, H., Li, Z., Qu, Z., Lewis, F.L.: On constructing Lyapunov functions for multi-agent systems. Automatica **58**, 39–42 (2015). https://doi.org/10.1016/j.automatica.2015.05.006
19. Zhao, Y., Duan, Z., Wen, G.: Finite-time consensus for second-order multi-agent systems with saturated control protocols. IET Control Theory & Applications **9**(3), 312–319 (2014)

Chapter 8
Time-Synchronized Spacecraft Control in Rendezvous and Docking

Abstract Spacecraft missions usually require precise operations under limited fuel consumption. In this chapter, for the short-range relative motion of on-orbit service spacecraft during rendezvous and docking missions, we will improve control accuracy, reducing energy consumption, and optimizing the travel distance through time-synchronized control. Finite-time and fixed-time-synchronized controllers are proposed to drive the relative attitude or orbit tracking error to the origin synchronously within the finite or fixed time. Time-synchronized convergence of all the state elements of the 6-DOF spacecraft is achieved, while the merits of shortening the travel length of the output and saving fuel consumption are revealed.

8.1 Problem Formulations

In this section, the relative attitude motion error model, the relative orbit motion error model, the 6-DOF coupling error model, and the control objectives are given.

To form the basis of the attitude and position of the on-orbit service spacecraft with respect to the target spacecraft, we use the standard definition of the Earth-Centered Inertial (ECI) frame $\mathfrak{F}_i$ as the essential reference coordinate frame [1] as in Fig. 8.1. It is attached to the center of the Earth with x_i axis towards spring equinox γ, z_i axis towards celestial north, y_i axis being mutually perpendicular to the x_i and z_i axis such that forms a right-hand coordinate frame.

The attitude movement of the target spacecraft and the on-orbit service spacecraft can be described by the rotation of their body-fixed reference frames relative to the $\mathfrak{F}_i$. The body-fixed Cartesian frame $\mathfrak{F}_t$ and $\mathfrak{F}_s$ are located at centroid on the target spacecraft and the on-orbit service spacecraft, respectively. Moreover, the relative attitude motion between them can be expressed by the relative rotation of $\mathfrak{F}_t$ and $\mathfrak{F}_s$.

The motion of an on-orbit service spacecraft can be described relative to the target spacecraft following a Keplerian orbit. The position of the service spacecraft relative to the target spacecraft is defined as $\mathbf{q}_p$. The vector $\mathbf{q}_p$ has components $[q_{px}, q_{py}, q_{pz}]^T$ in a local-vertical/local-horizontal (LVLH) reference frame

D. Li et al., *Time-Synchronized Control: Analysis and Design*,
https://doi.org/10.1007/978-981-16-3089-7_8

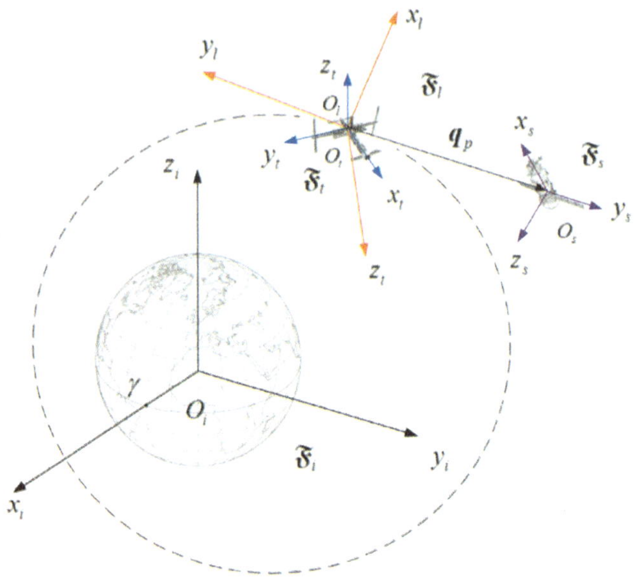

Fig. 8.1 Cartesian coordinate frames

$\mathfrak{F}_l$ attached to the target spacecraft with the x_l component pointing in the radial direction, the z_l component pointing in the direction of the reference orbit's angular momentum, and the y_l component obeying the right-hand coordinates [2]. Matrices representing rotation coordinate transformation between $\mathfrak{F}_s$ and $\mathfrak{F}_t$ are denoted as $\mathbf{R}_{ts}$, and $\mathbf{R}_{li}$, $\mathbf{R}_{si}$, $\mathbf{R}_{ti}$ are similar.

8.1.1 Relative Attitude Error Dynamics and Kinematics Model

Considering the problem of a rigid spacecraft with the body-fixed frame located at the center of mass, the attitude motion of the on-orbit service spacecraft, as well as the orientation of the body-fixed frame $\mathfrak{F}_s$ with respect to the inertial frame $\mathfrak{F}_i$, can be described by the unit quaternion $\mathbf{q}_a = [q_0, \mathbf{q}_v]^T \in \mathbb{R}^4$, which is subject to the constraint [3]

$$\mathbf{q}_v^T \mathbf{q}_v + q_0^2 = 1 \tag{8.1}$$

The dynamic model for the described rigid spacecraft can be expressed as follows [4]:

$$\mathbf{J}_s\omega_a = -\omega_a^{\times}\mathbf{J}_s\omega_a + \tau_a \tag{8.2}$$

$$\mathbf{q}_v = \frac{1}{2}\left(\mathbf{q}_v^{\times}\omega_a + q_0\omega_a\right) \tag{8.3}$$

$$\dot{q}_0 = -\frac{1}{2}\mathbf{q}_v^T\omega_a \tag{8.4}$$

where $\mathbf{J}_s \in \mathbb{R}^{3\times 3}$ is the constant, positive-definite, symmetric inertia matrix of the on-orbit service spacecraft, $\omega_a \in \mathbb{R}^3$ the angular velocity of the body-fixed frame $\mathfrak{F}_s$ with respect to the inertial reference frame $\mathfrak{F}_i$ expressed in $\mathfrak{F}_s$, $\tau_a \in \mathbb{R}^3$ a vector of control torques in $\mathfrak{F}_s$, and the notation $\mathbf{a}^{\times}$, $\forall \mathbf{a} = [a_1\ a_2\ a_3]^T$ denotes the following skew-symmetric matrix:

$$\mathbf{a}^{\times} \triangleq \begin{bmatrix} 0 & -a_3 & a_2 \\ a_3 & 0 & -a_1 \\ -a_2 & a_1 & 0 \end{bmatrix} \tag{8.5}$$

The rotation matrix that brings $\mathfrak{F}_i$ onto $\mathfrak{F}_s$, denoted by $\mathbf{R}_a(\mathbf{q}_v, q_0) \in SO(3)$, is defined as follows:

$$\mathbf{R}_a = (q_0^2 - \mathbf{q}_v^T\mathbf{q}_v)I_3 + 2\mathbf{q}_v\mathbf{q}_v^T - 2q_0\mathbf{q}_v^{\times} \tag{8.6}$$

where I_3 is the 3×3 identity matrix, and the angular velocity ω_a can be computed as follows:

$$\omega_a = 2(q_0\dot{\mathbf{q}}_v - \mathbf{q}_v\dot{q}_0) - 2\mathbf{q}_v^{\times}\dot{\mathbf{q}}_v \tag{8.7}$$

In order to obtain the attitude error model, we assume that the desired attitude of the on-orbit service spacecraft can be described by the desired unit quaternion $\mathbf{q}_{ad}(t) = [q_{d0}, \mathbf{q}_{dv}]^T \in \mathbb{R}^4$ and define the desired angular velocity $\omega_d \in \mathbb{R}^3$ through the following kinematic equations:

$$\mathbf{q}_{dv} = \frac{1}{2}\left(\mathbf{q}_{dv}^{\times}\omega_d + q_{d0}\omega_d\right) \tag{8.8}$$

$$\dot{q}_{d0} = -\frac{1}{2}\mathbf{q}_{dv}^T\omega_d \tag{8.9}$$

$$\omega_d = 2(q_{d0}\dot{\mathbf{q}}_{dv} - \mathbf{q}_{dv}\dot{q}_{d0}) - 2\mathbf{q}_{dv}^{\times}\dot{\mathbf{q}}_{dv} \tag{8.10}$$

The corresponding rotation matrix $\mathbf{R}_d(\mathbf{q}_{dv}, q_{d0}) \in SO(3)$ is defined as follows:

$$\mathbf{R}_d = (q_{d0}^2 - \mathbf{q}_{dv}^T\mathbf{q}_{dv})I_3 + 2\mathbf{q}_{dv}\mathbf{q}_{dv}^T - 2q_{d0}\mathbf{q}_{dv}^{\times} \tag{8.11}$$

To quantify the error between the actual and desired spacecraft attitudes, we define the rotation matrix $\mathbf{R}_e(\mathbf{q}_{ev}, q_{e0}) \in SO(3)$ and quaternion tracking error $\mathbf{q}_e(t) = [q_{e0}, \mathbf{q}_{ev}]^T \in \mathbb{R}^4$ as follows:

$$\mathbf{R}_e = \mathbf{R}_a\mathbf{R}_d^T = (q_{e0}^2 - \mathbf{q}_{ev}^T\mathbf{q}_{ev})I_3 + 2\mathbf{q}_{ev}\mathbf{q}_{ev}^T - 2q_{e0}\mathbf{q}_{ev}^{\times} \tag{8.12}$$

$$\mathbf{q}_{ev} = q_{d0}\mathbf{q}_v - q_0\mathbf{q}_{dv} + \mathbf{q}_v^{\times}\mathbf{q}_{dv} \tag{8.13}$$

$$q_{e0} = q_0 q_{d0} + \mathbf{q}_v^T\mathbf{q}_{dv} \tag{8.14}$$

Based on the preceding formulation, the tracking error of the angular velocity $\omega_e \in \mathbb{R}^3$ is defined as follows:

$$\omega_e = \omega_a - \mathbf{R}_e\omega_d \tag{8.15}$$

Computing (8.2)–(8.15), we obtain the open-loop tracking error dynamics as follows

$$\begin{aligned}\mathbf{J}_s\omega_e =& -(\omega_e + \mathbf{R}_e\omega_d)^{\times}\mathbf{J}_s(\omega_e + \mathbf{R}_e\omega_d)\\ &+ \mathbf{J}_s(\omega_e^{\times}\mathbf{R}_e\omega_d - \mathbf{R}_e\omega_d) + \tau_a\end{aligned} \tag{8.16}$$

$$\mathbf{q}_{ev} = \frac{1}{2}(\mathbf{q}_{ev}^{\times} + q_{e0}I_3)\omega_e \tag{8.17}$$

$$\dot{q}_{e0} = -\frac{1}{2}\mathbf{q}_{ev}^T\omega_e \tag{8.18}$$

where $\mathbf{R}_e = -\omega_e{}^{\times}\mathbf{R}_e$.

To express the open-loop tracking error dynamics, we rewrite (8.17) as follows:

$$\mathbf{q}_{ev} = \frac{1}{2}\mathbf{T}\omega_e \tag{8.19}$$

where $\mathbf{T} = \mathbf{q}_{ev}^{\times} + q_{e0}I_3$ is a Jacobian-type matrix.

After taking the time derivative of (8.19) and pre-multiplying both sides of the resulting expression by $\mathbf{T}^{-T}\mathbf{J}_s\mathbf{T}^{-1}$, we obtain,

$$\mathbf{M}_a\mathbf{q}_{ev} = \frac{1}{2}\mathbf{M}_a\mathbf{T}\omega_e + \frac{1}{2}\mathbf{P}^T\mathbf{J}_s\omega_e \tag{8.20}$$

where $\mathbf{M}_a = \mathbf{P}^T\mathbf{J}_s\mathbf{P}$ and $\mathbf{P} = \mathbf{T}^{-1}$.

After substituting (8.16) into the right-hand side of (8.20), we can obtain the following expression for the open-loop tracking error dynamics of the relative attitude motion

$$\mathbf{M}_a\mathbf{q}_{ev} + \mathbf{C}_a\mathbf{q}_{ev} + \mathbf{G}_a = \mathbf{Q}_a\tau_a \tag{8.21}$$

where

$$\mathbf{Q}_a = \frac{1}{2}\mathbf{P}^T \tag{8.22}$$

$$\mathbf{C}_a = -\mathbf{M}_a\mathbf{P}^{-1}\mathbf{P} - 2\mathbf{P}^T(\mathbf{J}_s\mathbf{P}\mathbf{q}_{ev})^{\times}\mathbf{P} \tag{8.23}$$

$$\begin{aligned}\mathbf{G}_a =& \mathbf{P}^T[(\mathbf{P}\mathbf{q}_{ev})^\times \mathbf{J}_s \mathbf{R}_e \omega_d] + \mathbf{P}^T[(\mathbf{R}_e \omega_d)^\times \mathbf{J}_s \mathbf{P}\mathbf{q}_{ev}] \\ &+ \frac{1}{2}\mathbf{P}^T[(\mathbf{R}_e \omega_d)^\times \mathbf{J}_s \mathbf{R}_e \omega_d] \\ &- \frac{1}{2}\mathbf{P}^T \mathbf{J}_s[(2\mathbf{P}\mathbf{q}_{ev})^\times \mathbf{R}_e \omega_d - \mathbf{R}_e \omega_d]\end{aligned} \tag{8.24}$$

8.1.2 Relative Orbit Error Dynamics and Kinematics Model

The relative orbit or position and relative velocity vectors from the target spacecraft to the on-orbit service spacecraft are defined as $\mathbf{q}_p = [q_{px}\ q_{py}\ q_{pz}]^T$ and $\mathbf{v}_p = [v_{px}\ v_{py}\ v_{pz}]^T$ with LVLH components $[x_l\ y_l\ z_l]^T$, respectively. Assuming that the target spacecraft follows a Keplerian orbit and that the on-orbit service spacecraft is in close proximity to the target, the relative motion model for the spacecraft in $\mathfrak{F}_l$ can be described by [1]

$$\begin{aligned}\mathbf{q}_p &= \mathbf{v}_p \\ \mathbf{M}_p \mathbf{v}_p + \mathbf{C}_p \mathbf{v}_p + \mathbf{G}_{pl} &= \tau_{pl}\end{aligned} \tag{8.25}$$

where $\mathbf{M}_p = \text{diag}(m_s, m_s, m_s)$, m_s denotes the mass of the service spacecraft, $\tau_{pl} \in \mathbb{R}^3$ represents the control force applied on the service spacecraft in LVLH. Also, $\mathbf{C}_p$ is a Coriolis-like anti-symmetric matrix and $\mathbf{G}_{pl}$ is a nonlinear term defined as

$$\mathbf{C}_p = 2m_s \begin{bmatrix} 0 & -\dot{\theta}_{pt} & 0 \\ \dot{\theta}_{pt} & 0 & 0 \\ 0 & 0 & 0 \end{bmatrix} \tag{8.26}$$

$$\mathbf{G}_{pl} = m_s \begin{bmatrix} -\ddot{\theta}_{pt} q_{py} - \dot{\theta}_{pt}^2 q_{px} + \dfrac{\mu_p (q_{px} + r_{et})}{r_{es}^3} - \mu_p / r_{et}^2 \\ \ddot{\theta}_{pt} q_{px} - \dot{\theta}_{pt}^2 q_{py} + \mu_p q_{py} / r_{es}^3 \\ \mu_p q_{pz} / r_{es}^3 \end{bmatrix} \tag{8.27}$$

$$r_{et} = a_{pt} \frac{1 - e_{pt}^2}{1 + e_{pt} \cos(\theta_{pt})} \tag{8.28}$$

$$r_{es} = \sqrt{(r_{et} + q_{px})^2 + q_{py}^2 + q_{pz}^2} \tag{8.29}$$

$$\dot{\theta}_{pt} = \sqrt{\mu / a_{pt}^3} \frac{[1 + e_{pt} \cos(\theta_{pt})]^2}{(1 - e_{pt}^2)^{3/2}} \tag{8.30}$$

$$\ddot{\theta}_{pt} = \frac{-2\mu e_{pt} [1 + e_{pt} \cos(\theta_{pt})]^3 \sin(\theta_{pt})}{a_{pt}^3 (1 - e_{pt}^2)^3} \tag{8.31}$$

where r_{et} and r_{es} denote the distance from the Earth center to the target spacecraft and the on-orbit service spacecraft, respectively. θ_{pt} is the true anomaly of the target spacecraft. a_{pt} and e_{pt} respectively represent the semi-major axis and the orbital eccentricity of the elliptical orbit of the target spacecraft, and μ_p represents the gravitational constant.

In order to obtain the relative position error model, the desired relative position vector from the target spacecraft to the on-orbit service spacecraft is defined as $\mathbf{q}_{pd} \in \mathbb{R}^3$ in $\mathfrak{F}_l$. Then the relative position tracking error vector $\mathbf{q}_{pe} \in \mathbb{R}^3$ is defined as $\mathbf{q}_{pe} = \mathbf{q}_p - \mathbf{q}_{pd}$. The first-order derivative and the second-order derivative of $\mathbf{q}_{pe}$ are as follows

$$\mathbf{q}_{pe} = \mathbf{q}_p - \mathbf{q}_{pd} \tag{8.32}$$

$$\mathbf{q}_{pe} = \mathbf{q}_p - \mathbf{q}_{pd} \tag{8.33}$$

Substituting (8.32) and (8.33) into (8.25) results in the following expression for the open-loop tracking error dynamics

$$\mathbf{M}_p\mathbf{q}_{pe} + \mathbf{C}_p\mathbf{q}_{pe} + \mathbf{G}_p = \tau_{pl} \tag{8.34}$$

where $\mathbf{G}_p = \mathbf{G}_{pl} + \mathbf{M}_p\mathbf{q}_{pd} + \mathbf{C}_p\mathbf{q}_{pd}$.

Considering the relative position motion of the on-orbit service spacecraft accompanied by the attitude motion, the control of the actuators installed in the body-fixed reference frame $\mathfrak{F}_s$ that provides body-fixed thrust is affected by the attitude motion. Therefore we need to convert actual control τ_p in $\mathfrak{F}_s$ into τ_{pl} in $\mathfrak{F}_l$ as follows with $\mathbf{Q}_p = \mathbf{R}_{li}\mathbf{R}_{bi}^T$.

$$\tau_{pl} = \mathbf{Q}_p\tau_p \tag{8.35}$$

Then, (8.34) can be written as the following form, which is the relative orbit model of the spacecraft coupled with the attitude motion.

$$\mathbf{M}_p\mathbf{q}_{pe} + \mathbf{C}_p\mathbf{q}_{pe} + \mathbf{G}_p = \mathbf{Q}_p\tau_p \tag{8.36}$$

8.1.3 6-DOF Coupling Spacecraft Error Model and Control Objectives

Combined with (8.21) and (8.36), the 6-DOF coupling model of the on-orbit service spacecraft can be described as follows

$$\mathbf{M}(\mathbf{q})\mathbf{q} + \mathbf{C}(\mathbf{q}, \mathbf{q})\mathbf{q} + \mathbf{G}(\mathbf{q}) = \mathbf{Q}\tau \tag{8.37}$$

where $\mathbf{q} = [\mathbf{q}_{ev}\mathbf{q}_{pe}]^T$, $\mathbf{M} = \text{diag}(\mathbf{M}_a, \mathbf{M}_p)$, $\mathbf{C} = \text{diag}(\mathbf{C}_a, \mathbf{C}_p)$,$\mathbf{G} = [\mathbf{G}_a\mathbf{G}_p]^T$, $\tau = [\tau_a\tau_p]^T$, $\mathbf{Q} = \text{diag}(\mathbf{Q}_a, \mathbf{Q}_p)$. We use i to express each dimension.

The control objective is given as follows.

Problem 8.1 For the short-range rendezvous and docking mission between the on-orbit service spacecraft and the target spacecraft, given the initial error system (8.37) and the desired system states $\mathbf{q}_d \in \mathbb{R}^n$ and $\mathbf{q}_{pd} \in \mathbb{R}^n$, design controllers, such that the closed-loop system is (fixed-) time-synchronized stable.

8.2 Control Design

8.2.1 Time-Synchronized Control Design for Spacecraft

We choose a sliding-mode manifold as follows so that the sliding surface $\mathbf{s}_1$ and the system related to the tracking error $\mathbf{q}$ satisfy the condition of ratio persistence.

$$\mathbf{s}_1 = \dot{\mathbf{q}} + \kappa_1 \mathrm{sig}_{\mathrm{n}}^{p_1}(\mathbf{q}) \tag{8.38}$$

For convenience, we calculate the derivative of $\mathbf{s}_1$ with respect to time as follows

$$\begin{aligned} \dot{\mathbf{s}}_1 &= \ddot{\mathbf{q}} + \kappa_1 \frac{\mathrm{d}}{\mathrm{d}t} \mathrm{sig}_{\mathrm{n}}^{p_1}(\mathbf{q}) = \ddot{\mathbf{q}} + \phi_1 \\ &= -\mathbf{M}^{-1}\left[\mathbf{C}\dot{\mathbf{q}} + \mathbf{G}\right] + \mathbf{M}^{-1}\mathbf{Q}\tau + \phi_1 \end{aligned} \tag{8.39}$$

where

$$\phi_1 = \kappa_1 (p_1 - 1) \|\mathbf{q}\|^{p_1 - 3} \mathbf{q}\mathbf{q}^T \cdot \dot{\mathbf{q}} + \kappa_1 \|\mathbf{q}\|^{p_1 - 1} I_n \cdot \dot{\mathbf{q}} \tag{8.40}$$

As already properly discussed in the previous chapters, there may exist a singularity problem in (8.40), that $\|\mathbf{q}\|^{-3}$ and $\|\mathbf{q}\|^{-1}$ are infinite when $\mathbf{q} = 0$ and $\dot{\mathbf{q}} \neq 0$. For the classical sign function $\mathrm{sig}_{\mathrm{c}}^{p_s}(x)$ (p_s is a positive constant), to avoid the singularity, a threshold value ϖ_s is set to tackle the singularity as follows

$$\frac{\mathrm{d}}{\mathrm{d}t}(\mathrm{sig}_{\mathrm{c}}^{p_s}(x)) = \begin{cases} p_s |x|^{p_s - 1} \dot{x}, \text{ if } |x| \geq \varpi_s, \dot{x} \neq 0 \\ p_s |\iota_s|^{p_s - 1} \dot{x}, \text{ if } |x| < \varpi_s, \dot{x} \neq 0 \\ 0, \text{ if } \dot{x} = 0 \end{cases} \tag{8.41}$$

where ϖ_s and ι_s are both small positive constants. For the norm-normalized sign function $\mathrm{sig}_{\mathrm{n}}(x)$, we modify the above switching law to

$$\frac{\mathrm{d}}{\mathrm{d}t}(\mathrm{sig}_{\mathrm{n}}^{p_N + 1}(x)) = \begin{cases} p_N \|x\|^{p_N - 2} x x^T \cdot \dot{x} + \|x\|^{p_N} I_n \cdot \dot{x}, \text{ if } \|x\| \geq \varpi_1, \dot{x} \neq 0 \\ p_N \|\iota_1\|^{p_N - 2} \iota_1 \iota_1^T \cdot \dot{x} + \|\iota_1\|^{p_N} I_n \cdot \dot{x}, \text{ if } \|x\| < \varpi_1, \dot{x} \neq 0 \\ 0, \text{ if } \dot{x} = 0 \end{cases} \tag{8.42}$$

where ϖ_1 and ι_1 are both small positive constants, p_N is the power exponent. Therefore, ϕ_1 is rewritten into the switching function form as follows to avoid singularity problems.

$$\phi_1 = \begin{cases} \kappa_1 (p_1 - 1) \|\mathbf{q}\|^{p_1-3}\mathbf{q}\mathbf{q}^T \cdot \dot{\mathbf{q}} + \kappa_1 \|\mathbf{q}\|^{p_1-1} I_n \cdot \dot{\mathbf{q}}, \text{ if } \|\mathbf{q}\| \geq \varpi_1, \dot{\mathbf{q}} \neq 0 \\ \kappa_1 (p_1 - 1) \|\iota_1\|^{p_1-3}\iota_1\iota_1^T \cdot \dot{\mathbf{q}} + \kappa_1 \|\iota_1\|^{p_1-1} I_n \cdot \dot{\mathbf{q}}, \text{ if } \|\mathbf{q}\| < \varpi_1, \dot{\mathbf{q}} \neq 0 \\ 0, \text{ if } \dot{\mathbf{q}} = 0 \end{cases} \tag{8.43}$$

Next, we propose a time-synchronized controller for spacecraft as

$$\tau_1 = -\kappa_2 \mathbf{Q}^{-1}\mathbf{M}\text{sig}_\text{n}^{p_2}(\mathbf{s}_1) - \mathbf{Q}^{-1}\mathbf{M}\phi_1 + \mathbf{Q}^{-1}(\mathbf{C}\dot{\mathbf{q}} + \mathbf{G}) \tag{8.44}$$

Remark 8.1 If the controller is only used for relative attitude synchronized control, then $\mathbf{q} = \mathbf{q}_{ev}$, $\mathbf{M} = \mathbf{M}_a$, $\mathbf{C} = \mathbf{C}_a$, $\mathbf{G} = \mathbf{G}_a$, and $\mathbf{Q} = \mathbf{Q}_a$ in (8.44). The same is true for relative orbit synchronized control.

The corresponding theorem is expressed as follows.

Theorem 8.1 Considering the spacecraft error model governed by (8.21) and (8.36), under the time-synchronized controller (8.44), the closed-loop system is time-synchronized stable. The sliding-mode surface $\mathbf{s}_1$ and the closed-loop system state $\mathbf{q}$ (when $\mathbf{s}_1 = 0$) are ratio persistent. The settling time of the state $\mathbf{q}$ is $T_{1m} \leq T_{11} + T_{12}$, where the synchronized settling time T_{11} for $\mathbf{s}_1$ and T_{12} for $\mathbf{q}$ (when $\mathbf{s}_1 = 0$) are respectively as follows

$$T_{11} \leq \frac{V_1^{\frac{1-p_2}{2}}(\mathbf{s}_1(0))}{2^{\frac{p_2-1}{2}}\kappa_2(1-p_2)} \tag{8.45}$$

$$T_{12} \leq \frac{V_2^{\frac{1-p_1}{2}}(\mathbf{q}_0)}{2^{\frac{p_1-1}{2}}\kappa_1(1-p_1)} \tag{8.46}$$

Proof Step 1. Analyze the convergence of the sliding mode variable $\mathbf{s}_1$.

The candidate Lyapunov function is selected as

$$V_1 = \frac{1}{2}\mathbf{s}_1^T\mathbf{s}_1 \tag{8.47}$$

Taking the derivative of (8.47) with respect to time and substituting controller (8.44), we have

$$\begin{aligned} \dot{V}_1 &= \mathbf{s}_1^T\dot{\mathbf{s}}_1 = \mathbf{s}_1^T\left[-\mathbf{M}^{-1}(\mathbf{C}\dot{\mathbf{q}} + \mathbf{G}) + \mathbf{M}^{-1}\mathbf{Q}\tau_1 + \phi_1\right] \\ &= -\kappa_2\mathbf{s}_1^T\text{sig}_\text{n}^{p_2}(\mathbf{s}_1) \\ &= -2^{\frac{1+p_2}{2}}\kappa_2 V_1^{\frac{1+p_2}{2}} \end{aligned} \tag{8.48}$$

Substituting (8.44) into (8.39), we have

$$\begin{aligned} \dot{\mathbf{s}}_1 &= -\mathbf{M}^{-1}\left[\mathbf{C}\dot{\mathbf{q}} + \mathbf{G}\right] + \mathbf{M}^{-1}\mathbf{Q}\tau_1 + \phi_1 \\ &= -\kappa_2\text{sig}_\text{n}^{p_2}(\mathbf{s}_1) \end{aligned} \tag{8.49}$$

We know that $\mathbf{s}_1$ is ratio persistent. According to Theorem 2.2, $\mathbf{s}_1$ is time-synchronized stable and the settling time is

$$T_{11} \leq \frac{V_1^{\frac{1-p_2}{2}}(\mathbf{s}_1(0))}{2^{\frac{p_2-1}{2}}\kappa_2(1-p_2)} \tag{8.50}$$

Step 2. Analyze the convergence $\mathbf{q}$ when $\mathbf{s}_1 = 0$.

The candidate Lyapunov function is selected as

$$V_2 = \frac{1}{2}\mathbf{q}^T\mathbf{q} \tag{8.51}$$

When $\mathbf{s}_1 = 0$, according to (8.38), we have

$$\dot{\mathbf{q}} = -\kappa_1 \mathrm{sig}_{\mathrm{n}}^{p_1}(\mathbf{q}) \tag{8.52}$$

Similarly, $\mathbf{q}$ is ratio persistent.

Taking the derivative of (8.51) with respect to time and substituting (8.52), we have

$$\begin{aligned} \dot{V}_2 &= \mathbf{q}^T\dot{\mathbf{q}} = -\kappa_1\mathbf{q}^T\mathrm{sig}_{\mathrm{n}}^{p_1}(\mathbf{q}) \\ &= -2^{\frac{1+p_1}{2}}\kappa_1 V_2^{\frac{1+p_1}{2}} \end{aligned} \tag{8.53}$$

According to Theorem 2.2, $\mathbf{q}$ is time-synchronized stable with the settling time

$$T_{12} \leq \frac{V_2^{\frac{1-p_1}{2}}(\mathbf{q}_0)}{2^{\frac{p_1-1}{2}}\kappa_1(1-p_1)} \tag{8.54}$$

The settling time of $\mathbf{q}$ is $T_{1m} \leq T_{11} + T_{12}$. Therefore, under controller (8.44), the closed-loop system is time-synchronized stable.

□

8.2.2 Fixed-Time-Synchronized Control Design for Spacecraft

We design the fixed-time-synchronized controller in this subsection. We choose a sliding-mode manifold $\mathbf{s}_2$ with the power exponents $p_3 < 1$ and $p_4 > 1$ as follows, so that the system related to the state $\mathbf{q}$ satisfies ratio persistence.

$$s_2 = \dot{\mathbf{q}} + \kappa_3\mathrm{sig}_{\mathrm{n}}^{p_3}(\mathbf{q}) + \kappa_4\mathrm{sig}_{\mathrm{n}}^{p_4}(\mathbf{q}) \tag{8.55}$$

To find the derivative of (8.55) with respect to time, we have

$$\begin{aligned}\dot{\mathbf{s}}_2 &= \ddot{\mathbf{q}} + \kappa_3 \frac{\mathrm{d}}{\mathrm{d}t}\mathrm{sig}_{\mathrm{n}}^{p_3}(\mathbf{q}) + \kappa_4 \frac{\mathrm{d}}{\mathrm{d}t}\mathrm{sig}_{\mathrm{n}}^{p_4}(\mathbf{q}) = \ddot{\mathbf{q}} + \phi_2 \\ &= -\mathbf{M}^{-1}\left[\mathbf{C}\dot{\mathbf{q}} + \mathbf{G}\right] + \mathbf{M}^{-1}\mathbf{Q}\tau_2 + \phi_2\end{aligned} \tag{8.56}$$

where

$$\begin{aligned}\phi_2 =& \kappa_3 (p_3 - 1) \|\mathbf{q}\|^{p_3-3}\mathbf{q}\mathbf{q}^T \cdot \dot{\mathbf{q}} + \kappa_3 \|\mathbf{q}\|^{p_3-1} I_n \cdot \dot{\mathbf{q}} \\ &+ \kappa_4 (p_4 - 1) \|\mathbf{q}\|^{p_4-3}\mathbf{q}\mathbf{q}^T \cdot \dot{\mathbf{q}} + \kappa_4 \|\mathbf{q}\|^{p_4-1} I_n \cdot \dot{\mathbf{q}}\end{aligned} \tag{8.57}$$

Similar to the previous subsection, there is also a singularity problem because $\|\mathbf{q}\|^{-3}$ and $\|\mathbf{q}\|^{-1}$ are infinite when $\mathbf{q} = 0$ and $\dot{\mathbf{q}} \neq 0$. Therefore, we replace ϕ_2 with a modified switching function as

$$\phi_2 = \begin{cases} \kappa_3 (p_3 - 1) \|\mathbf{q}\|^{p_3-3}\mathbf{q}\mathbf{q}^T \cdot \dot{\mathbf{q}} + \kappa_3 \|\mathbf{q}\|^{p_3-1} I_n \cdot \dot{\mathbf{q}} & \\ + \kappa_4 (p_4 - 1) \|\mathbf{q}\|^{p_4-3}\mathbf{q}\mathbf{q}^T \cdot \dot{\mathbf{q}} + \kappa_4 \|\mathbf{q}\|^{p_4-1} I_n \cdot \dot{\mathbf{q}}, & \text{if } \|\mathbf{q}\| \geq \varpi_2, \dot{\mathbf{q}} \neq 0 \\ \kappa_3 (p_3 - 1) \|\iota_2\|^{p_3-3}\iota_2\iota_2^T \cdot \dot{\mathbf{q}} + \kappa_3 \|\iota_2\|^{p_3-1} I_n \cdot \dot{\mathbf{q}} & \\ + \kappa_4 (p_4 - 1) \|\iota_2\|^{p_4-3}\iota_2\iota_2^T \cdot \dot{\mathbf{q}} + \kappa_4 \|\iota_2\|^{p_4-1} I_n \cdot \dot{\mathbf{q}}, & \text{if } \|\mathbf{q}\| < \varpi_2, \dot{\mathbf{q}} \neq 0 \\ 0, \text{ if } \dot{\mathbf{q}} = 0 & \end{cases} \tag{8.58}$$

where ϖ_2 and ι_2 are both small positive constants.

We propose a controller based on fixed-time-synchronized stability for spacecraft.

$$\begin{aligned}\tau_2 =& -\mathbf{Q}^{-1}\mathbf{M}\left(\kappa_5 \mathrm{sig}_{\mathrm{n}}^{p_5}(\mathbf{s}_2) + \kappa_6 \mathrm{sig}_{\mathrm{n}}^{p_6}(\mathbf{s}_2)\right) \\ &- \mathbf{Q}^{-1}\mathbf{M}\phi_2 + \mathbf{Q}^{-1}\left(\mathbf{C}\dot{\mathbf{q}} + \mathbf{G}\right)\end{aligned} \tag{8.59}$$

where the power exponents $p_5 < 1$ and $p_6 > 1$.

The corresponding theorem is expressed as follows.

Theorem 8.2 Considering the spacecraft error model governed by (8.21) and (8.36), under the fixed-time-synchronized controller (8.59), the closed-loop system is fixed-time-synchronized stable. The sliding-mode surface $\mathbf{s}_2$ and the closed-loop system state $\mathbf{q}$ (when $\mathbf{s}_2 = 0$) are ratio persistent. The total settling time of the state $\mathbf{q}$ is $T_{2m} \leq T_{21} + T_{22}$, where the synchronized settling time T_{21} for $\mathbf{s}_2$ and T_{22} for $\mathbf{q}$ (when $\mathbf{s}_2 = 0$) are respectively as follows

$$T_{21} \leq \frac{2^{\frac{1-p_5}{2}}}{\kappa_5(1-p_5)} + \frac{2^{\frac{1-p_6}{2}}}{\kappa_6(p_6-1)} \tag{8.60}$$

$$T_{22} \leq \frac{2^{\frac{1-p_3}{2}}}{\kappa_3(1-p_3)} + \frac{2^{\frac{1-p_4}{2}}}{\kappa_4(p_4-1)} \tag{8.61}$$

Proof Step 1. Analyze the convergence of the sliding mode variable $\mathbf{s}_2$.

The candidate Lyapunov function is selected as

$$V_3 = \frac{1}{2}\mathbf{s}_2^T\mathbf{s}_2 \tag{8.62}$$

Taking the derivative of (8.62) with respect to time and substituting controller (8.59), we have

$$\begin{aligned}\dot{V}_3 &= \mathbf{s}_2^T\dot{\mathbf{s}}_2 = \mathbf{s}_2^T\left[-\mathbf{M}^{-1}\left(\mathbf{C}\dot{q}+\mathbf{G}\right)+\mathbf{M}^{-1}\mathbf{Q}\tau_2+\phi_2\right] \\ &= -\mathbf{s}_2^T\left(\kappa_5\mathrm{sig}_{\mathrm{n}}^{p_5}(\mathbf{s}_2)+\kappa_6\mathrm{sig}_{\mathrm{n}}^{p_6}(\mathbf{s}_2)\right) \\ &= -2^{\frac{1+p_5}{2}}\kappa_5V_3^{\frac{1+p_5}{2}}-2^{\frac{1+p_6}{2}}\kappa_6V_3^{\frac{1+p_6}{2}}\end{aligned} \tag{8.63}$$

Substituting (8.59) into (8.56), we have

$$\dot{\mathbf{s}}_2 = -\kappa_5\mathrm{sig}_{\mathrm{n}}^{p_5}(\mathbf{s}_2)-\kappa_6\mathrm{sig}_{\mathrm{n}}^{p_6}(\mathbf{s}_2) \tag{8.64}$$

We know that $\mathbf{s}_2$ is ratio persistent. According to Theorem 2.2, the $\mathbf{s}_2$ is fixed-time-synchronized stable with the bounded settling time

$$T_{21} \le \frac{2^{\frac{1-p_5}{2}}}{\kappa_5(1-p_5)}+\frac{2^{\frac{1-p_6}{2}}}{\kappa_6(p_6-1)} \tag{8.65}$$

Step 2. Analyze the convergence of $\mathbf{q}$ when $\mathbf{s}_2 = 0$.

The candidate Lyapunov function is selected as

$$V_4 = \frac{1}{2}\mathbf{q}^T\mathbf{q} \tag{8.66}$$

When $\mathbf{s}_2 = 0$, according to (8.55), we have

$$\dot{\mathbf{q}} = -\kappa_3\mathrm{sig}_{\mathrm{n}}^{p_3}(\mathbf{q})-\kappa_4\mathrm{sig}_{\mathrm{n}}^{p_4}(\mathbf{q}) \tag{8.67}$$

Similarly, $\mathbf{q}$ is ratio persistent.

Taking the derivative of (8.66) with respect to time and substituting (8.55) with $\mathbf{s}_2 = 0$, we have

$$\begin{aligned}\dot{V}_4 &= \mathbf{q}^T\dot{\mathbf{q}} = -\kappa_3\mathbf{q}^T\mathrm{sig}_{\mathrm{n}}^{p_3}(\mathbf{q})-\kappa_4\mathbf{q}^T\mathrm{sig}_{\mathrm{n}}^{p_4}(\mathbf{q}) \\ &= -2^{\frac{1+p_3}{2}}\kappa_3V_4^{\frac{1+p_3}{2}}-2^{\frac{1+p_4}{2}}\kappa_4V_4^{\frac{1+p_4}{2}}\end{aligned} \tag{8.68}$$

According to Theorem 2.2, $\mathbf{q}$ is fixed-time-synchronized stable, and its bounded settling time is

$$T_{22} \le \frac{2^{\frac{1-p_3}{2}}}{\kappa_3(1-p_3)}+\frac{2^{\frac{1-p_4}{2}}}{\kappa_4(p_4-1)} \tag{8.69}$$

The settling time of $\mathbf{q}$ in the closed-loop system is $T_{2m} \leq T_{21} + T_{22}$. Therefore, under the controller (8.59), the closed-loop system is fixed-time-synchronized stable.

Remark 8.2 For the time-synchronized controller (8.44) and the fixed-time-synchronized controller (8.59), their innovation and uniqueness are mainly based on the $\text{sig}_\text{n}(\cdot)$ function. If they are replaced with the classical sign function $\text{sig}_\text{c}(\cdot)$, the conventional finite-time and fixed-time controllers can be obtained. The controllers are as follows:

$$\tau_{1c} = -\kappa_2 \mathbf{Q}^{-1}\mathbf{M}\text{sig}_\text{c}^{p_2}(\mathbf{s}_{1c}) - \mathbf{Q}^{-1}\mathbf{M}\phi_{1c} + \mathbf{Q}^{-1}(\mathbf{C}\dot{\mathbf{q}} + \mathbf{G}) \tag{8.70}$$

$$\begin{aligned}\tau_{2c} = &- \mathbf{Q}^{-1}\mathbf{M}\left(\kappa_5 \text{sig}_\text{c}^{p_5}(\mathbf{s}_{2c}) + \kappa_6 \text{sig}_\text{c}^{p_6}(\mathbf{s}_{2c})\right) - \mathbf{Q}^{-1}\mathbf{M}\phi_{2c} \\ &+ \mathbf{Q}^{-1}(\mathbf{C}\dot{\mathbf{q}} + \mathbf{G})\end{aligned} \tag{8.71}$$

where

$$\mathbf{s}_{1c} = \dot{\mathbf{q}} + \kappa_1 \text{sig}_\text{c}^{p_1}(\mathbf{q}) \tag{8.72}$$

$$\mathbf{s}_{2c} = \dot{\mathbf{q}} + \kappa_3 \text{sig}_\text{c}^{p_3}(\mathbf{q}) + \kappa_4 \text{sig}_\text{c}^{p_4}(\mathbf{q}) \tag{8.73}$$

Simultaneously, ϕ_{1c} and ϕ_{2c} are also switching functions suitable for $\text{sig}_\text{c}(\cdot)$ function. We uses these two controllers as a comparative controller to illustrate the effects of the proposed time-synchronized ones.

8.2.3 (Fixed-) Time-Synchronized Control Design for 6-DOF Spacecraft

Theorems 8.1 and 8.2 respectively design the time-synchronized and fixed-time-synchronized controllers for spacecraft. If we apply the above controllers to the attitude and orbit models, respectively, in this way, each dimension value of the relative attitude or relative orbit tracking error will converge at the same time. However, the relative attitude and relative orbit tracking errors do not necessarily converge at the same time in practical missions.

If we adopt the 6-DOF spacecraft tracking error model and apply the proposed controllers, then the 6-dimensional variables of the relative attitude-orbit tracking error will all converge synchronously. We propose the following theorem.

Theorem 8.3 Considering the 6-DOF spacecraft error model governed by (8.37), under the time-synchronized controller (8.44) and the fixed-time-synchronized controller (8.59), the closed-loop system is time-synchronized stable and fixed-time-synchronized stable, respectively. Simultaneously, all tracking errors of relative attitude and orbit converge synchronously.

Proof The proof of the time-synchronized controller for 6-DOF spacecraft is given by following that of Theorems 8.1 and 8.2.

8.3 Simulation Analysis

The numerical example is shown in this section to demonstrate the performance of the developed the time-synchronized and fixed-time-synchronized controllers. We will compare the proposed time-synchronized controllers with controllers (8.70) and (8.71) under similar parameters to yield a fair comparison.

8.3.1 Simulation Parameters

For the target spacecraft, the orbital parameters are given in Table 8.1. Since the service spacecraft is very close to the target spacecraft, they have the same orbital parameters.

Without loss of generality, the initial parameters of the service spacecraft are given in Table 8.2, and the parameters used for the simulation are all replaceable.

The desired signal is given by the following

Table 8.1 Orbital parameters of the target spacecraft

Parameters	Values
Semimajor axis	6.85×10^6m
Eccentricity	0.0073
Inclination	60°
Argument of periapsis	0°
Longitude of the ascending node	0°
Mean anomaly	0°

Table 8.2 Initial parameters of the on-orbit service spacecraft

Parm	Values	Parm	Values
$\mathbf{q}_{v0}$	$[0.6, 0.3, -0.3]^T$rad	$\mathbf{q}_{p0}$	$[32, 10, -20]^T$m
ω_{a0}	$[0, 0, 0]^T$rad/s	$\mathbf{v}_{p0}$	$[-0.3, -0.15, 0.2]^T$m/s
$\mathbf{J}_s$	$\begin{bmatrix} 100 & -10 & 5 \\ -10 & 120 & -3 \\ 5 & -3 & 100 \end{bmatrix}$ kg · m^2	$\mathbf{M}_p$	$\begin{bmatrix} 100 & 0 & 0 \\ 0 & 100 & 0 \\ 0 & 0 & 100 \end{bmatrix}$ kg

$$\mathbf{q}_d = \begin{bmatrix} \mathbf{q}_{dv} \\ \mathbf{q}_{pd} \end{bmatrix} = \begin{bmatrix} (r_{li-23} - r_{li-32})/(4q_{d0}) \\ (r_{li-31} - r_{li-13})/(4q_{d0}) \\ (r_{li-12} - r_{li-21})/(4q_{d0}) \\ \sin(0.001t) \\ \sin(0.001t + \pi/2) \\ \sin(0.001t + \pi/3) \end{bmatrix} \tag{8.74}$$

In this simulation, the service spacecraft tracks the attitude of the target spacecraft. We assume that the $\mathfrak{F}_t$ coincides with the $\mathfrak{F}_i$, i.e., it is inertially oriented. Therefore, $\mathbf{q}_{dv}$ is obtained from the quaternion of the relative motion of the $\mathfrak{F}_s$ and $\mathfrak{F}_i$. $r_{li-j,k}$ represents the element of the j th-row and k th-column in $\mathbf{R}_{li}$, $q_{d0} = 1/2\ (1 + r_{li-11} + r_{li-22} + r_{li-33})^{1/2}$.

Next, the controller parameters are provided in Tables 8.3 and 8.4. Here, 'A' stands for the attitude control, 'O' for the relative orbit control, and 'D' for the 6-DOF control.

In order to illustrate the energy consumption of the proposed controller, we use the energy index, which is related to the amount of energy (chemical or electrical energy) that actuators (thrusters or reaction flywheels) consume during closed-loop system stabilization, as $E = \sum_{i=1}^{n} \int_0^t |\tau_i(t)|^2 \mathrm{d}t$.

Table 8.3 The values of the finite-time controllers parameters

	(8.44)-A	(8.70)-A	(8.44)-O	(8.70)-O	(8.44)-D
κ_1	0.1	0.1	0.1	0.1	0.1
κ_2	0.06	0.06	0.06	0.06	0.01
p_1	0.6	0.6	0.6	0.6	0.6
p_2	0.9	0.9	0.9	0.9	0.9

Table 8.4 The values of the fixed-time controllers parameters

	(8.59)-A	(8.71)-A	(8.59)-O	(8.71)-O	(8.59)-D
κ_3	0.1	0.1	0.02	0.02	0.02
κ_4	0.1	0.1	0.008	0.01	0.008
κ_5	0.04	0.04	0.02	0.02	0.02
κ_6	0.04	0.04	0.012	0.015	0.012
p_4	1.1	1.1	1.5	1.5	1.5
p_5	0.6	0.6	0.6	0.6	0.6
p_6	1.1	1.1	1.5	1.5	1.5

8.3.2 Performance and Comparison of Controllers

For the tracking control of the on-orbit service spacecraft, we divide the simulations into three groups: (i) Figs. 8.2, 8.3, 8.4 and 8.6 show the control performance of the time-synchronized controller (8.44) compared with the finite-time controller (8.70); (ii) Figs. 8.7, 8.8, 8.9 and 8.11 present the control performance of the fixed-time-synchronized controller (8.59) compared with the fixed-time controller (8.71); and (iii) Figs. 8.12 and 8.13 reveal the control effects of the 6-DOF spacecraft synchronized convergence (Fig. 8.10).

In simulation group (i), we show the control performance of the time-synchronized controller (8.44) and compare it with the finite-time controller (8.70). For the controllers to be compared, we use exactly the same initial conditions and controller parameters. Our comparison focuses on three aspects: (i) the differences between the output trajectories of the closed-loop spacecraft system; (ii) time-synchronized convergence achieved by the control design; and (iii) the energy consumption under two controllers.

Regardless of the relative attitude or the relative orbital tracking trajectory, the time-synchronized controller (8.44) and the finite-time controller (8.70) show obvious differences, as shown in Fig. 8.2. The tracking trajectory is a straight line under (8.44) and a curve under (8.70). For an on-orbit service spacecraft, the time-synchronized control algorithm obviously yields the shortest length, while at the same time, avoids the adverse effects of possible overshoots, which is of substantial significance for the motion control of the on-orbit service spacecraft in rendezvous and docking.

Figures 8.3 and 8.4 compare the control effects of the two controllers in terms of the relative attitude and relative orbit tracking, respectively. It is found that under the proposed time-synchronized controller, each element of the tracking error can converge simultaneously, while under the finite-time controller, such phenomenon is not expected.

Figure 8.5 intuitively reflects the energy consumption under two controllers. It is observed that time-synchronized control substantially reduces the energy consumption compared with finite-time control. Fig. 8.6 shows that the control inputs are within a reasonable range under practical engineering conditions.

In simulation group (ii), we show the control performance of the controller (8.59) compared with the fixed-time controller (8.71). From Figs. 8.7, 8.8, 8.9 and 8.11, conclusions similar to that of simulation group (i) can be drawn, e.g., fixed-time synchronized control provides a shorter travel length, less energy consumption, precise simultaneous convergence.

In simulation group (iii), we emphatically consider the situation where the relative attitude and orbit all converge synchronously. In Fig. 8.12, under the time-synchronized controller (8.44), the 6-DOF spacecraft achieves the synchronous convergence of all relative attitude and orbit tracking errors in finite time. In Fig. 8.13, under the fixed-time-synchronized controller (8.59), the 6-DOF spacecraft achieves the synchronous convergence of all relative attitude and orbit tracking errors in fixed time.

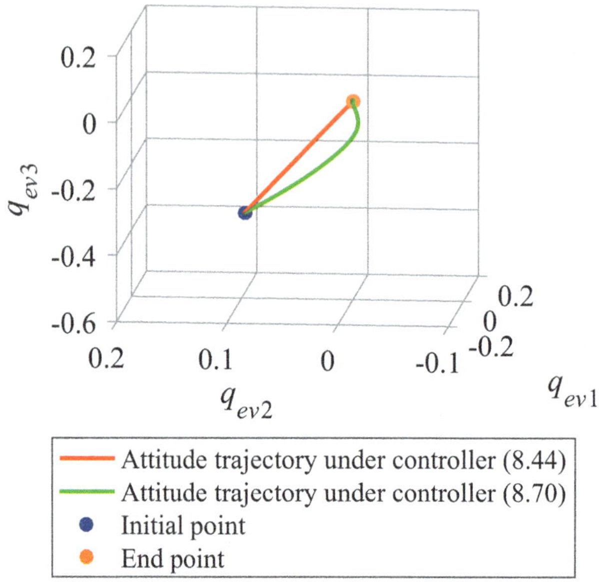

(a)

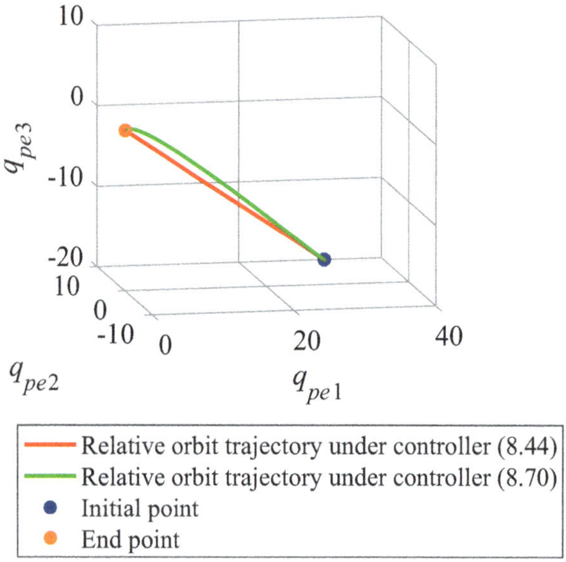

(b)

Fig. 8.2 The comparison chart of motion trajectory: **a** comparison of relative attitude motion trajectory under the time-synchronized controller (8.44) and finite-time controller (8.70); **b** comparison of relative orbit motion trajectory under the time-synchronized controller (8.44) and finite-time controller (8.70)

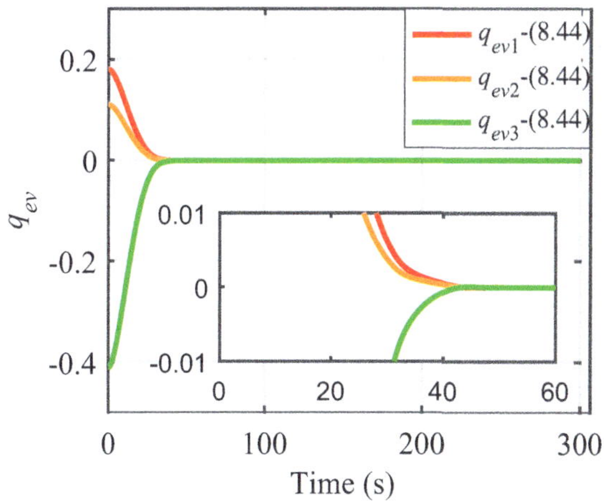

(a)

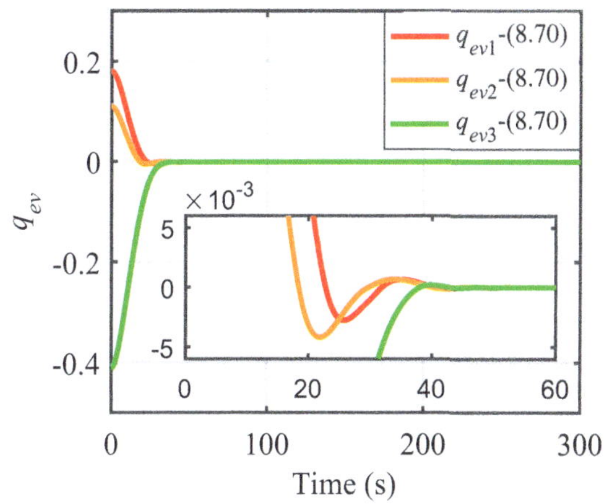

(b)

Fig. 8.3 The comparison of relative attitude tracking error under the time-synchronized controller (8.44) and finite-time controller (8.70): **a** tracking error under (8.44); **b** tracking error under (8.70)

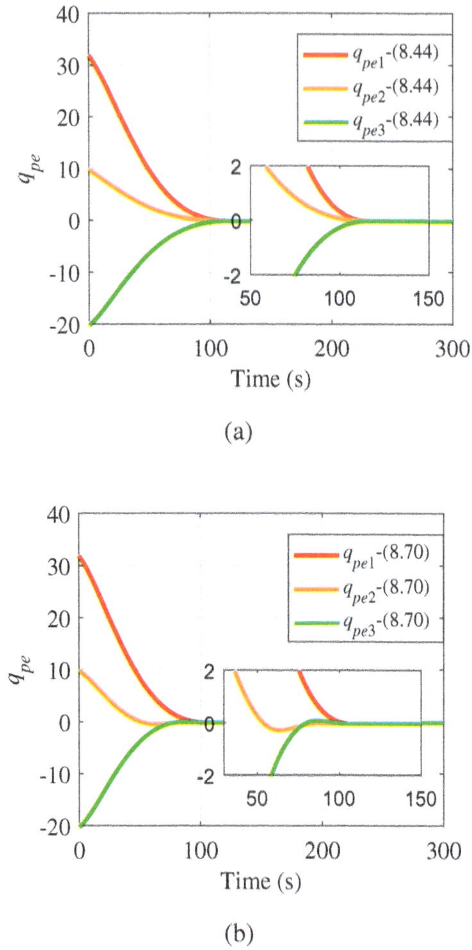

Fig. 8.4 The comparison of relative orbit tracking error under time-synchronized controller (8.44) and finite-time controller (8.70): **a** tracking error under (8.44); **b** tracking error under (8.70)

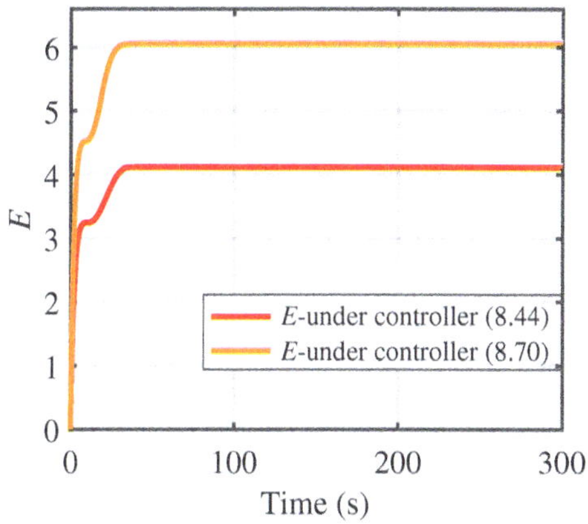

(a)

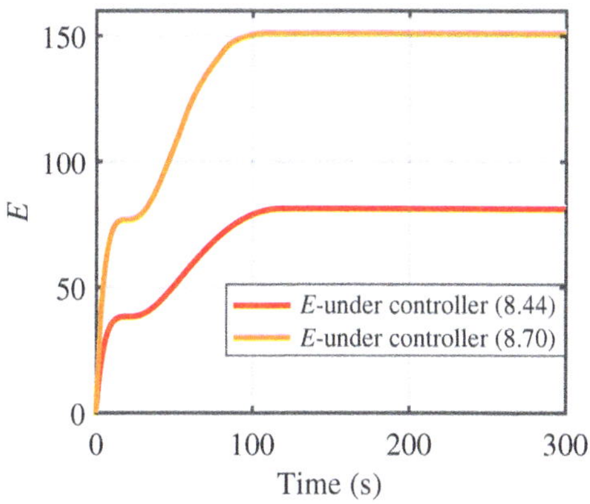

(b)

Fig. 8.5 The comparison of the energy consumption under the time-synchronized controller (8.44) and the finite-time controller (8.70): **a** the energy consumption of relative attitude tracking; **b** the energy consumption of relative orbit tracking

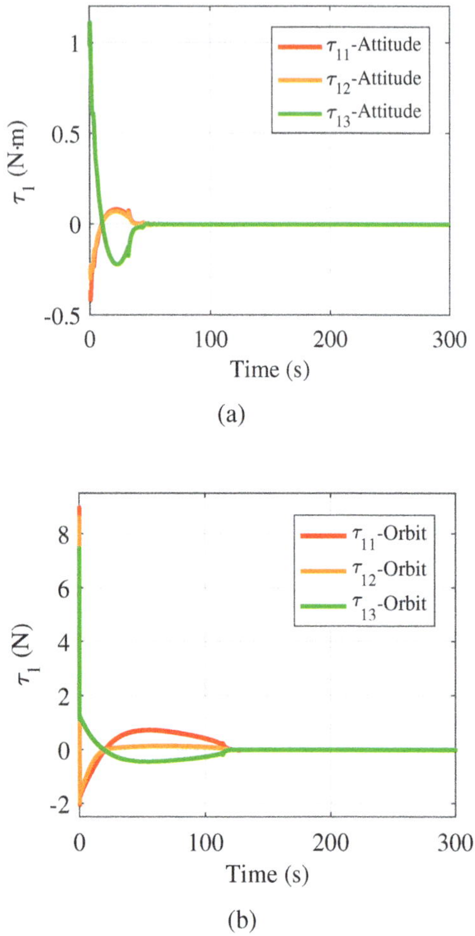

Fig. 8.6 The control input of the time-synchronized controller (8.44): **a** control input for the relative attitude tracking control; **b** control input for the relative orbit tracking control

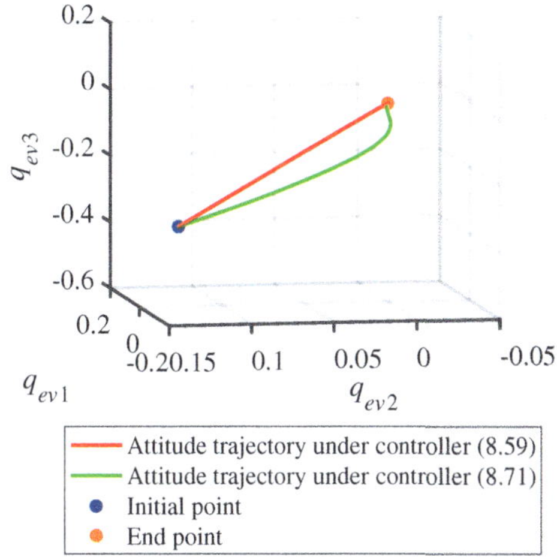

(a)

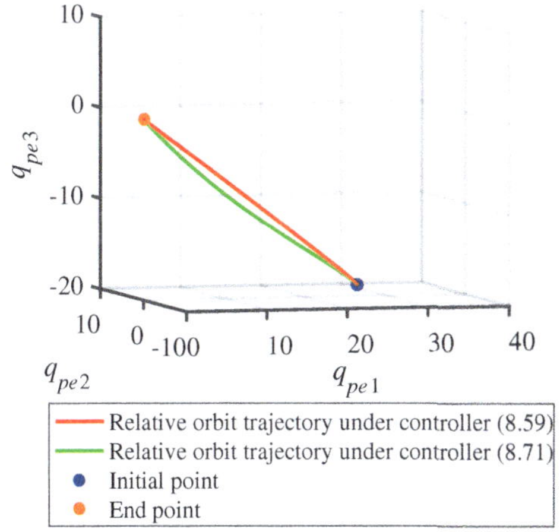

(b)

Fig. 8.7 The comparison chart of motion trajectory: **a** comparison of relative attitude motion trajectory under the fixed-time-synchronized controller (8.59) and fixed-time controller (8.71); **b** comparison of relative orbit motion trajectory under the time-synchronized controller (8.59) and finite-time controller (8.71)

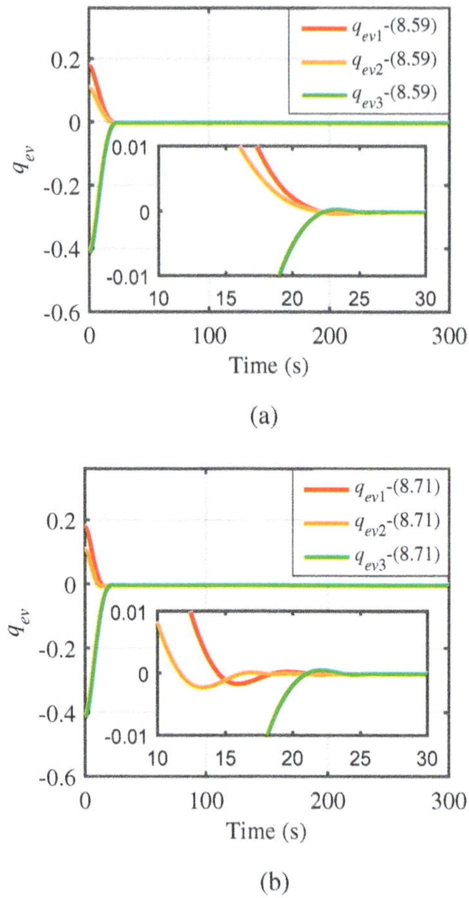

Fig. 8.8 The comparison of relative attitude tracking error under the fixed-time-synchronized controller (8.59) and fixed-time controller (8.71): **a** tracking error under (8.59); **b** tracking error under (8.71)

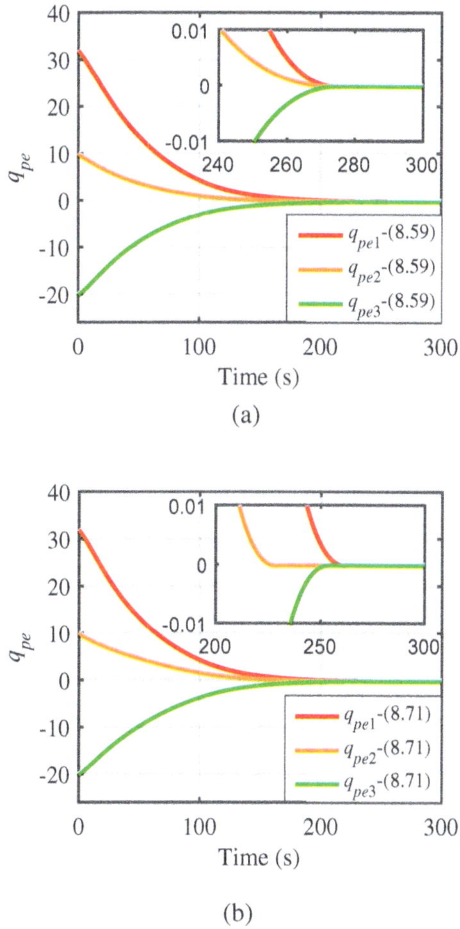

Fig. 8.9 The comparison of relative orbit tracking error under the fixed-time-synchronized controller (8.59) and fixed-time controller (8.71): **a** tracking error under (8.59); **b** tracking error under (8.71)

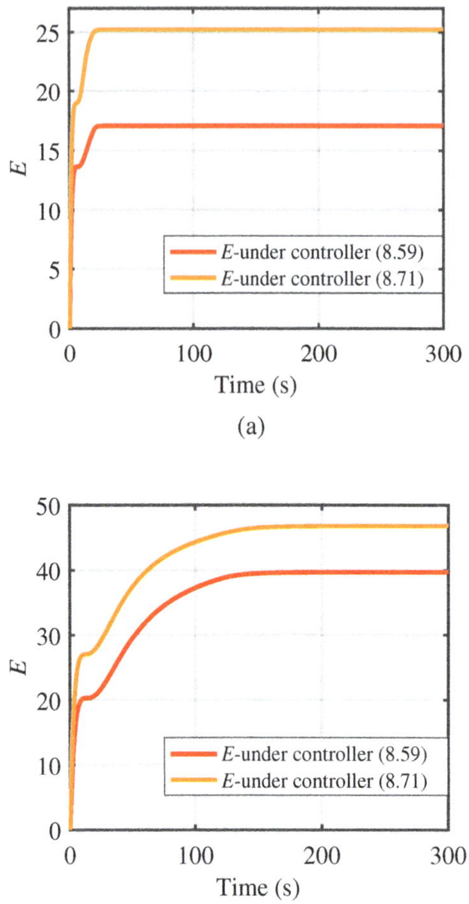

(b)

Fig. 8.10 The comparison of the energy consumption under the fixed-time-synchronized controller (8.59) and the fixed-time controller (8.71): **a** the energy consumption of relative attitude tracking; **b** the energy consumption of relative orbit tracking

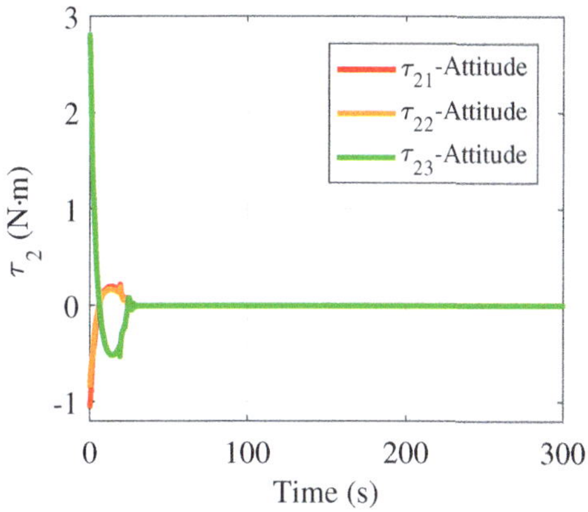

(a)

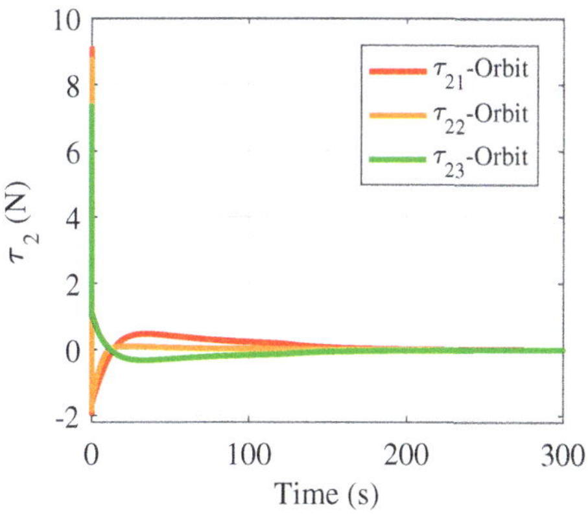

(b)

Fig. 8.11 The control input of the fixed-time-synchronized controller (8.59): **a** the input torque for the relative attitude tracking control; **b** the input force for the relative orbit tracking control

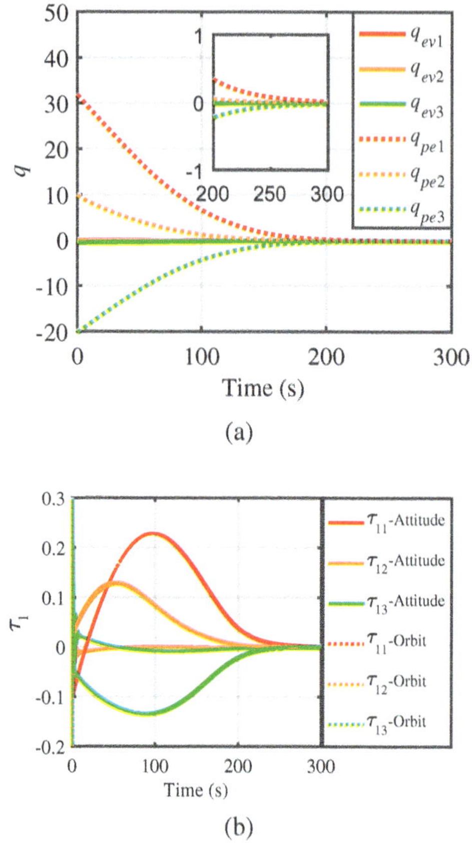

Fig. 8.12 For the display of 6-DOF-synchronized-convergence, according to Theorem 8.3, the relative attitude and relative orbit (6-DOF) tracking errors will simultaneously converge under the time-synchronized controller (8.44) in finite time: **a** 6-DOF tracking errors; **b** 6-DOF control inputs

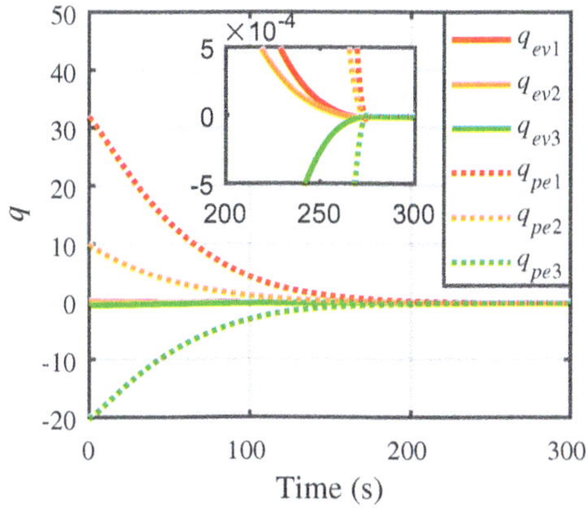

(a)

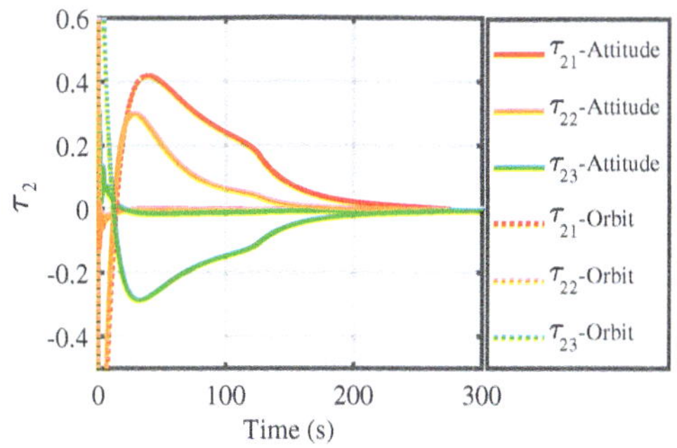

(b)

Fig. 8.13 For the display of 6-DOF-synchronized-convergence, according to Theorem 8.3, the relative attitude and relative orbit (6-DOF) tracking errors will simultaneously converge under the fixed-time-synchronized controller (8.59) in fixed time: **a** 6-DOF tracking errors; **b** 6-DOF control inputs

8.3.3 Hardware-in-Loop Experiment Results

As discussed in Lemma 4.3 (Chapter 4, Section 4.1.3), time-synchronized control can also be applied to a disturbed spacecraft system. Considering disturbed spacecraft attitude dynamics introduced in (4.35), experimental studies are conducted on a hardware-in-loop experimental platform, to further validate the practical applicability of the time-synchronized controller (4.22). The main consists of this hardware-in-loop experiment platform and experimental procedures are shown in Fig. 8.14 with the following components: 1) a high-performance real-time simulation computer, which runs the dynamics model on the VxWorks system; 2) a three-axis turntable for rotational motion simulation; 3) an angular velocity measurement unit comprising four fibre-optic gyroscopes (FOGs); 4) a reaction wheel assembly (RWA), which

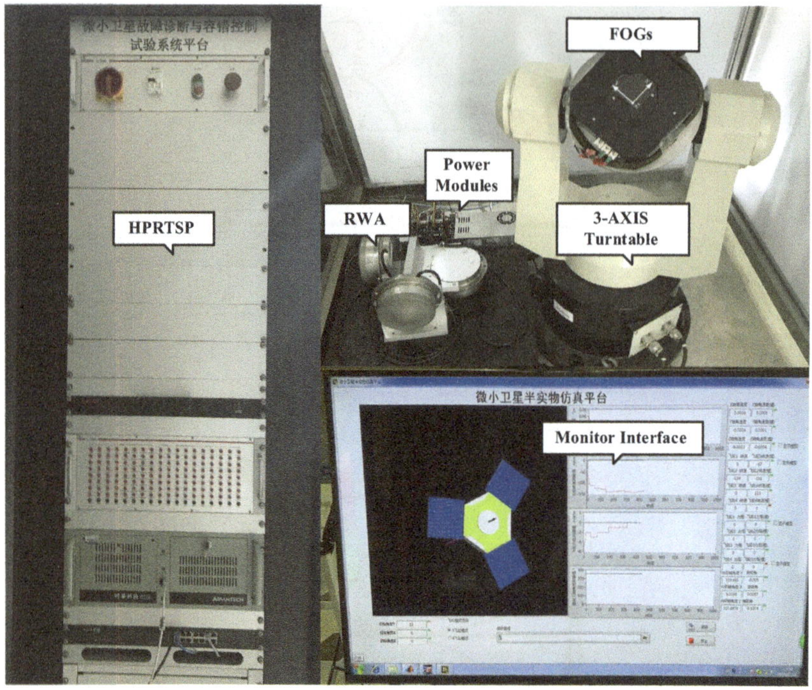

Fig. 8.14 Hardware-in-loop experiment platform

executes the control commands from the high-performance real-time simulation computer and feedbacks its output to itself; and 5) other hardware and software including power modules and monitor interface. The proposed controller of the spacecraft attitude reorientation can be tested and illustrated by this hardware-in-loop platform.

The experimental procedures are arranged as shown in Fig. 8.15. First, the attitude angle can be obtained from the turntable, and the angular velocity is measured by the four fibre-optic gyroscopes. Second, the motion information is sent to the high-performance real-time simulation computer as state feedback, prior to this, the attitude angle is converted into the form of the modified Rodrigues parameters (MRPs) to meet the requirement of the controller. Third, the control signal is calculated from the controller and implemented on the reaction wheel assembly. Note that, we add a saturation function to restrict the control signal within the maximum output torque of each reaction wheel (0.1 Nm). Besides, the output torque of reaction wheels can be measured by its inner sensor, then the misalignment simulator receives these signals and calculates the real torque applied to the spacecraft. Fourth, the attitude dynamics run in the high-performance real-time simulation computer, which received the torque and sent the attitude angle to the turntable for motion simulation.

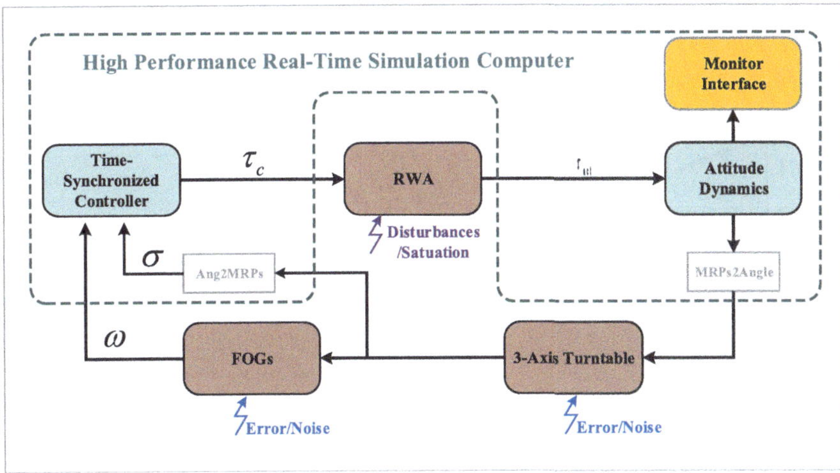

Fig. 8.15 Block diagram of hardware-in-loop experiment system

In addition, the monitor interface is used to display some key states and parameters during the experiment process.

Consider the inertia matrix of rigid spacecraft as

$$J = [12, 0.4, 0.2; 0.4, 10, 0.6; 0.2, 0.6, 11] \text{ kg} \cdot \text{m}^2. \tag{8.75}$$

Its initial attitude is $\sigma(0) = [0.1, 0.05, 0.01]^T$ with $\omega(0) = [0, 0, 0]^T$ rad/s. The simulated disturbances are assumed to occur at the beginning of simulation, and its form is defined as

$$\Delta = \begin{bmatrix} 0.0012 \\ 0.003 \sin(0.2\pi t) \\ 0.002 \cos(0.5\pi t) \end{bmatrix} \text{Nm} \tag{8.76}$$

The parameters of controller and observer are set as: $p_1 = 0.8$, $p_2 = 0.6$, $\gamma_2 = 0.01$, $\gamma_1 = 0.1$, $k_1 = 0.8$, $k_2 = 0.6$, $k_3 = 0.1$, $k_4 = 0.01$, $\varepsilon = 10^{-6}$.

From the experimental results shown in Figs. 8.16 and 8.17, we can see that all the attitude state of spacecraft synchronously stabilizes to the desired point within 50s, and the attitude accuracy arrives at the level of 10^{-4}. Although the attitude accuracy in the experiment is lower than that in simulation, due to the inevitable measurement noise and actuators' characteristics, this level also satisfies requirements in the practical applications. These experimental validation results show the application prospects in realistic missions.

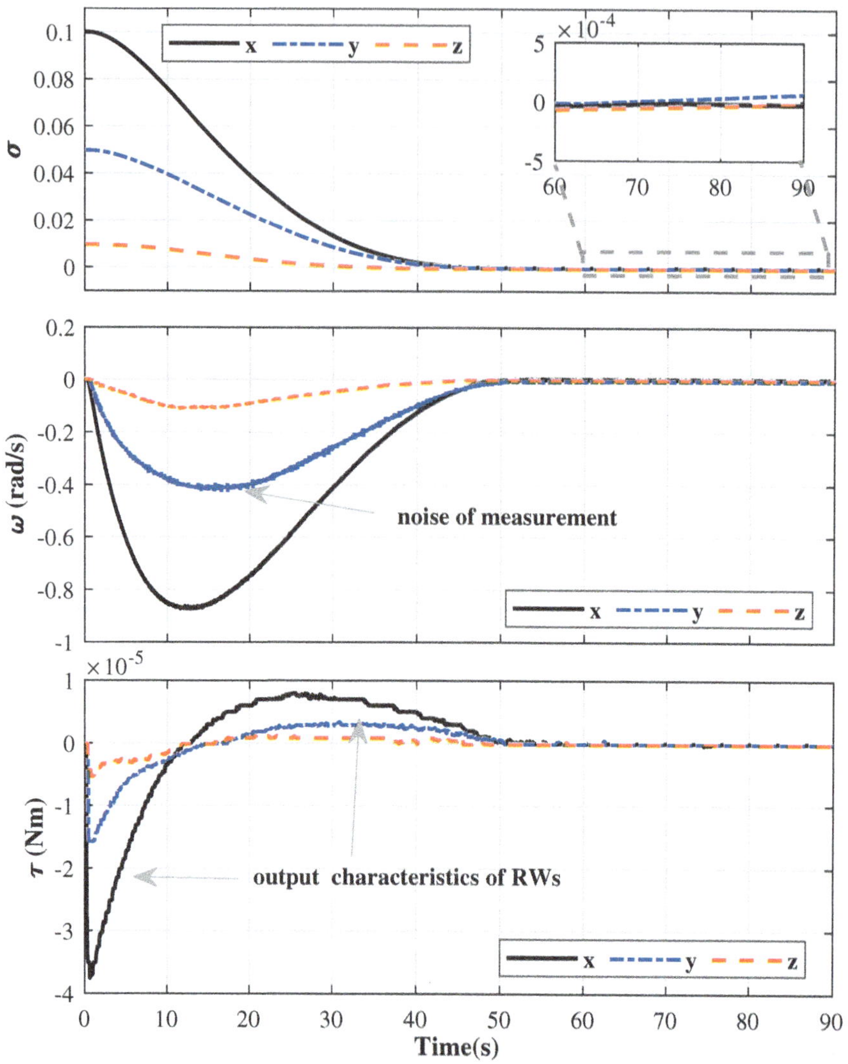

Fig. 8.16 Experimental results of the proposed method

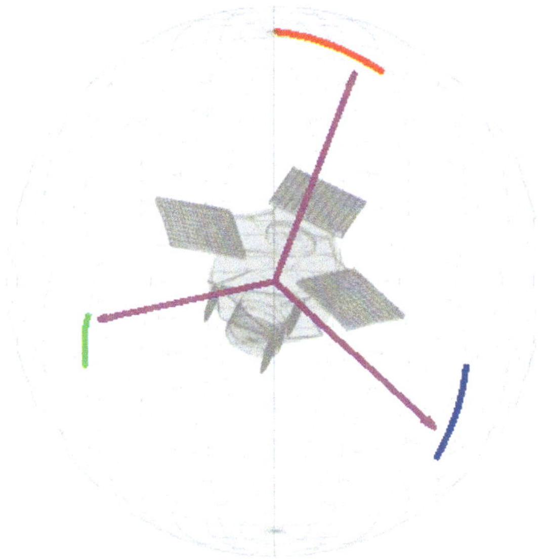

Fig. 8.17 Rotational trajectory of experimental results

8.4 Conclusion

This chapter focuses on the relative attitude and orbit control of the on-orbit service spacecraft during the close relative motion phase of rendezvous and docking. During the relative movement at close range, especially the docking phase, accurate and synchronized control performance are specifically required, otherwise possible harsh transient process may result in mission failure.

This chapter has designed several synchronized controllers for the 6-DOF spacecraft, which will provide a theoretical basis for more stringent scenarios that require synchronized convergence. In addition, we have showcased through numerical simulations and hardware-in-loop experiments that time-synchronized spacecraft control has the following advantages: (i) trajectories are optimal in terms of the travel length; (ii) possible unknown overshoot is overcome; (iii) precise control of relative motion is achieved; and (iv) energy consumption is significantly reduced.

References

1. J. L. Junkins and H. Schaub, *Analytical mechanics of space systems*. American Institute of Aeronautics and Astronautics, 2009.
2. P. Singla, K. Subbarao, and J. L. Junkins, "Adaptive output feedback control for spacecraft rendezvous and docking under measurement uncertainty," *Journal of guidance, control, and dynamics*, vol. 29, no. 4, pp. 892–902, 2006.

3. P. C. Hughes, *Spacecraft attitude dynamics*. Courier Corporation, 2012.
4. B. Costic, D. Dawson, M. De Queiroz, and V. Kapila, "Quaternion-based adaptive attitude tracking controller without velocity measurements," *Journal of Guidance, Control, and Dynamics*, vol. 24, no. 6, pp. 1214–1222, 2001.
5. H. Liu and T. Zhang, "Adaptive neural network finite-time control for uncertain robotic manipulators," *Journal of Intelligent & Robotic Systems*, vol. 75, no. 3, pp. 363–377, 2014.
6. B. Jiang, C. Li, and G. Ma, "Finite-time output feedback attitude control for spacecraft using "adding a power integrator" technique," *Aerospace Science and Technology*, vol. 66, pp. 342–354, 2017.

Chapter 9
Practical Time-Synchronized Attitude Control for Disturbed Spacecraft

Abstract In this chapter, we consider practical time-synchronized attitude control of the on-orbit service spacecraft under disturbances and input saturation, where transient performance (e.g., settling time, overshoot, trajectory), steady-state error, and fuel consumption are likewise evaluated herein. This chapter expands and enriches previous ratio persistence, defines a modified version of *ratio restriction* (slightly different from Definition 2.12 introduced in Chap. 2) and an extended time-synchronized stability concept—*practical time-synchronized stability*. These new techniques enable a control system to inherit most of the merits from standard ratio persistence and time-synchronized stability, but allow the closed-loop error to be within a small neighborhood of the origin rather than to be exactly at the equilibrium point. Practical time-synchronized controllers with command filter and low-pass filter for the spacecraft are designed, and the input saturation is considered. The effectiveness and advantages of the controllers proposed are verified by strict theoretical proof and sufficient comparative simulation. From a number of fair comparisons, the energy consumption is about 25–35% less than that of the state of the art controllers, e.g., finite-time control. This further approves the declared merit of saving energy consumption of (practical) time-synchronized control.

9.1 Discussion on the Direction of the Control Input

Consider a system

$$\dot{x} = u(x), \ x \in \mathbb{R}^n, \tag{9.1}$$

where $u(x) \in \mathbb{R}^n$ denotes the control input and $x \in \mathbb{R}^n$ the system state. From Fig. 9.1, we know if $u(x)$ is always in the opposite direction of the system state x, all the energy from the control input will be used to drive x back to the origin along the red straight arrow. Suppose there is a non-zero angle between $u(x)$ and $-x$ (the blue arrow). In this case, there are non-zero control components in the directions orthogonal to the negative direction of the state (the grey arrows). These control components (the grey arrows) have no positive effect on stabilizing x, which can be regarded as the ineffective control input under certain circumstances. In the

D. Li et al., *Time-Synchronized Control: Analysis and Design*,
https://doi.org/10.1007/978-981-16-3089-7_9

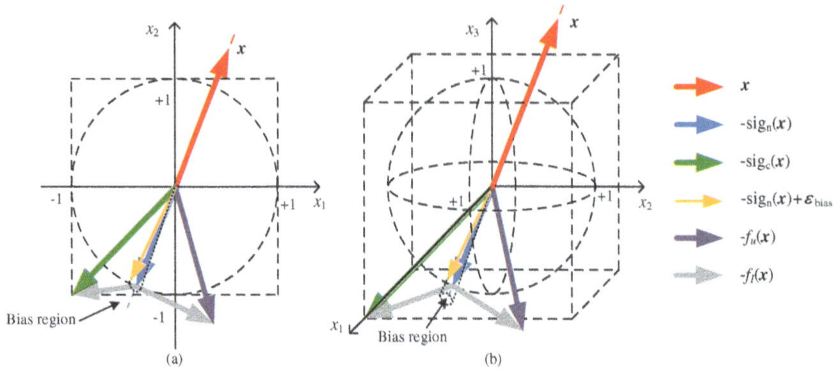

Fig. 9.1 The relationship between the system states and the control inputs, where x is the system state, $\mathrm{sig}_{\mathrm{n}}(\cdot)$ represents norm-normalized sig function, $\mathrm{sig}_{\mathrm{c}}(\cdot)$ indicates traditional sig function, $\varepsilon_{\mathrm{bias}}$ denotes any small bias, $f_u(x)$ can be any controller, $f_I(x)$ implies ineffective control input. **a** For the system with $x \in \mathbb{R}^2$, and **b** for the system with $x \in \mathbb{R}^3$

stabilization process, if the non-zero ineffective control input always exists, not only the closed-loop system trajectory (that is, the transient process) suffers from possible overshoot or jitter, but also energy consumption increases and valuable fuel is wasted (especially for spacecraft operations). The property that $u(x)$ is always in the negative/opposite direction of the system state x is called to have a negative direction of ratio persistence as in Definition 2.35. However, most controllers utilizing the classical sign function may yield a control input in the direction largely deviated from the negative direction of ratio persistence, e.g., the green or purple arrows.

For practical control system applications, steady errors are inevitable in most situations, which cannot be ideally zero as we demand. Thus there is a need to relax further and expand ratio persistence and time-synchronized stability to enable: (1) the closed-loop errors within a small neighborhood of the origin; and (2) the direction of the control input with a small bias of the negative direction of the system state x. As in Fig. 9.1, the yellow line represents the control input in the direction with a small bias of the negative direction of the system state.

Specifically, as shown in Fig. 9.1, the yellow line represents the equivalent control input in the direction with a small bias of the negative direction of the system state. The boundary of the small bias will form a region defined as *bias region* as shown in Fig. 9.1. When the bias is very small, the ineffective control component tends to be 0, and the system could be almost time-synchronized stable, which is called practical time-synchronized stable in this chapter.

9.2 Problem Formulations

Considering the kinematic description of the on-orbit service spacecraft attitude motion, in this section, we first give the corresponding coordinate system definition. As introduced in Chap. 8, the framework we adopt includes the earth-centered inertial (ECI) frame $\mathfrak{F}_i$ [6], the body-fixed Cartesian frame $\mathfrak{F}_s$ of the service spacecraft, and the body-fixed Cartesian frame $\mathfrak{F}_t$ of the target spacecraft. The attitude movement between the two spacecraft can be described by the rotation of their body-fixed reference frames relative to the $\mathfrak{F}_i$. The frames $\mathfrak{F}_t$ and $\mathfrak{F}_s$ are located at centroid on the two spacecraft, respectively. Moreover, the relative attitude motion between them can be expressed by the relative rotation of $\mathfrak{F}_s$ and $\mathfrak{F}_t$. Matrices $\mathbf{R}_{ts}$representing rotation coordinate transformation between $\mathfrak{F}_s$ and $\mathfrak{F}_t$, and $\mathbf{R}_{si}$, $\mathbf{R}_{ti}$ are defined in the same way.

Considering the relative attitude movement of the rigid spacecraft, we can use the unit quaternion $\mathbf{q}_a = [q_0 \; \mathbf{q}_v]^T \in \mathbb{R}^4$ to describe the rotation of the body-fixed frame $\mathfrak{F}_s$ with respect to the inertial frame $\mathfrak{F}_i$, which is subject to the constraint [5]

$$\mathbf{q}_v^T \mathbf{q}_v + q_0^2 = 1 \tag{9.2}$$

Then the dynamics and kinematics models of the spacecraft are governed as follows [1]

$$\mathbf{J}_s \dot{\boldsymbol{\omega}}_a = -\boldsymbol{\omega}_a^{\times} \mathbf{J}_s \boldsymbol{\omega}_a + \boldsymbol{\tau}_a + \mathbf{d} \tag{9.3}$$

$$\dot{\mathbf{q}}_v = \frac{1}{2} \left(\mathbf{q}_v^{\times} \boldsymbol{\omega}_a + q_0 \boldsymbol{\omega}_a \right) \tag{9.4}$$

$$\dot{q}_0 = -\frac{1}{2} \mathbf{q}_v^T \boldsymbol{\omega}_a \tag{9.5}$$

where $\mathbf{J}_s \in \mathbb{R}^{3\times3}$ is the constant, positive-definite, symmetric inertia matrix of the on-orbit service spacecraft, $\boldsymbol{\omega}_a \in \mathbb{R}^3$ is the angular velocity of the body-fixed frame $\mathfrak{F}_s$ with respect to the inertial reference frame $\mathfrak{F}_i$ expressed in $\mathfrak{F}_s$, $\boldsymbol{\tau}_a \in \mathbb{R}^3$ is a vector of control torques in $\mathfrak{F}_s$, and the notation $\mathbf{a}^{\times}$, $\forall \mathbf{a} = [a_1 \; a_2 \; a_3]^T$, denotes the following skew-symmetric matrix:

$$\mathbf{a}^{\times} \triangleq \begin{bmatrix} 0 & -a_3 & a_2 \\ a_3 & 0 & -a_1 \\ -a_2 & a_1 & 0 \end{bmatrix} \tag{9.6}$$

The rotation matrix that brings $\mathfrak{F}_i$ onto $\mathfrak{F}_s$, denoted by $\mathbf{R}_a(\mathbf{q}_v, q_0) \in SO(3)$, is defined as follows:

$$\mathbf{R}_a = (q_0^2 - \mathbf{q}_v^T \mathbf{q}_v) I_3 + 2\mathbf{q}_v \mathbf{q}_v^T - 2q_0 \mathbf{q}_v^{\times} \tag{9.7}$$

where I_3 is the 3×3 identity matrix, and the angular velocity ω_a can be computed as follows:

$$\boldsymbol{\omega}_a = 2(q_0\dot{\mathbf{q}}_v - \mathbf{q}_v\dot{q}_0) - 2\mathbf{q}_v^\times\dot{\mathbf{q}}_v \tag{9.8}$$

In order to obtain the attitude error model, we assume that the desired attitude of the on-orbit service spacecraft can be described by the desired unit quaternion $\mathbf{q}_{ad}(t) = [q_{d0}, \mathbf{q}_{dv}]^T \in \mathbb{R}^4$ and define the desired angular velocity $\boldsymbol{\omega}_d \in \mathbb{R}^3$ through the following dynamic equations:

$$\dot{\mathbf{q}}_{dv} = \frac{1}{2}\left(\mathbf{q}_{dv}^\times\boldsymbol{\omega}_d + q_{d0}\boldsymbol{\omega}_d\right) \tag{9.9}$$

$$\dot{q}_{d0} = -\frac{1}{2}\mathbf{q}_{dv}^T\boldsymbol{\omega}_d \tag{9.10}$$

$$\boldsymbol{\omega}_d = 2(q_{d0}\dot{\mathbf{q}}_{dv} - \mathbf{q}_{dv}\dot{q}_{d0}) - 2\mathbf{q}_{dv}^\times\dot{\mathbf{q}}_{dv} \tag{9.11}$$

The corresponding rotation matrix $\mathbf{R}_d(\mathbf{q}_{dv}, q_{d0}) \in SO(3)$ is defined as follows:

$$\mathbf{R}_d = (q_{d0}^2 - \mathbf{q}_{dv}^T\mathbf{q}_{dv})I_3 + 2\mathbf{q}_{dv}\mathbf{q}_{dv}^T - 2q_{d0}\mathbf{q}_{dv}^\times \tag{9.12}$$

To quantify the error between the actual and desired spacecraft attitudes, we define the rotation matrix $\mathbf{R}_e(\mathbf{q}_{ev}, q_{e0}) \in SO(3)$ and the quaternion tracking error $\mathbf{q}_e(t) = [q_{e0}, \mathbf{q}_{ev}]^T \in \mathbb{R}^4$ as follows:

$$\mathbf{R}_e = \mathbf{R}_a\mathbf{R}_d^T = (q_{e0}^2 - \mathbf{q}_{ev}^T\mathbf{q}_{ev})I_3 + 2\mathbf{q}_{ev}\mathbf{q}_{ev}^T - 2q_{e0}\mathbf{q}_{ev}^\times \tag{9.13}$$

$$\mathbf{q}_{ev} = q_{d0}\mathbf{q}_v - q_0\mathbf{q}_{dv} + \mathbf{q}_v^\times\mathbf{q}_{dv} \tag{9.14}$$

$$q_{e0} = q_0 q_{d0} + \mathbf{q}_v^T\mathbf{q}_{dv} \tag{9.15}$$

Based on the preceding formulation, the tracking error of the angular velocity $\boldsymbol{\omega}_e \in \mathbb{R}^3$ is defined as follows:

$$\boldsymbol{\omega}_e = \boldsymbol{\omega}_a - \mathbf{R}_e\boldsymbol{\omega}_d \tag{9.16}$$

Computing (9.3)–(9.16), we obtain the open-loop tracking error dynamics as follows

$$\begin{aligned}\mathbf{J}_s\dot{\boldsymbol{\omega}}_e &= -(\boldsymbol{\omega}_e + \mathbf{R}_e\boldsymbol{\omega}_d)^\times\mathbf{J}_s(\boldsymbol{\omega}_e + \mathbf{R}_e\boldsymbol{\omega}_d)\\ &+ \mathbf{J}_s(\boldsymbol{\omega}_e^\times\mathbf{R}_e\boldsymbol{\omega}_d - \mathbf{R}_e\dot{\boldsymbol{\omega}}_d) + \boldsymbol{\tau}_a + \mathbf{d}\end{aligned} \tag{9.17}$$

$$\dot{\mathbf{q}}_{ev} = \frac{1}{2}(\mathbf{q}_{ev}^\times + q_{e0}I_3)\boldsymbol{\omega}_e \tag{9.18}$$

$$\dot{q}_{e0} = -\frac{1}{2}\mathbf{q}_{ev}^T\boldsymbol{\omega}_e \tag{9.19}$$

where $\dot{\mathbf{R}}_e = -\boldsymbol{\omega}_e{}^{\times}\mathbf{R}_e$.

To express the open-loop tracking error dynamics, we rewrite (9.18) as follows:

$$\dot{\mathbf{q}}_{ev} = \frac{1}{2}\mathbf{T}\boldsymbol{\omega}_e \tag{9.20}$$

where $\mathbf{T} = \mathbf{q}_{ev}^{\times} + q_{e0}I_3$ is a Jacobian-type matrix.

After taking the time derivative of (9.20) and pre multiplying both sides of the resulting expression by $\mathbf{T}^{-T}\mathbf{J}_s\mathbf{T}^{-1}$, we obtain,

$$\mathbf{M}_a\ddot{\mathbf{q}}_{ev} = \frac{1}{2}\mathbf{M}_a\dot{\mathbf{T}}\boldsymbol{\omega}_e + \frac{1}{2}\mathbf{P}^T\mathbf{J}_s\dot{\boldsymbol{\omega}}_e \tag{9.21}$$

where $\mathbf{M}_a = \mathbf{P}^T\mathbf{J}_s\mathbf{P}$ and $\mathbf{P} = \mathbf{T}^{-1}$.

After substituting (9.17) into the right-hand side of (9.21), we can obtain the following expression for the open-loop tracking error dynamics of the relative attitude motion

$$\mathbf{M}_a\ddot{\mathbf{q}}_{ev} + \mathbf{C}_a\dot{\mathbf{q}}_{ev} + \mathbf{G}_a = \mathbf{Q}_a\boldsymbol{\tau}_a + \mathbf{Q}_a\mathbf{d} \tag{9.22}$$

where

$$\mathbf{Q}_a = \frac{1}{2}\mathbf{P}^T \tag{9.23}$$

$$\mathbf{C}_a = -\mathbf{M}_a\dot{\mathbf{P}}^{-1}\mathbf{P} - 2\mathbf{P}^T(\mathbf{J}_s\mathbf{P}\dot{\mathbf{q}}_{ev})^{\times}\mathbf{P} \tag{9.24}$$

$$\begin{aligned} \mathbf{G}_a =& \mathbf{P}^T[(\mathbf{P}\dot{\mathbf{q}}_{ev})^{\times}\mathbf{J}_s\mathbf{R}_e\boldsymbol{\omega}_d] + \mathbf{P}^T[(\mathbf{R}_e\boldsymbol{\omega}_d)^{\times}\mathbf{J}_s\mathbf{P}\dot{\mathbf{q}}_{ev}] \\ &+ \frac{1}{2}\mathbf{P}^T[(\mathbf{R}_e\boldsymbol{\omega}_d)^{\times}\mathbf{J}_s\mathbf{R}_e\boldsymbol{\omega}_d] \\ &- \frac{1}{2}\mathbf{P}^T\mathbf{J}_s[(2\mathbf{P}\dot{\mathbf{q}}_{ev})^{\times}\mathbf{R}_e\boldsymbol{\omega}_d - \mathbf{R}_e\dot{\boldsymbol{\omega}}_d] \end{aligned} \tag{9.25}$$

It is worth noting that $\mathbf{Q}_a$ is reversible in the above formula. Considering the input saturation, the above model has an additional condition, that is, $\tau_a = \delta(\tau)$ and $||\delta(\tau)|| \le \delta_{\max}$, where τ is the command input, $\delta(\tau)$ is the saturation function, and $\delta_{\max}$ is the positive definite constant.

After obtaining the model we need, we give the control objectives as follows. From the perspective of a second-order nonlinear system $\dot{x}_1 = x_2,\ \dot{x}_2 = f(x) + g(x)\tau + d$, where $x_1, x_2 \in \mathbb{R}^n$ are the states, $f(x) \in \mathbb{R}^n$ and $g(x) \in \mathbb{R}^{n\times n}$ is continuously differentiable functions, d represents disturbances, and $\tau \in \mathbb{R}^n$ denotes the control input.

We summarize the control objectives. For the relative attitude tracking control between the on-orbit service spacecraft and the target spacecraft, given an initial system state (9.22), and the desired state $\mathbf{q}_d \in \mathbb{R}^n$, design controllers, such that the following objectives are achieved:

(i) tracking errors converge time-synchronously within finite time;
(ii) the input saturation is handled so that the rate, amplitude, and bandwidth of the control command can be actively adjusted.
(iii) the output trajectory of the closed-loop system is shortened as much as possible considering the disturbances, while the energy consumption is reduced.

Since the command filter has the ability to limit the control command amplitude, it can be used to actively limit the amplitude of the control command to make it smaller than the saturation limit of the system input to achieve the objectives of overcoming input saturation.

9.3 Practical Time-Synchronized Stability and Command Filter

9.3.1 Ratio Persistence and Ratio Restriction

We first introduce the concept of ratio restriction, which is slightly different from Definition 2.12 introduced in Chap. 2).

Definition 9.1 (*Ratio Restriction*) The closed-loop state $x = [x_1, ..., x_i, ..., x_j, ..., x_n]^T$ of the system $\dot{x} = f(x)$, is *ratio restricted*, if for any $x_i, x_j \neq 0$, the *ratio bias* function of each pair of the state elements $R_{ij}(x_i, x_j) = |x_i| - |\kappa_{ij}||x_j|$ with the fixed non-zero *ratio constant* κ_{ij}, is constrained as $|R_{ij}(x_i, x_j)| \leq \varepsilon_{Rc}$, where ε_{Rc} can be any positive small constant that is independent of x.

Theorem 9.1 *The closed-loop state x of the system $\dot{z} = f(z)$, $z = x + \varepsilon(t)$ is ratio restricted, if for $x \neq 0$,*
(i) $f(z)$ and $\varepsilon(t)$ are bounded;
and (ii) z is bounded and ratio persistent.

Proof Considering that $\dot{z} = f(z)$ satisfies ratio persistence, $\varepsilon = [\varepsilon_1, ..., \varepsilon_i, ..., \varepsilon_j, ..., \varepsilon_n]^T$, $\forall i, j$, $z_i(t) \neq 0$, $z_j(t) \neq 0$, we have

$$z_i(t)/z_j(t) = z_i(0)/z_j(0) = c_{ij} \tag{9.26}$$

where c_{ij} is the non-zero ratio constant of $z_i(t)$ and $z_j(t)$. Then, we have

$$\frac{z_i}{z_j} = \frac{x_i + \varepsilon_i}{x_j + \varepsilon_j} = c_{ij} \tag{9.27}$$

Convert the above formula to

$$x_j = c_{ij}^{-1} x_i + c_{ij}^{-1} \varepsilon_i - \varepsilon_j \tag{9.28}$$

Then, we have

$$\begin{aligned} |x_j| &= |c_{ij}^{-1}x_i + c_{ij}^{-1}\varepsilon_i - \varepsilon_j| \\ &\le |c_{ij}^{-1}||x_i| + |c_{ij}^{-1}\varepsilon_i - \varepsilon_j| \end{aligned} \tag{9.29}$$

and

$$|x_i| - |c_{ij}||x_j| \ge -|c_{ij}||c_{ij}^{-1}\varepsilon_i - \varepsilon_j| \ge -\varepsilon_{Rc1} \tag{9.30}$$

where ε_{Rc1} is a positive constant and represents the maximum value of $|c_{ij}||c_{ij}^{-1}\varepsilon_i - \varepsilon_j|$ due to boundedness of ε_i and ε_j.

On the other hand, considering

$$\frac{|z_i|}{|z_j|} = \frac{|x_i + \varepsilon_i|}{|x_j + \varepsilon_j|} = |c_{ij}| \tag{9.31}$$

we have

$$\begin{aligned} |x_i| - |\varepsilon_i| &\le |x_i + \varepsilon_i| = |c_{ij}||x_j + \varepsilon_j| \\ &\le |c_{ij}||x_j| + |c_{ij}||\varepsilon_j| \end{aligned} \tag{9.32}$$

Then,

$$|x_i| - |c_{ij}||x_j| \le |\varepsilon_i| + |c_{ij}||\varepsilon_j| \le \varepsilon_{Rc2} \tag{9.33}$$

where ε_{Rc2} is a positive constant and represents the maximum value of $|\varepsilon_i| + |c_{ij}||\varepsilon_j|$ due to boundedness of ε_i and ε_j. Therefore, we have $||x_i| - |c_{ij}||x_j|| \le \varepsilon_{Rc}$, where $\varepsilon_{Rc} = \max\{\varepsilon_{Rc1}, \varepsilon_{Rc2}\}$. Thus, x is ratio restricted.

The closed-loop system satisfies ratio restriction, which means that:
(i) the direction of the control input is always in the small neighborhood of the state negative direction;
(ii) all state elements converge nearly or practically synchronously;
(iii) the trajectory of the system tends to a straight line from the initial point to the equilibrium point.
The above analysis shows that ratio restriction requires that the ratio between each state element is fixed with a bounded small bias. When the boundary value of the ratio bias function is small, the system performance is similar to that under ratio persistence, which refers that the system trajectory will tend to be a straight line. When small bias $\varepsilon(t) \to 0$, ratio restriction degrades to ratio persistence.

Remark 9.1 R_{ij} and κ_{ij} are not related to $x(t)$ but related to the initial state. The ratio bias function means that during the entire process from the initial state to the equilibrium point, the ratio between any pair of non-zero elements is bounded and fixed with a small bias. Therefore, the trajectory of the closed-loop system that satisfies ratio restriction will tend to be a straight line.

Remark 9.2 The physical meaning of ratio restriction needs to be clarified. If the closed-loop system satisfies ratio persistence, then every element of the closed-loop system state converges to the origin synchronously. However, there are inevitable measurement noise and disturbances in the practical system, and the practical closed-loop system cannot ideally satisfy ratio persistence. In this chapter, we try to make the ratio fluctuates within a small range by making the control system ratio restricted. This fluctuation range is independent of the system state. Therefore, the behavior of the systems satisfying ratio restriction is similar to ones satisfying ratio persistence.

9.3.2 Practical Time-Synchronized Stability

In this subsection, based on ratio restriction, we will show that for the system $\dot{z} = f(z)$, $z = x + \varepsilon(t)$, if z is time-synchronized stable, then x will be practical time-synchronized stable.

Lemma 9.1 (Finite-Time Uniformly Ultimately Boundedness [4]) *Consider system $\dot{x} = f(x)$, $f(0) = 0$, $x \in \mathbb{R}^n$, where $f(\cdot) : \mathbb{R}^n \to \mathbb{R}^n$ is a continuous function. Suppose there is a Lyapunov function $V(x) : \mathbb{U} \to \mathbb{R}^+$ with the initial value $V(x_0)$, positive constants $p_1 \in (0, 1)$, $p_2 < p_1$, $c_1 > 0$, $c_2 > 0$, $\varepsilon > 0$. If $\dot{V}(x) \leq -c_1 V^{p_1}(x) - c_2 V(x) + \varepsilon V^{p_2}(x)$, then the trajectory of the system is finite-time uniformly ultimately bounded. The states x can converge to compact set Ω(a region of equilibrium point) within finite time T as follows:*

$$\Omega = \{x : c_3 V^{p_1 - p_2}(x) + c_4 V^{1-p_2}(x) < \varepsilon\} \tag{9.34}$$

$$T \leq \frac{1}{(c_2 - c_4)(1 - p_1)} \ln\left(1 + \frac{c_2 - c_4}{c_1 - c_3} V^{1-p_1}(x_0)\right) \tag{9.35}$$

where c_3 and c_4 are arbitrary positive constants satisfying $c_3 \in (0, c_1)$ and $c_4 \in (0, c_2)$.

Definition 9.2 (*Practical Time-Synchronized Stability*) Considering the system $\dot{x} = f(x)$, $x \in \mathbb{R}^n$, for an open neighborhood $N_0 \subseteq D_0$ of the origin, the origin of a Lyapunov function $V(x)$ is *practical time-synchronized stable* with the *practical synchronized settling time* $T_x : N_0 \backslash \{0\} \to (0, \infty)$ and the neighborhood of the origin $\Omega_x = \{x(t) : ||x(t)|| \leq \varepsilon_x, \forall t \geq T_x\}$ if:

(i) it is finite-time uniformly ultimately boundedness within the settling time T_x and the neighborhood of the origin Ω_x;
(ii) the state x is ratio restricted.
If the equilibrium of the system is practical time-synchronized stable with $N_0 = D_0 = \mathbb{R}^n$, it is globally practical time synchronized stable.

Theorem 9.2 *Considering the non-zero state $x(t)$ of the system $\dot{z} = f(z)$, $z = x + \varepsilon(t)$, where $\varepsilon(t)$ and $\dot{\varepsilon}(t)$ are bounded, the state $x(t)$ is ratio restricted and the origin of a Lyapunov function $V(x)$ is practical time-synchronized stable with the settling time T_x, if the Lyapunov function $V(z)$ is time-synchronized stable.*

Proof We select the candidate Lyapunov functions for x and z as $V_x = \frac{1}{2}x^T x$, $V_z = \frac{1}{2}z^T z$, respectively. Considering $x = z - \varepsilon$, we have

$$\begin{aligned} V_x &= \frac{1}{2}x^T x = \frac{1}{2}z^T z - z^T \varepsilon + \frac{1}{2}\varepsilon^T \varepsilon \\ &\leq \frac{1}{2}z^T z + \frac{1}{2}\varepsilon^T \varepsilon + \|z^T \varepsilon\| \end{aligned} \tag{9.36}$$

Note that for $a > 0$, $b > 0$ and $0 < p < 1$, the inequality $(a+b)^p \leq a^p + b^p$ holds. Thus, introducing the power $p \in (0, 1)$ and we have

$$\begin{aligned} V_x^p &\leq \left(\frac{1}{2}z^T z + \frac{1}{2}\varepsilon^T \varepsilon + \|z^T \varepsilon\|\right)^p \\ &\leq V_z^p + \left(\frac{1}{2}\varepsilon^T \varepsilon\right)^p + \|z^T \varepsilon\|^p \end{aligned} \tag{9.37}$$

Find the derivative of V_x with respect to time, we have

$$\dot{V}_x = x^T \dot{x} = z^T \dot{z} - \dot{z}^T \varepsilon - z^T \dot{\varepsilon} + \varepsilon^T \dot{\varepsilon} \tag{9.38}$$

Combining the above equations with constants $c_1 > 0$, $c_2 > 0$ we have

$$\dot{V}_x + c_1 V_x^p + c_2 V_x \leq \dot{V}_z + c_1 V_z^p + \varepsilon_z \tag{9.39}$$

where

$$\begin{aligned} \varepsilon_z = &+ c_2 V_z - \dot{z}^T \varepsilon - z^T \dot{\varepsilon} + \varepsilon^T \dot{\varepsilon} + c_1 \left(\frac{1}{2}\varepsilon^T \varepsilon\right)^p \\ &+ c_1 \|z^T \varepsilon\|^p + c_2 \frac{1}{2}\varepsilon^T \varepsilon + c_2 \|z^T \varepsilon\| \end{aligned} \tag{9.40}$$

Considering that z is time-synchronized stable, we have $\dot{V}_z + c_1 V_z^p \leq 0$ with synchronized settling time

$$T_z(z_0) \leq \frac{V_z^{1-p}(z_0)}{c_1(1-p)}, \tag{9.41}$$

and $z = \dot{z} = 0$ for $\forall t \geq T_z(z_0)$.

Considering $\|\varepsilon\| \leq \varepsilon_{\max}$ and $\|\dot{\varepsilon}\| \leq \varepsilon_{d\max}$, for $\forall t \geq T_z(z_0)$, we have

$$\dot{V}_x + c_1 V_x^p + c_2 V_x \leq \varepsilon_{z\max} \tag{9.42}$$

and

$$\begin{aligned} \varepsilon_z &= +\varepsilon^T \dot{\varepsilon} + c_1 \left(\frac{1}{2} \varepsilon^T \varepsilon \right)^p + c_2 \frac{1}{2} \varepsilon^T \varepsilon \\ &\leq \varepsilon_{z\max} = \varepsilon_{\max} \varepsilon_{d\max} + 2^{-p} c_1 \varepsilon_{\max}^{2p} + \frac{1}{2} c_2 \varepsilon_{\max}^2 \end{aligned} \tag{9.43}$$

According to the Lemma 9.1, the states x can converge to compact set Ω_x with the maximum radius $\varepsilon_{\Omega x \max}$ within finite time T_x as follows:

$$\Omega_x = \{x : c_3 V_x^p + c_4 V_x < \varepsilon_{z\max}\} \tag{9.44}$$

$$T_x \leq \frac{1}{(c_2 - c_4)(1 - p)} \ln \left(1 + \frac{c_2 - c_4}{c_1 - c_3} V_x^{1-p}(x_0) \right). \tag{9.45}$$

where c_3 and c_4 are arbitrary positive constants satisfying $c_3 \in (0, c_1)$ and $c_4 \in (0, c_2)$.

The above process proves that x is finite-time uniformly ultimately bounded. Considering that z satisfies ratio persistence, according to Theorem 9.1, x is ratio restricted. Thus, V_x is practical time-synchronized stable.

Consider any pair of non-zero elements x_i and x_j of the state. Then, we have $x_j = c_{ij}^{-1} x_i + d_\varepsilon$, $d_\varepsilon = c_{ij}^{-1} \varepsilon_i - \varepsilon_j$ as in (9.28). d_ε is not related to the state. Consider any time interval $[t_k, t_{k+1}]$ in $[0, T_x]$. We define t_k as the time when x_i converges to a certain set, and t_{k+1} is the time when x_j converges to this set. If there is finite time t_k so that $\lim_{t \to t_k} |x_i| \leq \varepsilon_{xik}$, then there must be $\lim_{t \to t_k} |x_j| \leq |c_{ij}^{-1}| \varepsilon_{xik} + |d_\varepsilon(t_k)|$. This means that x_j will converge to a neighborhood related to the neighborhood where x_i converges in t_k. When $\lim_{t \to t_{k+1}} |x_i| \leq \varepsilon_{xik+1}$, we have $\lim_{t \to t_{k+1}} |x_j| \leq |c_{ij}^{-1}| \varepsilon_{xik+1} + |d_\varepsilon(t_{k+1})|$. Considering the definition of t_{k+1}, we have $\varepsilon_{xik+1} = |c_{ij}| \varepsilon_{xik} - |c_{ij}| |d_\varepsilon(t_{k+1})|$. We can see that for any time interval, the neighborhoods where x_i and x_j converge are in fixed ratios with a small range of bias. Therefore, in the transient process, the convergence of x_i and x_j is synchronized with small bias $d_\varepsilon(t)$. If the bias $d_\varepsilon(t) = 0$, the system converges synchronously. When the bias $d_\varepsilon(t)$ tends to 0, the system tends to converge synchronously, that is, the practical synchronized stability.

For steady state, when $t_{k+1} = T_x$, x_j converges to set Ω_x before or after x_i converges to set Ω_x at t_k. If $d_\varepsilon(t) = 0$, then x_i and x_j enter the set Ω_x synchronously. When $d_\varepsilon(t) \neq 0$, within a short time interval $[t_k, t_{k+1}]$, x_i and x_j enter set Ω_x, which is practical time-synchronized convergence of x_i and x_j with small biases of the steady-state error and synchronized settling time. If x_i and x_j are the first and last elements of the system to enter set Ω_x, then the state x is practical time-synchronized stable.

Remark 9.3 Practical time-synchronized stability is different from traditional finite-time stability. For finite-time uniformly ultimately boundedness, it requires that within finite time T, all states converge to the neighborhood of the equilibrium point. For practical time-synchronized stability, the boundary of ratio restriction is not related to the time-varying x state but only related to its initial state. Practical time-synchronized stability further emphasizes the relationship between each non-zero element of the state. When the ratio between any pair of elements is kept to be fixed but allowed to be varying within a certain small range, the closed-loop system will exhibit time-synchronized-like properties.

Corollary 9.1 *Considering the system* $\dot{x} = f(x)$ *and an open neighborhood* $D_0 \subseteq \mathbb{R}^n$ *of the origin, where* $f : D_0 \rightarrow \mathbb{R}^n$ *is continuous with* $f(0) = 0$*, and there exist a unique solution for any initial state* $x(0) = x_0$*, the origin of a Lyapunov function* $V(x)$ *is practical time-synchronized stable, with the compact set* Ω *and the synchronized settling time* T *as*

$$\Omega = \{x : c_3 V^{p_1 - p_2}(x) + c_4 V^{1-p_2}(x) < \varepsilon\} \tag{9.46}$$

$$T \leq \frac{1}{(c_2 - c_4)(1 - p_1)} \ln\left(1 + \frac{c_2 - c_4}{c_1 - c_3} V^{1-p_1}(x_0)\right), \tag{9.47}$$

if the following conditions hold:

(i) there exist any real constants $p_1 \in (0, 1)$, $p_2 < p_1$, $c_1 > 0$, $c_2 > 0$, $c_3 \in (0, c_1)$, $c_4 \in (0, c_2)$, *and* $\varepsilon > 0$, *such that*

$$\dot{V}(x) + c_1 V^{p_1}(x) + c_2 V(x) \leq \varepsilon V^{p_2}(x) \tag{9.48}$$

(ii) the state x *is ratio restricted.*

Proof According to (9.48), we know that x is finite-time uniformly ultimately bounded. Through the analysis of the proof of Theorem 9.1, we can verify that the system is practical time-synchronized stable.

In Corollary 9.1, we can construct the auxiliary system $z = x + \varepsilon(t)$, and obtain ratio restriction of x by proving that z is ratio persistent. For the systems whose information of $\varepsilon(t)$ in $z = x + \varepsilon(t)$ can be obtained, Theorem 9.2 is suitable for designing the practical time-synchronized controller. For the systems whose information of $\varepsilon(t)$ cannot be directly obtained, Corollary 9.1 is suitable for designing the practical time-synchronized controller.

For synchronized control systems with unknown disturbances, we hope to accurately observe the disturbances through observers. We uses multivariable super-twisting sliding mode observers to solve this problem.

Lemma 9.2 (Multivariable Super-Twisting Sliding Mode Observer [7]) *For the system*

$$\dot{\sigma} = -k_1 \frac{\sigma}{||\sigma||^{1/2}} + z - k_2\sigma + \gamma(t, \sigma) \tag{9.49}$$

$$\dot{z} = -k_3 \frac{\sigma}{||\sigma||} - k_4\sigma + \phi(t) \tag{9.50}$$

where $\sigma \in \mathbb{R}^n$ *and* $z \in \mathbb{R}^n$ *denote the state vector, and the terms* $\gamma(t, \sigma) \in \mathbb{R}^n$ *and* $\phi(t) \in \mathbb{R}^n$ *are assumed to satisfy* $||\gamma(t, \sigma)|| \leq \delta_1||\sigma||$ *and* $||\phi(t)|| \leq \delta_2$ *for known* *scalar bounds $\delta_1, \delta_2 > 0$, there exist a range of values for the gains k_1, k_2, k_3, k_4, such that the variables σ and $\dot{\sigma}$ are forced to zero in finite time and remain zero for all subsequent time.*

For the system $\dot{x} = f + gv + d$ in $\mathbb{R}^n$, where $x \in \mathbb{R}^n$ denotes the state vector, $f \in \mathbb{R}^n$ indicates a known nonlinear term, $g \in \mathbb{R}^{n\times n}$ represents the nonsingular known part, $v \in \mathbb{R}^n$ indicates the control input, $d \in \mathbb{R}^n$ is the unknown disturbance and uncertainties, and $\dot{d}$ is bounded. We define $\tilde{y}_0 = y_0 - x$ and $\tilde{y}_1 = y_1 - d$. According to Lemma 1, we can design the finite-time observer to observe d as follows

$$\dot{\tilde{y}}_0 = -k_1 \frac{\tilde{y}_0}{||\tilde{y}_0||^{1/2}} + \tilde{y}_1 - k_2 \tilde{y}_0 \tag{9.51}$$

$$\dot{\tilde{y}}_1 = -k_3 \frac{\tilde{y}_0}{||\tilde{y}_0||} - k_4 \tilde{y}_0 - \dot{d} \tag{9.52}$$

Substituting $\tilde{y}_0$ and $\tilde{y}_1$ into the above formula and simplifying, we can obtain the general form of the observer for system $\dot{x} = f + gv + d$ as follows

$$\dot{y}_0 = -k_1 \frac{y_0 - x}{||y_0 - x||^{1/2}} + y_1 - k_2 (y_0 - x) + f + gv \tag{9.53}$$

$$\dot{y}_1 = -k_3 \frac{y_0 - x}{||y_0 - x||} - k_4 (y_0 - x) \tag{9.54}$$

where k_1, k_2, k_3, and k_4 represent positive gains, y_0 and y_1 are the approximate values of the state vector x and the unknown disturbance d, respectively. In the following, we will use (9.53) and (9.54) to design the corresponding observers.

9.3.3 Command Filter Technique

The command filter-based controller is actually composed of many filters and compensators. In this structure, many signals have to pass through the filter alone to obtain new signals. These command filters on the command signal flow form a chain [3]. Here, we first introduce the structure of a single filter.

The second-order nonlinear command filter that can adjust the magnitude, rate, and bandwidth of the command signal is designed as follows [8]

$$\begin{aligned} \dot{\varsigma}_1 &= \varsigma_2 \\ \dot{\varsigma}_2 &= 2\eta\omega_n \left[\Upsilon_R \left(\frac{\omega_n^2}{2\eta\omega_n} \left(\Upsilon_M (\chi_d) - \varsigma_1 \right) \right) - \varsigma_2 \right] \end{aligned} \tag{9.55}$$

where χ_d is the signal to be filtered, the output filtered-limited signals χ_c and $\dot{\chi}_c$ are defined as

$$\begin{aligned} \chi_c &= \varsigma_1 \\ \dot{\chi}_c &= \varsigma_2 \end{aligned}, \tag{9.56}$$

where the state variables ς_1, ς_2 satisfy $\varsigma_1(0) = \varsigma_2(0) = 0$, $\eta \in (0, 1]$ represents the damping ratio, $\omega_n > 0$ is the natural frequency, and $\Upsilon_R(\cdot)$ and $\Upsilon_M(\cdot)$ governed by the following formula are the rate and magnitude constraint functions, respectively, where $\chi_{\max}$ and $\chi_{\min}$ are constants.

$$\Upsilon_{R/M}(\chi) = \begin{cases} \chi_{\max}, & \chi \geq \chi_{\max} \\ \chi, & \chi_{\min} < \chi < \chi_{\max} \\ \chi_{\min}, & \chi \leq \chi_{\min} \end{cases} \tag{9.57}$$

In the second-order nonlinear filter (9.55), by inputting the bounded signal χ_d, we can obtain the continuous and bounded signal χ_c and its derivative $\dot{\chi}_c$ after it has been filtered. According to the characteristics of the command filter, the output χ_c is modified in magnitude, rate, and bandwidth relative to the input χ_d. Therefore, we can use these properties of the filter to deal with a series of practical problems of the system, such as excessive peaks, control input saturation, and signal delay. This also means that we can use filters to keep the various characteristics of the commands obtained by the spacecraft's actuators within a reasonable range. The convergence analysis of the command filter is presented in [3].

A traditional command filter-based controller structure is obtained when we introduce several filters into the controller step by step through the backstepping method. In traditional command filter-based algorithms and a large amount of the latest research works, three filters and two compensators are generally required for a second-order system. In this chapter, a *simplified command filter* structure is designed, using only one filter and two compensator structures. So that the simplified command filter-based controller not only has the advantages of the filter but also alleviates the complicated chain structure.

To show the role of the filter clearly, we have made some necessary explanations for the corresponding signal in advance. Here, $\boldsymbol{\tau}_a$ represents the control input obtained by the closed-loop system or actuators, which is also the constrained and modified control command signal after filtering. $\boldsymbol{\tau}_d$ is the control signal we designed theoretically. Furthermore, $\boldsymbol{\tau}_a$ is obtained when $\boldsymbol{\tau}_d$ passes through the filter. $(\boldsymbol{\tau}_a - \boldsymbol{\tau}_d)$ is the error caused by filtering, and we need to define and design the corresponding signals $\boldsymbol{\xi}_1$ and $\boldsymbol{\xi}_2$ to compensate for the loss.

Due to the existence of compensators, the command filter itself does not affect the design of the finite-time controller. Because as long as the controller and compensator converge within finite time, the steady-state error of the filter will only affect the error boundary of the closed-loop system.

9.4 Controller Design

9.4.1 Command-Filter-Based Practical Time-Synchronized Control for Spacecraft

Considering the error caused by the filter, we need to design the compensators. We design two compensators $\boldsymbol{\xi}_1$ and $\boldsymbol{\xi}_2$, and the auxiliary variables are as follows

$$\mathbf{z}_1 = \mathbf{q}_{e1} - \boldsymbol{\xi}_1 \tag{9.58}$$

$$\boldsymbol{\alpha} = \mathbf{q}_{e2} - \boldsymbol{\xi}_2 \tag{9.59}$$

where $\mathbf{q}_{e1} = \mathbf{q}_{ev}$, $\mathbf{q}_{e2} = \dot{\mathbf{q}}_{ev}$.

The derivative of (9.58) is given as

$$\begin{aligned} \dot{\mathbf{z}}_1 =& \mathbf{q}_{e2} - \dot{\boldsymbol{\xi}}_1 \\ =& \boldsymbol{\alpha} + \boldsymbol{\xi}_2 - \dot{\boldsymbol{\xi}}_1 - \kappa_{\xi 1}\boldsymbol{\xi}_1 + \kappa_{\xi 1}\mathbf{q}_{e1} - \kappa_{\xi 1}\mathbf{z}_1 \\ & - \kappa_{\xi 2}\mathrm{sig}_{\mathrm{c}}^{p_{\xi 1}}\left(\boldsymbol{\xi}_1\right) + \kappa_{\xi 2}\mathrm{sig}_{\mathrm{c}}^{p_{\xi 1}}\left(\mathbf{q}_{e1} - \mathbf{z}_1\right) \end{aligned} \tag{9.60}$$

where $\kappa_{\xi 1} \geq 1/2$ and $\kappa_{\xi 2} \geq 0$ are constants, $p_{\xi 1} \in (0, 1)$ is a power number.

Observing (9.60), we design the adaptive law of finite-time compensator $\boldsymbol{\xi}_1$ as

$$\dot{\boldsymbol{\xi}}_1 = -\kappa_{\xi 1}\boldsymbol{\xi}_1 - \kappa_{\xi 2}\mathrm{sig}_{\mathrm{c}}^{p_{\xi 1}}\left(\boldsymbol{\xi}_1\right) + \boldsymbol{\xi}_2 \tag{9.61}$$

In order to simplify the expression, we use simple symbol to replace complex function items as

$$\boldsymbol{\Psi}_1 = \kappa_{\xi 1}\mathbf{q}_{e1} - \kappa_{\xi 1}\mathbf{z}_1 + \kappa_{\xi 2}\mathrm{sig}_{\mathrm{c}}^{p_{\xi 1}}\left(\mathbf{q}_{e1} - \mathbf{z}_1\right) \tag{9.62}$$

Then, we have

$$\dot{\mathbf{z}}_1 = \boldsymbol{\Psi}_1 + \boldsymbol{\alpha} \tag{9.63}$$

Taking the derivative of (9.59) with respect to time, we have

$$\begin{aligned} \dot{\boldsymbol{\alpha}} =& \dot{\mathbf{q}}_{e2} - \dot{\boldsymbol{\xi}}_2 \\ =& -\mathbf{M}_a^{-1}\left(\mathbf{C}_a\dot{\mathbf{q}}_{ev} + \mathbf{G}_a\right) + \mathbf{M}_a^{-1}\mathbf{Q}_a\boldsymbol{\tau}_a \\ & - \mathbf{M}_a^{-1}\mathbf{Q}_a\boldsymbol{\tau}_d + \mathbf{M}_a^{-1}\mathbf{Q}_a\boldsymbol{\tau}_d + \mathbf{M}_a^{-1}\mathbf{Q}_a\mathbf{d} - \dot{\boldsymbol{\xi}}_2 \\ =& -\mathbf{M}_a^{-1}\left(\mathbf{C}_a\dot{\mathbf{q}}_{ev} + \mathbf{G}_a\right) + \mathbf{M}_a^{-1}\mathbf{Q}_a\left(\boldsymbol{\tau}_a - \boldsymbol{\tau}_d\right) \\ & + \mathbf{M}_a^{-1}\mathbf{Q}_a\boldsymbol{\tau}_d + \mathbf{M}_a^{-1}\mathbf{Q}_a\mathbf{d} \\ & - \boldsymbol{\xi}_2 - \kappa_{\xi 3}\boldsymbol{\xi}_2 - \kappa_{\xi 4}\mathrm{sig}_{\mathrm{c}}^{p_{\xi 1}}\left(\boldsymbol{\xi}_2\right) \\ & + \kappa_{\xi 3}\left(\mathbf{q}_{e2} - \boldsymbol{\alpha}\right) + \kappa_{\xi 4}\mathrm{sig}_{\mathrm{c}}^{p_{\xi 1}}\left(\mathbf{q}_{e2} - \boldsymbol{\alpha}\right) \end{aligned} \tag{9.64}$$

where $\kappa_{\xi 3} \geq 1$ and $\kappa_{\xi 4} \geq 0$ are constants.

Similarly, we design a finite-time adaptive law for $\boldsymbol{\xi}_2$ as follows

$$\dot{\boldsymbol{\xi}}_2 = -\kappa_{\xi 3}\boldsymbol{\xi}_2 - \kappa_{\xi 4}\text{sig}_c^{p_{\xi 1}}\left(\boldsymbol{\xi}_2\right) + \mathbf{M}_a^{-1}\mathbf{Q}_a\left(\boldsymbol{\tau}_a - \boldsymbol{\tau}_d\right) \tag{9.65}$$

Intuitively, $\boldsymbol{\xi}_2$ has the ability to compensate for the loss caused by $(\boldsymbol{\tau} - \boldsymbol{\tau}_d)$.

Then, (9.64) can be expressed as

$$\dot{\boldsymbol{\alpha}} = \boldsymbol{\Psi}_2 + \mathbf{M}_a^{-1}\mathbf{Q}_a\boldsymbol{\tau}_d + \mathbf{M}_a^{-1}\mathbf{Q}_a\mathbf{d} \tag{9.66}$$

where

$$\begin{aligned}\boldsymbol{\Psi}_2 = &- \mathbf{M}_a^{-1}\left(\mathbf{C}_a\dot{\mathbf{q}}_{ev} + \mathbf{G}_a\right) + \kappa_{\xi 3}\left(\mathbf{q}_{e2} - \boldsymbol{\alpha}\right)\\ &+ \kappa_{\xi 4}\text{sig}_c^{p_{\xi 1}}\left(\mathbf{q}_{e2} - \boldsymbol{\alpha}\right)\end{aligned} \tag{9.67}$$

We further perform a coordinate transformation as $\mathbf{z}_2 = \dot{\mathbf{z}}_1 = \boldsymbol{\Psi}_1 + \boldsymbol{\alpha}$ and the derivative of $\mathbf{z}_2$ is

$$\dot{\mathbf{z}}_2 = \dot{\boldsymbol{\Psi}}_1 + \dot{\boldsymbol{\alpha}} = \boldsymbol{\Psi}_3 + \mathbf{M}_a^{-1}\mathbf{Q}_a\boldsymbol{\tau}_d + \mathbf{M}_a^{-1}\mathbf{Q}_a\mathbf{d} \tag{9.68}$$

where $\boldsymbol{\Psi}_3 = \dot{\boldsymbol{\Psi}}_1 + \boldsymbol{\Psi}_2$.

Remark 9.4 There is a singularity problem in $\dot{\boldsymbol{\Psi}}_1$ because the derivative of $\text{sig}_c^{p_{\xi 1}}(\boldsymbol{\beta})$ is infinite when $\beta_i = 0$ and $\dot{\beta}_i \neq 0$ in $\text{d}(\kappa_{\xi 2}\text{sig}_c^{p_{\xi 1}}(\boldsymbol{\beta}))/\text{d}t$, $\boldsymbol{\beta} = \mathbf{q}_{e1} - \mathbf{z}_1$. The threshold value ϖ_ξ is set to avoid this singularity problem, and the following definition is given

$$\frac{\text{d}}{\text{d}t}(\kappa_{\xi 2}\text{sig}_c^{p_{\xi 1}}(\beta_i)) = \begin{cases}\kappa_{\xi 2}p_{\xi 1}|\beta_i|^{p_{\xi 1}-1}\dot{\beta}_i, \text{ if } |\beta_i| \geq \varpi_\xi, \dot{\beta}_i \neq 0\\ \kappa_{\xi 2}p_{\xi 1}|\iota_\xi|^{p_{\xi 1}-1}\dot{\beta}_i, \text{ if } |\beta_i| < \varpi_\xi, \dot{\beta}_i \neq 0\\ 0\,, \text{ if } \dot{\beta}_i = 0\end{cases} \tag{9.69}$$

where ϖ_ξ and ι_ξ are both small positive constants.

For system $\dot{\mathbf{z}}_2 = \dot{\boldsymbol{\Psi}}_1 + \dot{\boldsymbol{\alpha}} = \boldsymbol{\Psi}_3 + \mathbf{M}_a^{-1}\mathbf{Q}_a\boldsymbol{\tau}_d + \mathbf{M}_a^{-1}\mathbf{Q}_a\mathbf{d}$, considering the unknown bounded disturbances, we use the super-twisting sliding mode observers (9.53) and (9.54) to observe $\mathbf{M}_a^{-1}\mathbf{Q}_a\mathbf{d}$. Design the observers as

$$\begin{aligned}\dot{\vartheta}_1 = &- \kappa_{\vartheta 1}\frac{\vartheta_1 - \mathbf{z}_2}{\|\vartheta_1 - \mathbf{z}_2\|^{1/2}} - \kappa_{\vartheta 2}(\vartheta_1 - \mathbf{z}_2)\\ &+ \boldsymbol{\Psi}_3 + \mathbf{M}_a^{-1}\mathbf{Q}_a\boldsymbol{\tau}_d + \vartheta_2\end{aligned} \tag{9.70}$$

$$\dot{\vartheta}_2 = - \kappa_{\vartheta 3}\frac{\vartheta_1 - \mathbf{z}_2}{\|\vartheta_1 - \mathbf{z}_2\|} - \kappa_{\vartheta 4}(\vartheta_1 - \mathbf{z}_2) \tag{9.71}$$

where $\kappa_{\vartheta 1}$, $\kappa_{\vartheta 2}$, $\kappa_{\vartheta 3}$, and $\kappa_{\vartheta 4}$ are positive gains. According to Lemma 9.2 and the properties of the observer, there is finite time $T_{\vartheta 1}$, when $t \geq T_{\vartheta 1}$, $\vartheta_2 = \mathbf{M}_a^{-1}\mathbf{Q}_a\mathbf{d}$

holds. After this, we have

$$\dot{\mathbf{z}}_2 = \boldsymbol{\Psi}_3 + \vartheta_2 + \mathbf{M}_a^{-1}\mathbf{Q}_a\boldsymbol{\tau}_d \tag{9.72}$$

To make the closed-loop system of spacecraft satisfy ratio restriction or ratio persistence, we use the norm-normalized sign function to map $\mathbf{z}_1$. We choose a sliding-mode variable as follows, so that $\mathbf{z}_1$ satisfies ratio persistence when ζ_1 converges to 0.

$$\zeta_1 = \dot{\mathbf{z}}_1 + \kappa_1 \mathrm{sig}_{\mathrm{n}}^{p_1}(\mathbf{z}_1) \tag{9.73}$$

The derivative of ζ_1 with respect to time is calculated as follows

$$\dot{\zeta}_1 = \dot{\mathbf{z}}_2 + \kappa_1 \frac{\mathrm{d}}{\mathrm{d}t}\mathrm{sig}_{\mathrm{n}}^{p_1}(\mathbf{z}_1) = \dot{\mathbf{z}}_2 + \phi_1 \tag{9.74}$$

where

$$\phi_1 = \kappa_1 (p_1 - 1) \|\mathbf{z}_1\|^{p_1-3}\mathbf{z}_1\mathbf{z}_1^T \cdot \dot{\mathbf{z}}_1 + \kappa_1 \|\mathbf{z}_1\|^{p_1-1} I_n \cdot \dot{\mathbf{z}}_1. \tag{9.75}$$

Considering the existence of singularities in ϕ_1, we refer to the method of Remark 9.4 and design the following modified switching function,

$$\frac{\mathrm{d}}{\mathrm{d}t}(\mathrm{sig}_{\mathrm{n}}^{p_N+1}(x)) = \begin{cases} p_N \|x\|^{p_N-2} x x^T \cdot \dot{x} + \|x\|^{p_N} I_n \cdot \dot{x}, \\ \text{if } \|x\| \geq \varpi_1, \dot{x} \neq 0 \\ p_N \|\iota_1\|^{p_N-2} \iota_1 \iota_1^T \cdot \dot{x} + \|\iota_1\|^{p_N} I_n \cdot \dot{x}, \\ \text{if } \|x\| < \varpi_1, \dot{x} \neq 0 \\ 0, \ \text{if } \dot{x} = 0 \end{cases} \tag{9.76}$$

where ϖ_1 and ι_1 are both small positive constants, p_N is the power exponent. Then, ϕ_1 is modified as

$$\phi_1 = \begin{cases} \kappa_1 (p_1 - 1) \|\mathbf{z}_1\|^{p_1-3}\mathbf{z}_1\mathbf{z}_1^T \cdot \dot{\mathbf{z}}_1 + \kappa_1 \|\mathbf{z}_1\|^{p_1-1} I_n \cdot \dot{\mathbf{z}}_1, \\ \text{if } \|\mathbf{z}_1\| \geq \varpi_1, \dot{\mathbf{z}}_1 \neq 0 \\ \kappa_1 (p_1 - 1) \|\iota_1\|^{p_1-3}\iota_1\iota_1^T \cdot \dot{\mathbf{z}}_1 + \kappa_1 \|\iota_1\|^{p_1-1} I_n \cdot \dot{\mathbf{z}}_1, \\ \text{if } \|\mathbf{z}_1\| < \varpi_1, \dot{\mathbf{z}}_1 \neq 0 \\ 0, \ \text{if } \dot{\mathbf{z}}_1 = 0 \end{cases} \tag{9.77}$$

Next, we propose a practical time-synchronized controller for spacecraft as

$$\boldsymbol{\tau}_d = -\kappa_2 \mathbf{Q}_a^{-1}\mathbf{M}_a \mathrm{sig}_{\mathrm{n}}^{p_2}(\zeta_1) - \mathbf{Q}_a^{-1}\mathbf{M}_a (\boldsymbol{\Psi}_3 + \vartheta_2 + \phi_1) \tag{9.78}$$

Theorem 9.3 *Considering the spacecraft attitude error model governed by (9.22) with input saturation and unknown bounded disturbances, under the command filter-based practical time-synchronized controller (9.78), adaptive compensators (9.61, 9.65), and uncertain nonlinear observers (9.70, 9.71), the sliding mode surface* ζ_1 *satisfies ratio persistence and is time-synchronized stable within finite time* $T_{\zeta 1}$*, the auxiliary variable* $\mathbf{z}_1$ *satisfies ratio persistence and is time-synchronized stable within finite time* T_{z1}*, the adaptive compensators are finite-time uniformly ultimately bounded within finite time* $T_{\xi 1}$*, and the closed-loop system state* $\mathbf{q}_{ev}$ *is ratio restricted and practical time-synchronized stable within finite time* $T_{\mathbf{q}_{ev}} \leq \max\{T_{\vartheta 1}, T_{\zeta 1}, T_{z1}, T_{\xi 1}\}$. *The relevant time parameters will be defined later.*

Proof Focusing on the entire proof process, under the proposed controller, we first prove that the sliding mode surface ζ_1 is time-synchronized stable, and secondly prove that the auxiliary variable $\mathbf{z}_1$ is time-synchronized stable, and then explain that the adaptive compensators of the filters are finite-time uniformly ultimately bounded. Finally, from Theorem 9.2, we prove that the spacecraft tracking error $\mathbf{q}_{ev}$ is practical time-synchronized stable.

We first select the candidate Lyapunov function for ζ_1 as follows

$$V_1 = \frac{1}{2}\zeta_1^T \zeta_1 \tag{9.79}$$

Find the derivative of (9.79) with respect to time, and substitute the controller (9.78), then we have

$$\begin{aligned} \dot{V}_1 &= \zeta_1^T \dot{\zeta}_1 = \zeta_1^T \left(\boldsymbol{\Psi}_3 + \vartheta_2 + \mathbf{M}_a^{-1}\mathbf{Q}_a \boldsymbol{\tau}_d + \phi_1\right) \\ &= \zeta_1^T \left(\boldsymbol{\Psi}_3 + \vartheta_2 - \kappa_2 \mathrm{sig}_{\mathrm{n}}^{p_2}(\zeta_1) - (\boldsymbol{\Psi}_3 + \vartheta_2 + \phi_1) + \phi_1\right) \\ &= -\kappa_2 \zeta_1^T \mathrm{sig}_{\mathrm{n}}^{p_2}(\zeta_1) \\ &= -2^{\frac{1+p_2}{2}} \kappa_2 V_1^{\frac{1+p_2}{2}} \end{aligned} \tag{9.80}$$

According to Theorem 2.2, V_1 is time-synchronized stable with the settling time

$$T_1 \leq \frac{V_1^{\frac{1-p_2}{2}}(\zeta_1(0))}{2^{\frac{p_2-1}{2}} \kappa_2 (1-p_2)} \tag{9.81}$$

Considering that the observer converges in finite time $T_{\vartheta 1}$, the actual settling time of ζ_1 is $T_{\zeta 1} \leq \max\{T_1, T_{\vartheta 1}\}$.

When $\zeta_1 = 0$, $\zeta_1 = \dot{\mathbf{z}}_1 + \kappa_1 \mathrm{sig}_{\mathrm{n}}^{p_1}(\mathbf{z}_1) = 0$ holds, so we have $\dot{\mathbf{z}}_1 = -\kappa_1 \mathrm{sig}_{\mathrm{n}}^{p_1}(\mathbf{z}_1)$. We choose the candidate Lyapunov function for $\mathbf{z}_1$ as follows

$$V_2 = \frac{1}{2}\mathbf{z}_1^T \mathbf{z}_1 \tag{9.82}$$

with its derivative with respect to time

$$\dot{V}_2 = \mathbf{z}_1^T \dot{\mathbf{z}}_1 = -\kappa_1 \mathbf{z}_1^T \mathrm{sig}_{\mathrm{n}}^{p_1}(\mathbf{z}_1) = -2^{\frac{1+p_1}{2}} \kappa_1 V_2^{\frac{1+p_1}{2}} \tag{9.83}$$

According to Theorem 2.2, V_2 is time-synchronized stable with the settling time

$$T_2 \leq \frac{V_2^{\frac{1-p_1}{2}}(\mathbf{z}_1(0))}{2^{\frac{p_1-1}{2}} \kappa_1 (1-p_1)} \tag{9.84}$$

Taking into account the settling time of the observer, sliding mode surface ζ_1, and the compensators, the settling time of $\mathbf{z}_1$ is $T_{z1} \leq \max\{T_{\zeta 1} + T_2, T_3\}$, where the settling time T_3 of the compensators is described next.

Consider the candidate Lyapunov function for $\boldsymbol{\xi}_1$ and $\boldsymbol{\xi}_2$ as follows

$$V_3 = \frac{1}{2} \boldsymbol{\xi}_1^T \boldsymbol{\xi}_1 + \frac{1}{2} \boldsymbol{\xi}_2^T \boldsymbol{\xi}_2 \tag{9.85}$$

We can obtain the derivative of the above formula with respect to time as

$$\begin{aligned} \dot{V}_3 =& \boldsymbol{\xi}_1^T \dot{\boldsymbol{\xi}}_1 + \boldsymbol{\xi}_2^T \dot{\boldsymbol{\xi}}_2 = -\kappa_{\xi 1} \boldsymbol{\xi}_1^T \boldsymbol{\xi}_1 - \kappa_{\xi 3} \boldsymbol{\xi}_2^T \boldsymbol{\xi}_2 + \boldsymbol{\xi}_1^T \boldsymbol{\xi}_2 \\ & - \kappa_{\xi 2} \boldsymbol{\xi}_1^T \mathrm{sig}_{\mathrm{c}}^{p_{\xi 1}}(\boldsymbol{\xi}_1) - \kappa_{\xi 4} \boldsymbol{\xi}_2^T \mathrm{sig}_{\mathrm{c}}^{p_{\xi 1}}(\boldsymbol{\xi}_2) \\ & + \boldsymbol{\xi}_2^T \mathbf{M}_a^{-1} \mathbf{Q}_a (\boldsymbol{\tau}_a - \boldsymbol{\tau}_d) \end{aligned} \tag{9.86}$$

Considering the nature of the command filter itself, we assume that $||\boldsymbol{\tau}_a - \boldsymbol{\tau}_d|| \leq \varepsilon_\tau$, and $||\mathbf{M}_a^{-1} \mathbf{Q}_a||$ is bounded in the actual spacecraft system, that is, $||\mathbf{M}_a^{-1} \mathbf{Q}_a|| \leq \varepsilon_M$, where ε_τ and ε_M are constants. So we have

$$\boldsymbol{\xi}_2^T \mathbf{M}_a^{-1} \mathbf{Q}_a (\boldsymbol{\tau}_a - \boldsymbol{\tau}_d) \leq \frac{1}{2} ||\boldsymbol{\xi}_2||^2 + \frac{1}{2} (\varepsilon_\tau \varepsilon_M)^2 \tag{9.87}$$

Considering $\boldsymbol{\xi}_1^T \boldsymbol{\xi}_2 \leq \frac{1}{2} ||\boldsymbol{\xi}_1||^2 + \frac{1}{2} ||\boldsymbol{\xi}_2||^2$, we have

$$\begin{aligned} \dot{V}_3 \leq & -(\kappa_{\xi 1} - \frac{1}{2}) \boldsymbol{\xi}_1^T \boldsymbol{\xi}_1 - (\kappa_{\xi 3} - 1) \boldsymbol{\xi}_2^T \boldsymbol{\xi}_2 + \frac{1}{2} (\varepsilon_\tau \varepsilon_M)^2 \\ & - \kappa_{\xi 2} \boldsymbol{\xi}_1^T \mathrm{sig}_{\mathrm{c}}^{p_{\xi 1}}(\boldsymbol{\xi}_1) - \kappa_{\xi 4} \boldsymbol{\xi}_2^T \mathrm{sig}_{\mathrm{c}}^{p_{\xi 1}}(\boldsymbol{\xi}_2) \\ \leq & -\kappa_{\xi 5} V_3 - \kappa_{\xi 6} V_3^{\frac{1+p_{\xi 1}}{2}} + \kappa_{\xi 7} V_3^0 \end{aligned} \tag{9.88}$$

where $\kappa_{\xi 5} = \min\{(\kappa_{\xi 1} - 1/2), (\kappa_{\xi 3} - 1)\}$, $\kappa_{\xi 6} = \min\{2^{\frac{1+p_{\xi 1}}{2}} \kappa_{\xi 2}, 2^{\frac{1+p_{\xi 1}}{2}} \kappa_{\xi 4}\}$, and $\kappa_{\xi 7} = \frac{1}{2} (\varepsilon_\tau \varepsilon_M)^2$.

According to Lemma 9.1, V_3 is finite-time uniformly ultimately bounded within the compact set $\Omega_{\xi 1} = \{\boldsymbol{\xi}_1, \boldsymbol{\xi}_2 : \kappa_{\xi 8} V_3^{\frac{1+p_{\xi 1}}{2}} + \kappa_{\xi 9} V_3 < \kappa_{\xi 7}\}$ with the settling time

$$T_3 \leq \frac{2\ln\left(1+\frac{\kappa_{\xi 5}-\kappa_{\xi 9}}{\kappa_{\xi 6}-\kappa_{\xi 8}} V_3^{\frac{1-p_{\xi 1}}{2}}\left(\boldsymbol{\xi}_1(0), \boldsymbol{\xi}_2(0)\right)\right)}{(\kappa_{\xi 5}-\kappa_{\xi 9})(1-p_{\xi 1})} \tag{9.89}$$

where $\kappa_{\xi 8} \in (0, \kappa_{\xi 6})$, $\kappa_{\xi 9} \in (0, \kappa_{\xi 5})$. So $\boldsymbol{\xi}_1$, $\boldsymbol{\xi}_2$ are bounded with $T_{\xi 1} \leq T_3$. Further, we explain the boundedness of the derivative of $\boldsymbol{\xi}_1$ with respect to time.

$$\begin{aligned} \left\|\dot{\boldsymbol{\xi}}_1\right\| &\leq \left\|-\kappa_{\xi 1}\boldsymbol{\xi}_1 - \kappa_{\xi 2}\mathrm{sig}_{\mathrm{c}}^{p_{\xi 1}}\left(\boldsymbol{\xi}_1\right) + \boldsymbol{\xi}_2\right\| \\ &\leq +\kappa_{\xi 1}\left\|\boldsymbol{\xi}_1\right\| + \kappa_{\xi 2}\left\|\boldsymbol{\xi}_1\right\|^{p_{\xi 1}} + \left\|\boldsymbol{\xi}_2\right\| \end{aligned} \tag{9.90}$$

Therefore, in the system $\mathbf{z}_1 = \mathbf{q}_{e1} - \boldsymbol{\xi}_1$, both $\boldsymbol{\xi}_1$ and $\dot{\boldsymbol{\xi}}_1$ are bounded. Considering that $\mathbf{z}_1$ is time-synchronized stable, according to Theorem 9.2, the closed-loop system tracking error $\mathbf{q}_{ev} = \mathbf{q}_{e1}$ is practical time-synchronized stable, and the settling time is $T_{\mathbf{q}_{ev}} \leq \max\{T_{\vartheta 1}, T_{\zeta 1}, T_{z1}, T_{\xi 1}\}$.

9.4.2 *Improved Command-Filter-Based Practical Time-Synchronized Control for Spacecraft*

Considering that the controller designed in the previous subsection requires all the information of $\boldsymbol{\tau}_a$, it is necessary to add corresponding measurement equipment on the spacecraft, which will increase the burden on the spacecraft. At the same time, we consider that the controller has low-pass characteristics. We use a low-pass filter to improve the command filter-based practical time-synchronized controller. In this way, we can achieve the control objectives without actually measuring $\boldsymbol{\tau}_a$.

We improve the adaptive law on the compensation signal $\boldsymbol{\xi}_2$ as follows,

$$\dot{\boldsymbol{\xi}}_2 = -\kappa_{\xi 3}\boldsymbol{\xi}_2 - \kappa_{\xi 4}\mathrm{sig}_{\mathrm{c}}^{p_{\xi 1}}\left(\boldsymbol{\xi}_2\right) + \mathbf{M}_a^{-1}\mathbf{Q}_a\left(\mathrm{Fil}(\boldsymbol{\tau}_d) - \boldsymbol{\tau}_d\right) \tag{9.91}$$

where $\mathrm{Fil}(\boldsymbol{\tau}_d)$ represents the signal that $\boldsymbol{\tau}_d$ passes through the low-pass filter. For this, we recalculate the relevant variables as

$$\begin{aligned} \dot{\boldsymbol{\alpha}} = &-\mathbf{M}_a^{-1}\left(\mathbf{C}_a\dot{\mathbf{q}}_{ev} + \mathbf{G}_a\right) + \mathbf{M}_a^{-1}\mathbf{Q}_a\boldsymbol{\tau}_d + \mathbf{M}_a^{-1}\mathbf{Q}_a\mathbf{d} \\ &+ \kappa_{\xi 3}\left(\mathbf{q}_{e2} - \boldsymbol{\alpha}\right) + \kappa_{\xi 4}\mathrm{sig}_{\mathrm{c}}^{p_{\xi 1}}\left(\mathbf{q}_{e2} - \boldsymbol{\alpha}\right) \\ &+ \mathbf{M}_a^{-1}\mathbf{Q}_a\left(\boldsymbol{\tau}_a - \mathrm{Fil}(\boldsymbol{\tau}_d)\right) - \dot{\boldsymbol{\xi}}_2 - \kappa_{\xi 3}\boldsymbol{\xi}_2 \\ &- \kappa_{\xi 4}\mathrm{sig}_{\mathrm{c}}^{p_{\xi 1}}\left(\boldsymbol{\xi}_2\right) + \mathbf{M}_a^{-1}\mathbf{Q}_a\left(\mathrm{Fil}(\boldsymbol{\tau}_d) - \boldsymbol{\tau}_d\right) \end{aligned} \tag{9.92}$$

Substituting the improved compensators (9.91) and (9.92) into $\dot{\mathbf{z}}_2$, we have

$$\begin{aligned} \dot{\mathbf{z}}_2 = &\dot{\boldsymbol{\Psi}}_1 + \dot{\boldsymbol{\alpha}} = \boldsymbol{\Psi}_4 + \mathbf{M}_a^{-1}\mathbf{Q}_a\boldsymbol{\tau}_d + \mathbf{M}_a^{-1}\mathbf{Q}_a\mathbf{d} \\ &+ \mathbf{M}_a^{-1}\mathbf{Q}_a\left(\boldsymbol{\tau}_a - \mathrm{Fil}(\boldsymbol{\tau}_d)\right) \end{aligned} \tag{9.93}$$

where

$$\begin{aligned}\boldsymbol{\Psi}_4 =& \dot{\boldsymbol{\Psi}}_1 - \mathbf{M}_a^{-1}\left(\mathbf{C}_a\dot{\mathbf{q}}_{ev} + \mathbf{G}_a\right) + \kappa_{\xi 3}\left(\mathbf{q}_{e2} - \boldsymbol{\alpha}\right) \\ &+ \kappa_{\xi 4}\mathrm{sig}_{\mathrm{c}}^{p_{\xi 1}}\left(\mathbf{q}_{e2} - \boldsymbol{\alpha}\right)\end{aligned} \tag{9.94}$$

For the unknown disturbance $\mathbf{M}_a^{-1}\mathbf{Q}_a\mathbf{d}$ and the unobtainable parameter $\mathbf{M}_a^{-1}\mathbf{Q}_a$ $(\boldsymbol{\tau}_a - \mathrm{Fil}(\boldsymbol{\tau}_d))$, we design the following observer according to Lemma 9.2.

$$\begin{aligned}\dot{\vartheta}_3 =& -\kappa_{\vartheta 5}\frac{\vartheta_3 - \mathbf{z}_2}{\|\vartheta_3 - \mathbf{z}_2\|^{1/2}} - \kappa_{\vartheta 6}(\vartheta_3 - \mathbf{z}_2) \\ &+ \boldsymbol{\Psi}_4 + \mathbf{M}_a^{-1}\mathbf{Q}_a\boldsymbol{\tau}_d + \vartheta_4\end{aligned} \tag{9.95}$$

$$\dot{\vartheta}_4 = -\kappa_{\vartheta 7}\frac{\vartheta_3 - \mathbf{z}_2}{\|\vartheta_3 - \mathbf{z}_2\|} - \kappa_{\vartheta 8}(\vartheta_3 - \mathbf{z}_2) \tag{9.96}$$

where $\kappa_{\vartheta 5}$, $\kappa_{\vartheta 6}$, $\kappa_{\vartheta 7}$, and $\kappa_{\vartheta 8}$ are positive gains. According to Lemma 9.2 and the properties of the observer, there is finite time $T_{\vartheta 2}$, when $t \geq T_{\vartheta 2}$, $\vartheta_4 = (\mathbf{M}_a^{-1}\mathbf{Q}_a$ $(\boldsymbol{\tau}_a - \mathrm{Fil}(\boldsymbol{\tau}_d)) + \mathbf{M}_a^{-1}\mathbf{Q}_a\mathbf{d})$ holds. After that, $\dot{\mathbf{z}}_2$ is rewritten as

$$\dot{\mathbf{z}}_2 = \boldsymbol{\Psi}_4 + \mathbf{M}_a^{-1}\mathbf{Q}_a\boldsymbol{\tau}_d + \vartheta_4 \tag{9.97}$$

Similar to the previous subsection, we design the sliding surface and the controller as follows

$$\zeta_2 = \dot{\mathbf{z}}_1 + \kappa_1\mathrm{sig}_{\mathrm{n}}^{p_1}(\mathbf{z}_1) \tag{9.98}$$

$$\begin{aligned}\boldsymbol{\tau}_d =& -\kappa_2\mathbf{Q}_a^{-1}\mathbf{M}_a\mathrm{sig}_{\mathrm{n}}^{p_2}(\zeta_2) \\ &- \mathbf{Q}_a^{-1}\mathbf{M}_a\left(+\boldsymbol{\Psi}_4 + \vartheta_4 + \phi_1\right)\end{aligned} \tag{9.99}$$

Theorem 9.4 *Considering the spacecraft attitude error model governed by (9.22) with input saturation and unknown bounded disturbances, under the improved command filter-based practical time-synchronized controller (9.99), adaptive compensators (9.61, 9.91), and uncertain nonlinear observers (9.95, 9.96), the sliding mode surface ζ_2 satisfies ratio persistence and is time-synchronized stable within finite time $T_{\zeta 2}$, the auxiliary variable $\mathbf{z}_1$ satisfies ratio persistence and is time-synchronized stable within finite time T_{z2}, the adaptive compensators are finite-time uniformly ultimately bounded within finite time $T_{\xi 2}$, and the closed-loop system state $\mathbf{q}_{ev}$ is ratio restricted and practical time-synchronized stable within finite time $T_{\mathbf{q}_{ev}} \leq \max\{T_{\vartheta 2}, T_{\zeta 2}, T_{z2}, T_{\xi 2}\}$. The relevant time parameters will be defined later.*

Proof Considering the candidate Lyapunov function $V_4 = \frac{1}{2}\zeta_2^T\zeta_2$ for ζ_2, substituting the controller (9.99) into the derivative of V_4, we have

$$\dot{V}_4 = \zeta_2^T\dot{\zeta}_2 = -2^{\frac{1+p_2}{2}}\kappa_2 V_4^{\frac{1+p_2}{2}}. \tag{9.100}$$

It can be seen that V_4 is time-synchronized stable within finite time

$$T_4 \leq \left(2^{\frac{p_2-1}{2}} \kappa_2(1-p_2)\right)^{-1} V_4^{\frac{1-p_2}{2}}(\zeta_2(0)). \tag{9.101}$$

Further, considering the observer, ζ_2 is time-synchronized stable within finite time

$$T_{\zeta 2} \leq \max\{T_4, T_{\vartheta 2}\}. \tag{9.102}$$

When $\zeta_2 = 0$, we have $\dot{\mathbf{z}}_1 = -\kappa_1 \mathrm{sig}_{\mathrm{n}}^{p_1}(\mathbf{z}_1)$. So that, $\mathbf{z}_1$ is time-synchronized stable with the settling time $T_{z2} \leq \max\{T_{\zeta 2} + T_2, T_5\}$, where the settling time T_5 of the improved compensators is described next.

Consider the candidate Lyapunov function $V_5 = \frac{1}{2}\boldsymbol{\xi}_1^T\boldsymbol{\xi}_1 + \frac{1}{2}\boldsymbol{\xi}_2^T\boldsymbol{\xi}_2$ of $\boldsymbol{\xi}_1$ and $\boldsymbol{\xi}_2$, its derivative is as follows

$$\begin{aligned}\dot{V}_5 =& \boldsymbol{\xi}_1^T\dot{\boldsymbol{\xi}}_1 + \boldsymbol{\xi}_2^T\dot{\boldsymbol{\xi}}_2 \\ =& -\kappa_{\xi 1}\boldsymbol{\xi}_1^T\boldsymbol{\xi}_1 - \kappa_{\xi 3}\boldsymbol{\xi}_2^T\boldsymbol{\xi}_2 + \boldsymbol{\xi}_1^T\boldsymbol{\xi}_2 \\ & - \kappa_{\xi 2}\boldsymbol{\xi}_1^T \mathrm{sig}_{\mathrm{c}}^{p_{\xi 1}}\left(\boldsymbol{\xi}_1\right) - \kappa_{\xi 4}\boldsymbol{\xi}_2^T \mathrm{sig}_{\mathrm{c}}^{p_{\xi 1}}\left(\boldsymbol{\xi}_2\right) \\ & + \boldsymbol{\xi}_2^T\mathbf{M}_a^{-1}\mathbf{Q}_a\left(\mathrm{Fil}(\boldsymbol{\tau}_d) - \boldsymbol{\tau}_d\right)\end{aligned} \tag{9.103}$$

Considering the nature of the command filter itself, we assume $||\mathrm{Fil}(\boldsymbol{\tau}_d) - \boldsymbol{\tau}_d|| \leq \varepsilon_{f\tau}$, where $\varepsilon_{f\tau}$ is a constant. Then we have

$$\boldsymbol{\xi}_2^T\mathbf{M}_a^{-1}\mathbf{Q}_a\left(\boldsymbol{\tau}_a - \boldsymbol{\tau}_d\right) \leq \frac{1}{2}||\boldsymbol{\xi}_2||^2 + \frac{1}{2}(\varepsilon_{f\tau}\varepsilon_M)^2 \tag{9.104}$$

Further, we have

$$\dot{V}_5 \leq -\kappa_{\xi 5}V_5 - \kappa_{\xi 6}V_5^{\frac{1+p_{\xi 1}}{2}} + \kappa_{\xi 10}V_5^0 \tag{9.105}$$

where $\kappa_{\xi 10} = \frac{1}{2}(\varepsilon_{f\tau}\varepsilon_M)^2$.

According to Lemma 9.1, V_5 is finite-time uniformly ultimately bounded within the compact set $\Omega_{\xi 2} = \{\boldsymbol{\xi}_1, \boldsymbol{\xi}_2 : \kappa_{\xi 8}V_5^{\frac{1+p_{\xi 1}}{2}} + \kappa_{\xi 9}V_5 < \kappa_{\xi 10}\}$ with the settling time

$$T_5 \leq \frac{2\ln\left(1 + \frac{\kappa_{\xi 5}-\kappa_{\xi 9}}{\kappa_{\xi 6}-\kappa_{\xi 8}}V_5^{\frac{1-p_{\xi 1}}{2}}(\boldsymbol{\xi}_1(0), \boldsymbol{\xi}_2(0))\right)}{(\kappa_{\xi 5} - \kappa_{\xi 9})(1 - p_{\xi 1})} \tag{9.106}$$

So $\boldsymbol{\xi}_1$, $\boldsymbol{\xi}_2$ are bounded with $T_{\xi 2} \leq T_5$. Similar to the previous subsection, the error $\mathbf{q}_{ev}$ is practical time-synchronized stable, and the settling time is $T_{\mathbf{q}_{ev}} \leq \max\{T_{\vartheta 2}, T_{\zeta 2}, T_{z2}, T_{\xi 2}\}$.

9.5 Simulation Analysis

The numerical example is shown in this section to demonstrate the performance of the developed command filter-based and improved command filter-based practical time-synchronized controllers. For the convenience of distinguishing from the comparison controllers, we label them as 'A' for (9.78) and 'B' for (9.99). To illustrate the advantages of the $\mathrm{sig_n}(\cdot)$ function, we replaced the $\mathrm{sig_n}(\cdot)$ in (9.78) and (9.99) with $\mathrm{sig_c}(\cdot)$, and got the controllers marked by 'C' and 'D' for comparison. These can also be considered as traditional finite-time controllers based on command filters under the same framework. To compare the controllers based on the command filter and improved command filter, we refer to the command filter backstepping controller in [8] marked as 'E' and the improved command filter backstepping controller via the stable first-order low-pass filter in [8] marked as 'F'. In order to reduce the interference of non-essential variables to the comparison simulation, we only refer to the controllers with command filter and improved command filter in [8]. Finally, we also give the conventional finite-time controller in [2] marked as 'G' as a comparison.

9.5.1 Simulation Parameters

For the target spacecraft, the orbital parameters are given in Table 9.1. Since the service spacecraft is very close to the target spacecraft, it has the same orbital parameters.

Without loss of generality, we directly give the initial parameters of the service spacecraft simulation model here, $\mathbf{q}_{v0} = [0.6, 0.3, -0.3]^T$, $q_0(0) = \sqrt{1 - \mathbf{q}_{v0}^T \mathbf{q}_{v0}}$, $\boldsymbol{\omega}_{a0} = [0, 0, 0]^T \mathrm{rad/s}$,

$$\mathbf{J}_s = \begin{bmatrix} 100 & -10 & 5 \\ -10 & 120 & -3 \\ 5 & -3 & 100 \end{bmatrix} \mathrm{kg \cdot m^2}.$$

The parameters used for the simulation are all replaceable.

Table 9.1 Orbital parameters of the target spacecraft

Parameters	Values
Semimajor axis	6.85×10^6m
Eccentricity	0.0073
Inclination	60 deg
Argument of periapsis	0 deg
Longitude of the ascending node	0 deg
Mean anomaly	0 deg

The desired signal is given by the following equation. In this simulation, the service spacecraft tracks the attitude of the target spacecraft. We assume that the $\mathfrak{F}_t$ coincides with the $\mathfrak{F}_i$, i.e., it is inertially oriented. Therefore, $\mathbf{q}_{dv}$ is obtained from the quaternion of the relative motion of the $\mathfrak{F}_s$ and $\mathfrak{F}_i$. $r_{li-j,k}$ represents the element of the j th-row and k th-column in $\mathbf{R}_{li}$, which represents the rotation matrix of the target spacecraft orbit frame and the ECI frame, $q_{d0} = 1/2\ (1 + r_{li-11} + r_{li-22} + r_{li-33})^{1/2}$.

$$\mathbf{q}_{dv} = \begin{bmatrix} (r_{li-23} - r_{li-32})/(4q_{d0}) \\ (r_{li-31} - r_{li-13})/(4q_{d0}) \\ (r_{li-12} - r_{li-21})/(4q_{d0}) \end{bmatrix} \tag{9.107}$$

In this simulation, we limit the minimum value -1 N $\cdot$ m and the maximum value 1 N $\cdot$ m of the control input, which can be accepted by the spacecraft actuators. We choose the disturbances used in the simulation as follows

$$\mathbf{d} = \begin{bmatrix} 0.09\sin(0.5t) \\ 0.018\sin(0.5t + \pi/2\ ) \\ 0.12\sin(0.5t) \end{bmatrix} \mathrm{N}\cdot\mathrm{m} \tag{9.108}$$

To show the disturbance rejection capability of the controllers, the disturbances we used are more extensive than those suffered by the spacecraft in the actual situation.

We only change the sign function used in the controllers 'A' and 'C', 'B' and 'D'. To illustrate the superiority of the proposed controllers, we use exactly the same controller parameters in 'A' and 'C', and the same for 'B' and 'D'. These reflect the considerable improvement of the control performance by the change of the function items. In controllers 'E', 'F', and 'G', because the controller structures are not entirely similar, we will adjust the appropriate parameters for comparison in terms of control performance and energy consumption. All the parameters used in controllers 'A', 'B', 'C', and 'D' are given in Table 9.2.

To show the tracking error in the whole simulation process, we define the error parameter for the controllers as $I = ||\mathbf{q}_{ev}|| + ||\boldsymbol{\omega}_e||$. For illustrating the energy consumption of the proposed controllers, we use the energy index, which is related to the amount of energy (chemical or electrical energy) that actuators (thrusters or reaction flywheels) consume during closed-loop system stabilization, as $E = \sum_{i=1}^{n} \int_0^t |\boldsymbol{\tau}_i(t)|^2 \mathrm{d}t$.

9.5.2 *Performance and Comparison of Controllers*

From the perspective of the original design intention, we hope to obtain a time-synchronized stable or practical time-synchronized stable system through ratio persistence or ratio restriction. This shows that the trajectories of the closed-loop systems tend to be in a straight line from the initial point to the equilibrium point. Figure 9.2 reflects this property. Compared with similar controllers 'C', 'D', 'E', 'F', and 'G',

Table 9.2 The values of the controllers parameters

Parameters	A	B	C	D
p_1	0.60	0.60	0.60	0.60
p_2	0.90	0.90	0.90	0.90
$p_{\xi 1}$	0.99	0.99	0.99	0.99
κ_1	0.10	0.10	0.10	0.10
κ_2	0.06	0.06	0.06	0.06
$\kappa_{\xi 1}$	1.00	1.00	1.00	1.00
$\kappa_{\xi 2}$	1.00	1.00	1.00	1.00
$\kappa_{\xi 3}$	1.00	1.00	1.00	1.00
$\kappa_{\xi 4}$	1.00	1.00	1.00	1.00
$\kappa_{\vartheta 1(5)}$	0.10	0.10	0.10	0.10
$\kappa_{\vartheta 2(6)}$	0.10	0.10	0.10	0.10
$\kappa_{\vartheta 3(7)}$	0.10	0.01	0.10	0.01
$\kappa_{\vartheta 4(8)}$	0.10	0.01	0.10	0.01
$\chi_{\max}$	1.00	1.00	1.00	1.00
$\chi_{\min}$	−1.00	−1.00	−1.00	−1.00

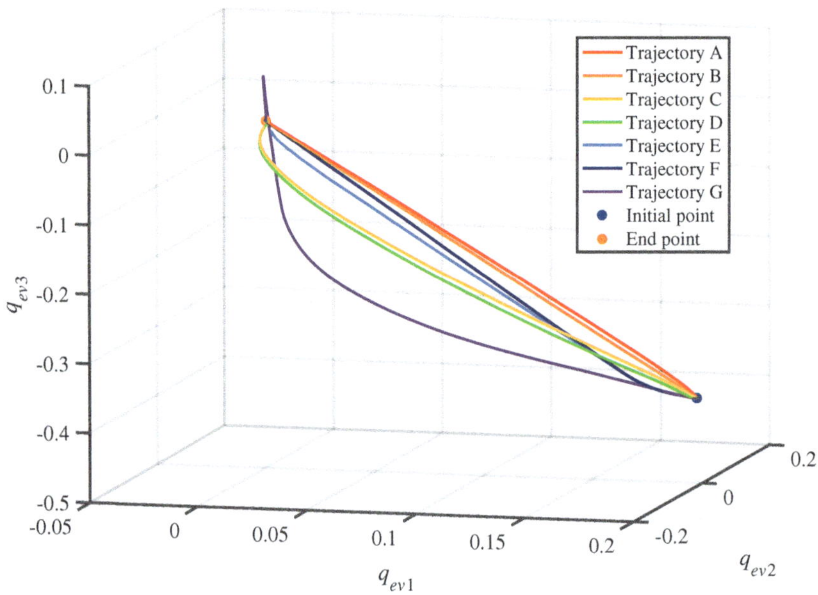

Fig. 9.2 The trajectories of relative attitude tracking errors under the controllers 'A', 'B', 'C', 'D', 'E', 'F', and 'G'

under the proposed controllers 'A' and 'B', the trajectories of the closed-loop systems are approximately a straight line directly linking the initial point and the equilibrium point. Under other controllers, the trajectories have a large deviation, which manifest as unnecessary overshoot and negative transient behaviors. Figure 9.3 shows the tracking errors of the closed-loop systems under controllers 'A', 'C', 'E', and 'G'. Compared with 'C', under controller 'A', the tracking error of the closed-loop system can synchronously converge to the neighborhood of the equilibrium point within finite time, and the steady-state error is small. In contrast, 'C' does not converge synchronously and has overshoot. This pair of comparisons reflects the critical effect and characteristics of the $\mathrm{sig}_{\mathrm{n}}(\cdot)$ function. Compared with 'E', under the controller 'A', the closed-loop system can converge faster in finite time, and the steady-state error is smaller. Compared with 'G', under controller 'A', the transient performance is better, including no overshoot, faster convergence rate, and smoother trajectory.

For the improved command filter-based practical time-synchronized controller, Fig. 9.4 shows the different control effects under the controllers 'B', 'D', and 'F'. Compared 'B' with 'D', based on the theoretical design, the simulation examples further show that the $\mathrm{sig}_{\mathrm{n}}(\cdot)$ function used can indeed make the trajectory tend to the shortest straight line. Compared with 'F', under the proposed controller 'B', the convergence rate is faster, and the steady-state error is smaller. Besides, the controller 'B' does not need to obtain the actual control information τ_a and can also achieve transient and steady-state performance similar to that of 'A'.

Figures 9.5 and 9.6 respectively show the relationship between τ_d and τ_a under the controllers 'A' and 'B'. Due to the existence of the command filter, the actual input τ_a will obtain the signals that satisfy the requirements of amplitude, rate, and bandwidth. The input saturation problem is solved.

Figure 9.7 shows the Euclidean norms of the tracking errors $I = ||\mathbf{q}_{ev}|| + ||\boldsymbol{\omega}_e||$ and the energy consumption of the seven controllers. Additionally, the energy consumption caused by disturbances is also shown in Fig. 9.7b. We can see that under the proposed command filter-based practical time-synchronized controller 'A' and the improved command filter-based practical time-synchronized controller 'B', the Euclidean norms of the tracking errors are smaller, and the energy consumptions are less, and even reduced at the level of 35%. The authors have likewise conducted a large number of simulations under other initial status and control parameters. From these simulations, the proposed controllers are able to reduce the energy consumption by 25–35% compared with the traditional finite-time controllers. These are due to the closed-loop system satisfying ratio restriction. In addition, the trajectory of the closed-loop system tends to the shortest path. Therefore, better transient and steady-state performance are obtained by (practical) time-synchronized controllers.

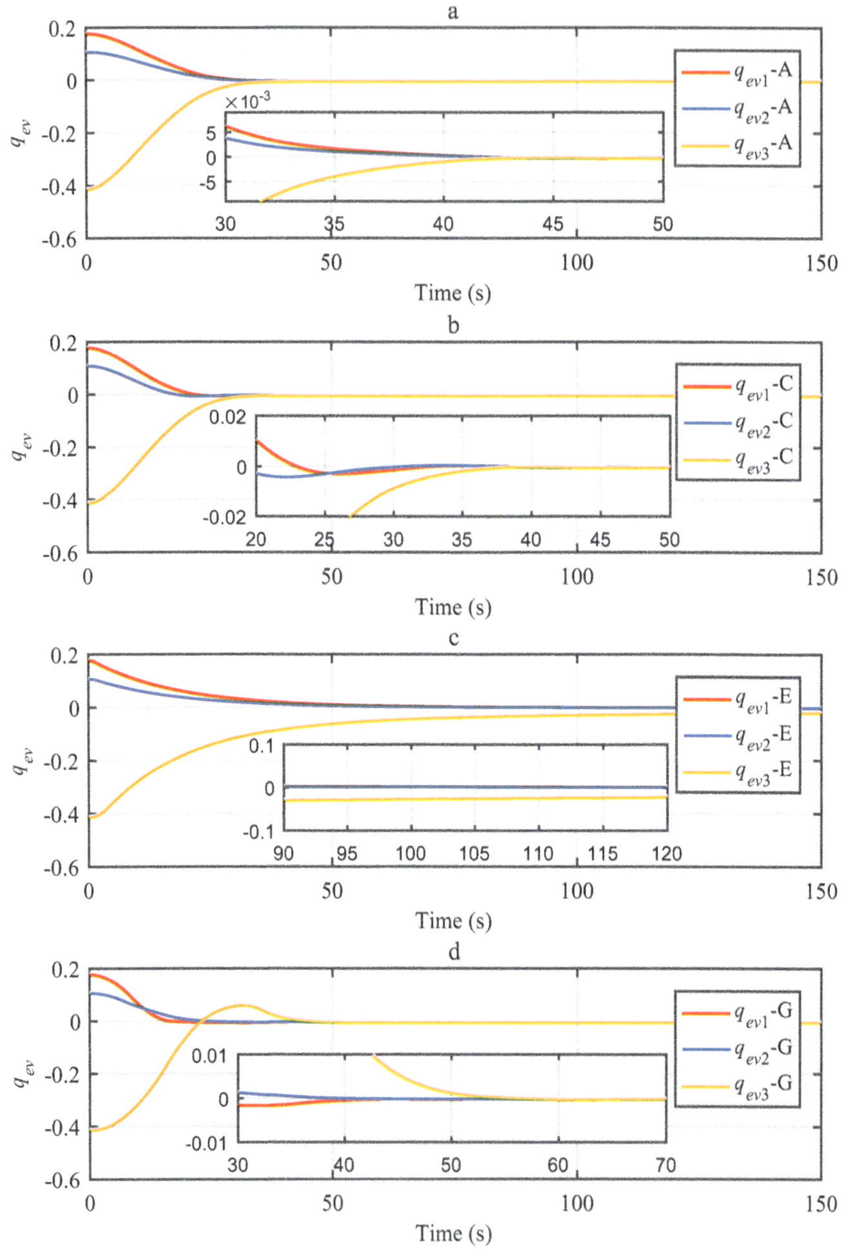

Fig. 9.3 The comparison of relative attitude tracking errors under **a** the command filter-based practical time-synchronized controller 'A', **b** the command filter-based finite-time controller 'C', **c** the command filter backstepping controller 'E', and **d** the conventional finite-time controller 'G'

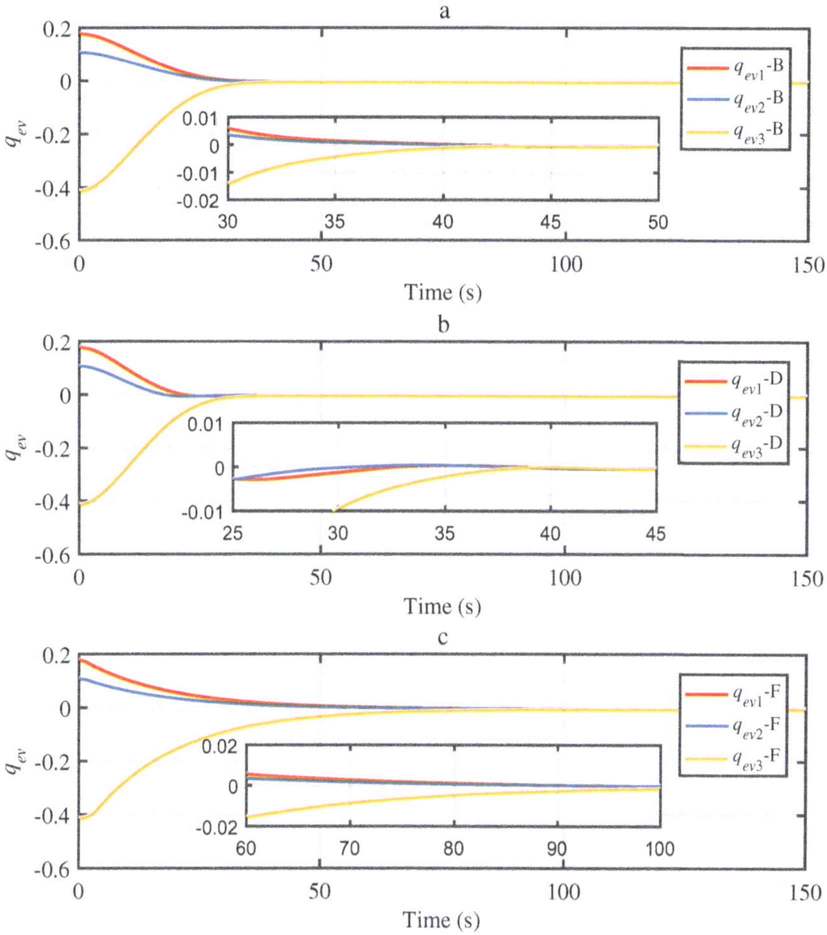

Fig. 9.4 The comparison of relative attitude tracking errors under **a** the improved command filter-based practical time-synchronized controller 'B', **b** the improved command filter-based finite-time controller 'D', and **c** the improved command filter backstepping controller 'F'

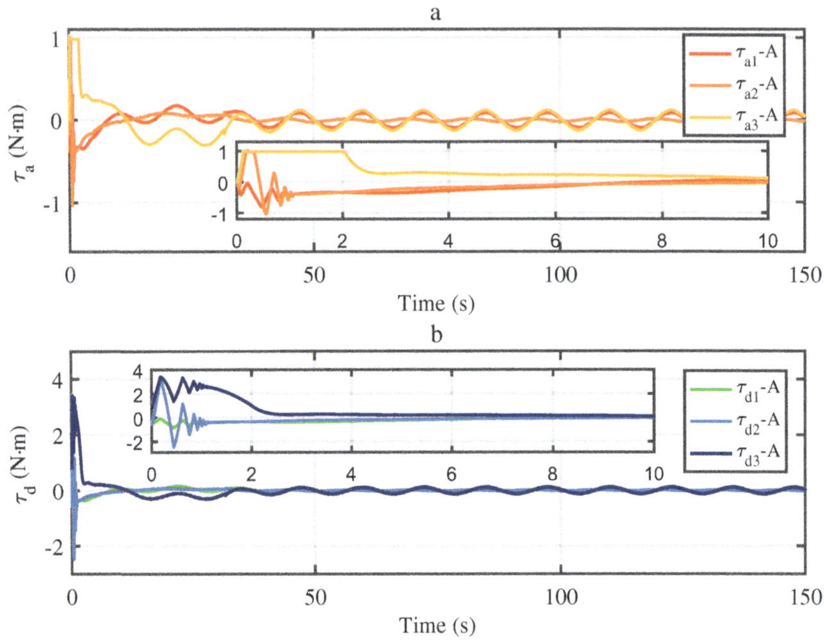

Fig. 9.5 The control input under the command filter-based practical time-synchronized controller 'A': **a** τ_a, the control input actually obtained by the actuator; **b** τ_d, the control input obtained by the command filter

9.6 Conclusion

Considering the disturbed on-orbit service spacecraft attitude tracking control, we achieve a comprehensive improvement in control performance and fuel consumption by defining and utilizing ratio restriction and practical time-synchronized stability.

Based on these techniques, practical time-synchronized controllers integrated with the command filter and observer are proposed. Among them, the command filter solves the input saturation, and the observer deals with the problem of uncertain signals. The application of the low-pass filter needs no measuring equipment of the actual control input.

This chapter further extends the application of time-synchronized techniques, where (i) the state elements are allowed to converge to the neighborhood of the equilibrium point synchronously rather than an exact equilibrium; (ii) the trajectory of the closed-loop system tends to be a straight line from the initial point to the equilibrium point; (iii) the problems of the input saturation and external disturbances are appropriately handled; and (iv) the energy consumption is significantly reduced.

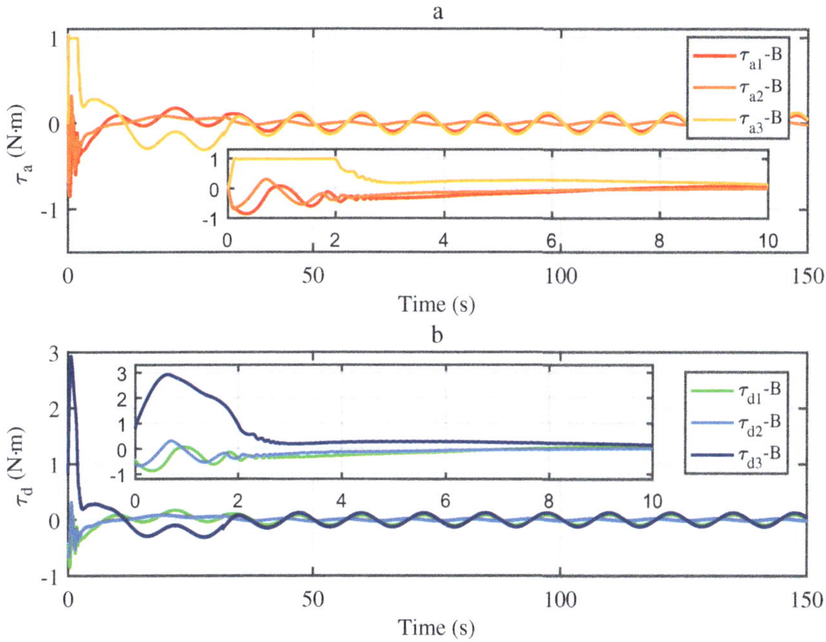

Fig. 9.6 The control input under the improved command filter-based practical time-synchronized controller 'B': **a** τ_a, the control input actually obtained by the actuator; **b** τ_d, the control input obtained by the command filter

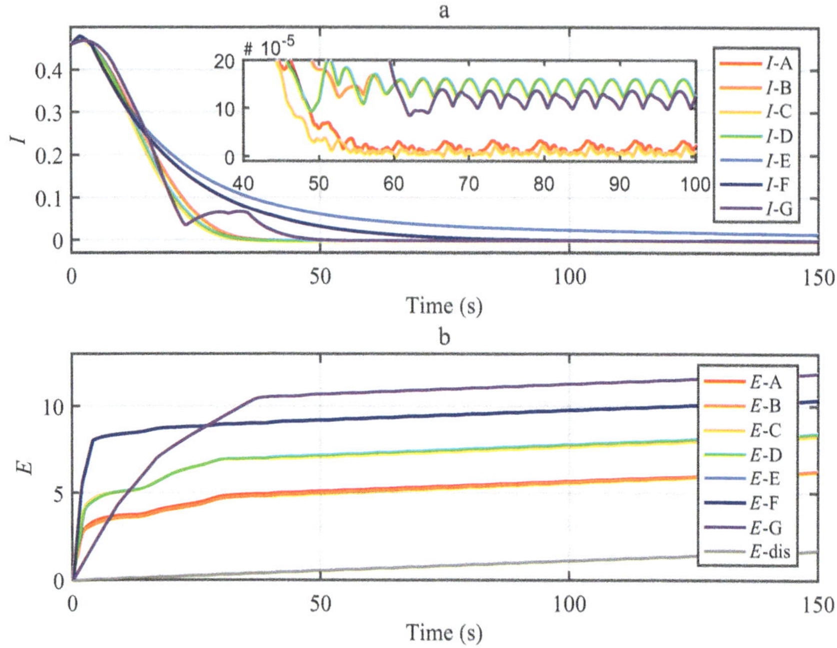

Fig. 9.7 The Euclidean norms of the tracking errors and the energy consumption under the controllers 'A', 'B', 'C', 'D', 'E', 'F', and 'G'. **a** The Euclidean norms, and **b** the energy consumption under the controllers and the energy index of the disturbances marked by 'E-dis'

References

1. Costic, B., Dawson, D., De Queiroz, M., Kapila, V.: Quaternion-based adaptive attitude tracking controller without velocity measurements. Journal of Guidance, Control, and Dynamics **24**(6), 1214–1222 (2001)
2. Du, H., Li, S., Qian, C.: Finite-time attitude tracking control of spacecraft with application to attitude synchronization. IEEE Transactions on Automatic Control **56**(11), 2711–2717 (2011)
3. Farrell, J.A., Polycarpou, M., Sharma, M., Dong, W.: Command filtered backstepping. IEEE Transactions on Automatic Control **54**(6), 1391–1395 (2009)
4. Hu, Q., Jiang, B.: Continuous finite-time attitude control for rigid spacecraft based on angular velocity observer. IEEE Transactions on Aerospace and Electronic Systems **54**(3), 1082–1092 (2017)
5. Hughes, P.C.: Spacecraft attitude dynamics. Courier Corporation (2012)
6. Junkins, J.L., Schaub, H.: Analytical mechanics of space systems. American Institute of Aeronautics and Astronautics (2009)
7. Nagesh, I., Edwards, C.: A multivariable super-twisting sliding mode approach. Automatica **50**(3), 984–988 (2014)
8. Wang, Y., Chen, W., Zhang, S., Zhu, J., Cao, L.: Command-filtered incremental backstepping controller for small unmanned aerial vehicles. Journal of Guidance, Control, and Dynamics **41**(4), 954–967 (2018)

Chapter 10
Conclusions

In this book, to meet the requirements of intelligent system autonomy, advanced manufacturing, network system cooperation, and exploration in space, air, ground and oceans, time-synchronized stability and control have been formulated and investigated, under which all the system state elements converge to the equilibrium at exactly the same time. The relevant time-synchronized controllers not only provide the specific time-synchronized convergence property to a target system, but also hold advantages in terms of smoothing the system output, shortening the system travel length, and reducing the energy consumption.

In general, time-synchronized stability has been built on finite-time stability. As a unique type of time-synchronized stability, fixed-time stability has been established where the synchronized setting time is independent of any initial state. Further, predefined-time-synchronized stability has been likewise developed based on fixed-time-synchronized stability. If a control system is predefined-time-synchronized stable, it meets all the requirements of fixed-time-synchronized stability and its control parameters explicitly contain the upper bound of the synchronized settling time. According to different properties of the predefined synchronized setting time, predefined-time-synchronized stability has been classified into three specific types: (1) weak predefined-time-synchronized stability; (2) strong predefined-time-synchronized stability; and (3) true predefined-time-synchronized stability.

To achieve time-synchronized stability, one effective method is to design relevant time-synchronized sliding modes. The system state elements on such a sliding surface are driven to the origin time-synchronously. Various time-synchronized sliding modes have been introduced with different properties on the continuity, singularity, and convergence speed. By this way, time-synchronized controllers can be constructed based on time-synchronized sliding modes by back-stepping methods.

Time-synchronized control for disturbed systems is more challenging, since the external disturbances or system uncertainties may impede the achievement of time-synchronized convergence. To solve this problem, the multi-variable super-twisting

D. Li et al., *Time-Synchronized Control: Analysis and Design*,
https://doi.org/10.1007/978-981-16-3089-7_10

sliding-mode observer and full-order sliding mode techniques have been introduced and integrated into time-synchronized control. In addition, for a class of more general MIMO systems under matched and unmatched disturbances, we have classified the target system into two representative types: (1) the input-dimension-dominant MIMO system; and (2) the state-dimension-dominant MIMO system. Time-synchronized controllers are proposed for these two classes of systems, respectively.

The design procedure of fixed-time-synchronized control needs more attention due to its inherent complicity in stability analysis. One important matter consists in avoiding the possible singularity in the designed control laws. For the first-order and second-order affine systems, singularity-free fixed-time-synchronized controllers have been proposed. For the more complicated general MIMO systems, a fixed-time-synchronized controller has been proposed for the input-dimension-dominant MIMO system, and a semi-fixed-time-synchronized controller for the state-dimension-dominant MIMO system, where the singularity has been eliminated by modified sliding modes.

The estimation of the upper bound of the synchronized settling time for classical fixed-time-synchronized control is usually significantly larger than the real convergence time. By utilizing the beta function and its related mathematical tools, the least upper bound of the synchronized settling time has been defined and calculated. This least upper bound substantially alleviates the overestimation issue widely existing in the traditional fixed-time control methods. As a suitable extension, predefined-time-synchronized control has been investigated. Its control parameters contain explicitly the upper bound of the synchronized setting time. For simple first-order systems, proper predefined-time-synchronized controllers have been constructed to achieve strong predefined-time-synchronized convergence and true predefined-time-synchronized convergence.

Time-synchronized consensus has been explored, where all the state elements of all the agents achieve consensus time-synchronously. This convergence property offers crucial advantages to multi-agent systems during the cooperative transport, formation flying, and simultaneous attack missions. Sufficient communication topology has been established to achieve time-synchronized consensus, based on which time-synchronized and fixed-time-synchronized consensus controllers have been designed for the multi-agent system under undirected and directed graphs. To time-synchronously obtain an unknown global state by distributed information flow, a time-synchronized distributed observer has been proposed for the multi-agent observation mission.

Finally, time-synchronized control is studied for spacecraft rendezvous and docking with or without external disturbances. It is noteworthy that the definitions of practical time-synchronized stability and ratio restrictions are introduced, which help to facilitate a more generalized time-synchronized convergence in practical applications as there always exists noise, disturbance, and uncertainty preventing the steady error from the origin. The hardware-in-loop experiments, as well as numerical simulations, show that the proposed algorithms achieved improved control accuracy, travel trajectory, and energy consumption compared with other control designs.

Despite the results introduced in this book, there exist more interesting but challenging problems that remain open. We here list these possible future directions as follows:

i. *Practical time-synchronized stability*: Practical finite-time stability (which has been proved to be essentially the same as semi-global uniform ultimate boundedness) allows the final convergence error to be contained in a small neighborhood of the origin. However, at this moment, such a concept has not been fully introduced to time-synchronized stability although Chap. 9 has carried out a preliminary exploration of practical time-synchronized stability. Certainly, more efforts should be taken on the mathematical formulation of "practical time-synchronized stability".
ii. *Time-synchronized tracking control*: The controllers proposed in this book mainly focus on the stabilization of a target MIMO system (including the multi-agent system). Apparently, the tracking control problem with the consideration of time-synchronized convergence is more difficult. The system is required to track a dynamic target signal time-synchronously, making the extension to time-synchronized tracking non-trivial.
iii *Time-synchronized control for more complex systems*: In this book, time-synchronized control has been studied for affine systems, disturbed systems, and general MIMO systems. In fact, there exist more complex control systems to be considered. How to realize time-synchronized control on these systems is an important problem to be further addressed.
iv. *Time-synchronized formation control*: Time-synchronized consensus has been investigated in this book. However, as a strongly related multi-agent control problem, the time-synchronized formation control has not been touched on. It requires all the state elements of all the agents to reach target values at the same time, to form a spacial configuration. There are abundant methods to describe the nominal formation to be achieved in the existing literature, e.g., the position-based, displacement-based, bearing-based, and angle-based techniques. Choosing a proper method to draw the nominal formation and designing the corresponding time-synchronized formation controller is appealing.
v. *Time-synchronized control for state-dimension-dominant MIMO systems*: In this book, in the case of time-synchronized control for state-dimension-dominant MIMO systems, the outcome of the system convergence becomes complicated and undesired, where only semi-time-synchronized convergence is expected to be achieved. The specific challenge preventing "full" time-synchronized convergence lies in that the right inverse of a crucial system sub-matrix no longer exit in the case of state-dimension-dominant MIMO systems. This difficult but interesting problem is to be solved in possible future work.

The authors hope this book will also catalyze efforts towards solving the above interesting problems.

The manufacturer's authorised representative in the EU is Springer Nature Customer Service Centre GmbH, Europaplatz 3, 69115 Heidelberg, Germany. If you have any concerns regarding our products, please contact ProductSafety@springernature.com

Printed and bound by CPI Group (UK) Ltd, Croydon, CR0 4YY

30/09/2026

02225293-0006